The Hidden Chapters of Human Origins

ANCIENT CIVILIZATIONS AND MYSTERIES, Volume 7

Alistair Ravenhurst

Published by AJ CARMICHAEL, 2026.

THE HIDDEN CHAPTERS OF HUMAN ORIGINS

First edition. March 8, 2026.

Written by Alistair Ravenhurst.

Also by Alistair Ravenhurst

Ancient Civilizations and Enigmas

Egypt Before Egypt, the World Before the Pharaohs

ANCIENT CIVILIZATIONS AND MYSTERIES

After the Flood, Cataclysm and the Forgotten Origins of Civilization

Megalithic Science, Architecture and Astronomy in the Ancient World

Skywatchers of Prehistory, the Hidden Origins of Ancient Shamanic Astronomy

Beneath the Sands: The Lost Worlds of Earth's Forgotten Deserts

Empire of the Ocean: The Lost Maritime Civilization of the Pacific

The Serpent Code: The Hidden Language of Ancient Myth

The Hidden Chapters of Human Origins

Watch for more at https://www.youtube.com/@AncientCivilization.

Table of Contents

INTRODUCTION

Every civilization carries an origin story. Sometimes it is written in sacred language, where gods form humanity from clay or breathe life into figures shaped from the earth. Sometimes it appears in scientific language, where species emerge through natural processes across immense spans of time. At first glance these two styles of explanation seem very different, yet they share a common purpose. Both attempt to answer the same question: *How did we become what we are?* And both shape the way people understand themselves long after the story has been told.

In the modern world, the dominant explanation of human origins is framed through evolutionary science. It describes a deep past populated by ancient ancestors who gradually changed over time, producing new species and new forms of life. Eventually, through a long chain of biological transitions, a particular lineage emerged that we recognize today as **Homo sapiens**. This narrative has extraordinary explanatory power. It connects humanity to the wider web of life, situates our existence within the history of the Earth, and demonstrates how small biological changes can accumulate into profound transformations.

Yet the public version of this story is often far simpler than the evidence itself. It tends to appear as a smooth progression from primitive forms to modern humans, a clear line moving steadily toward greater intelligence, greater capability, and finally the cultural complexity of the present. This version is easy to remember and easy to teach. It fits neatly into textbooks, documentaries, and museum exhibits. But the deeper researchers examine the evidence—from fossils, archaeology, genetics, and climate science—the more complicated the story becomes.

Human origins do not resemble a straight line. They resemble a tangled web of populations, migrations, and encounters unfolding across shifting landscapes. Multiple human groups existed at the same time. Some adapted to cold steppe environments, others to tropical forests, deserts, or mountain plateaus. Some populations remained isolated for long periods, while others met and exchanged genes or cultural knowledge. Rather than a solitary ascent, the deep human past increasingly appears as a dynamic ecological system filled with related beings whose lives overlapped in ways that are only now becoming visible.

This complexity does not weaken evolutionary science. In fact, it strengthens it. Scientific knowledge grows precisely by confronting evidence that challenges older assumptions. As new tools emerge—especially the powerful analytical methods of genetics—previously invisible chapters of the human story begin to appear. Genetic data can reveal ancient population mixtures, migrations that left few archaeological traces, and even the presence of human groups that are barely known from fossils. These discoveries do not overturn the basic framework of evolution; instead, they show that the process was richer, more interconnected, and more surprising than earlier generations imagined.

At the same time, the story of human origins is shaped not only by evidence but by the ways societies choose to interpret it. People naturally prefer narratives that are clear and coherent. A simple path from primitive to advanced feels emotionally satisfying because it places modern humanity at the culmination of a long developmental arc. In such a narrative, the past becomes a staircase and we stand at the highest step.

But nature rarely constructs staircases. Evolution is not guided by a predetermined goal. It does not aim toward intelligence, civilization, or technological mastery. It is a process of adaptation within changing environments. Species survive not because they are universally superior, but because they fit particular ecological circumstances at particular moments in time. When those circumstances shift, even highly successful species can vanish.

Understanding this point transforms how we interpret the human story. Instead of imagining a single heroic lineage climbing toward destiny, we begin to see a landscape of experimentation. Different human groups explored different ecological strategies. Some succeeded for tens of thousands of years in environments that would challenge modern populations. Others developed cultural traditions and survival skills that allowed them to thrive under specific climatic conditions. In this sense, the human genus was once far more diverse than it is today.

The disappearance of these other human groups is one of the most intriguing mysteries in our past. Fossil and genetic evidence now confirm that populations such as Neanderthals and Denisovans existed alongside early modern humans. These were not mythical creatures or evolutionary curiosities. They were human relatives who lived, hunted, formed families, and navigated the challenges of their worlds. In some cases they also interacted with the ancestors of modern populations, leaving small but detectable genetic traces that survive today.

This recognition has profound implications. It means that humanity's history was not isolated. The ancestors of modern humans did not evolve in splendid solitude. They emerged within a broader community of human lineages, some of which contributed to the genetic and cultural landscape from which our species eventually spread. The human story, therefore, is not simply the rise of one species. It is also the story of encounters, exchanges, and disappearances among several.

Another factor that complicates our understanding of the past is the nature of the evidence itself. The archaeological and fossil records are incomplete archives. Most living beings leave no trace after death. Bones decay, landscapes erode, and entire habitats vanish beneath oceans or shifting sediments. What survives for scientists to study is only a tiny fraction of what once existed. As a result, the record of human evolution is not a continuous documentary but a patchwork of fragments scattered across time and geography.

Each fragment—whether a fossil bone, a stone tool, or a strand of ancient DNA—offers a glimpse into a vanished world. When combined carefully, these glimpses reveal patterns of migration, adaptation, and contact. Yet they also remind us how much remains unknown. Entire populations may have lived and disappeared without leaving clear traces. Cultural practices that shaped daily life—language, music, ritual, storytelling—rarely fossilize at all. The past we reconstruct is therefore both real and partial, a reconstruction assembled from the pieces that survived.

The recognition of these limitations does not diminish the search for human origins. Instead, it deepens it. It encourages a more careful reading of the evidence and a greater openness to complexity. Rather than asking only when modern humans appeared, researchers increasingly ask how different populations interacted, how environmental pressures shaped biological change, and how cultural innovations spread across regions and generations.

Within this expanding framework, anomalies and puzzles become especially valuable. Unexpected fossil dates, unusual genetic signals, and cultural innovations that appear suddenly in the archaeological record all invite new questions. Sometimes these anomalies dissolve under further study. At other times they reveal important features of the past that were previously overlooked. In this way, scientific inquiry advances not only through confirmation but also through the careful examination of what does not fit easily into established models.

The chapters that follow explore these questions from several angles. They examine the complex structure of the human family, the surprising revelations of genetic research, and the cultural transformations that shaped the emergence of modern societies. They also address the tension between evidence and interpretation—the space where scientific data, cultural memory, and human imagination sometimes intersect.

Approaching this subject requires balance. On one side lies the risk of oversimplification, where the deep human past is reduced to a tidy narrative that ignores complexity. On the other lies the risk of speculation untethered from evidence. The goal of this work is neither to dismiss established science nor to replace it with sensational claims. Instead, it aims to widen the frame through which the human story is viewed, acknowledging both what we know and what remains uncertain.

Human origins are not a closed case. They are an unfolding investigation shaped by new discoveries and new technologies. With every decade, additional pieces of the puzzle emerge—from remote caves, ancient sediments, and the genomes carried by living populations. Each discovery adds detail to a story that is already extraordinary: the story of a species that emerged from a broader human family, survived immense environmental challenges, and developed the capacity to reflect on its own beginnings.

In that sense, the study of human origins is not merely about the distant past. It is about identity. To ask where we came from is to ask what kind of beings we are and how we fit within the natural world. The deeper the inquiry goes, the clearer it becomes that humanity is not the endpoint of a simple evolutionary ladder but the product of a complex history of survival, interaction, and adaptation.

Recognizing that complexity does not diminish our place in the world. If anything, it makes our existence more remarkable. We are the living heirs of a vast and intricate past—one that includes not only our direct ancestors but a wider community of human relatives whose lives helped shape the world we inherited.

Understanding that story is the beginning of understanding ourselves.

If the popular story of human evolution presents a straight line, the real evidence suggests something closer to a living landscape. Instead of a single lineage moving steadily toward modern humanity, the deep past reveals populations spreading, separating, reconnecting, and sometimes disappearing. These populations did not exist in isolation from one another. They occupied overlapping regions, shared ecological pressures, and in some cases encountered each other directly. The emerging picture is not one of simple succession but of coexistence.

For much of the twentieth century, the dominant image of human evolution was shaped largely by fossils. This was understandable. Fossils are tangible and dramatic. A skull emerging from ancient sediments feels like a direct window into prehistory. When early researchers assembled these discoveries, they naturally attempted to organize them into a sequence. A jaw from one region, a skull from another, and a stone tool tradition nearby could be placed along a timeline that suggested gradual transformation.

This approach produced a coherent narrative that helped establish the foundations of paleoanthropology. Yet the fossil record itself is extraordinarily selective. Fossilization requires rare conditions: rapid burial, stable geological environments, and long-term preservation without disturbance. The majority of living organisms leave no fossil remains at all. Even when fossils are preserved, they are often incomplete. A single tooth or fragmentary bone may represent an entire population that once thrived across a large region.

Because of this uneven preservation, the fossil record cannot be read as a complete historical account. It is closer to a scattered archive in which most of the pages are missing. Entire landscapes that once supported human populations are now submerged beneath oceans or buried under later sediments. Organic materials such as wood, fiber, or skin—objects that may have formed a large part of daily life—almost never survive across tens of thousands of years. What remains are fragments: bones, stone tools, traces of fire, and occasional preserved artifacts that offer glimpses into vanished worlds.

These fragments, however, are enough to reveal an important pattern. At various points in the past, more than one human lineage existed simultaneously. Fossil discoveries across Africa, Europe, and Asia demonstrate that different populations developed distinct anatomical characteristics while occupying different environments. Some possessed robust skeletons adapted to cold climates. Others had features suited to warmer regions or varied ecological niches.

For a long time, these differences were interpreted as stages in a ladder of progress. Early humans appeared primitive, later forms appeared more advanced, and modern humans seemed to represent the final stage of the process. Yet this interpretation relied heavily on arranging discoveries into a chronological order. As more fossils accumulated, it became increasingly difficult to maintain the image of a simple progression.

Instead of a ladder, researchers began to see branching patterns. Some populations appeared to diverge from earlier ancestors and then develop along separate trajectories. Others seemed to retain certain archaic features while adopting new traits. In several regions, the timelines of these populations overlapped, indicating that different human groups inhabited the same broad geographic areas during the same periods.

This realization transformed the metaphor used to describe human origins. The “human family tree” replaced the ladder, suggesting that multiple branches emerged from shared ancestral roots. In this framework, Homo sapiens represented one branch among several rather than the inevitable outcome of a linear path.

Yet even the tree metaphor has limitations. Trees imply clean divisions: a branch splits and then grows independently from the others. Real biological populations do not always behave this way. Groups that separate can later encounter each other again. When they do, gene flow can occur between them. Traits may pass from one population to another through interbreeding or cultural exchange. Under such conditions, the evolutionary pattern becomes less like a tree and more like a network.

This network model has gained increasing support as new lines of evidence have emerged. Genetic research, in particular, has revealed that the boundaries between ancient human populations were sometimes more permeable than previously assumed. Certain modern populations carry small portions of DNA inherited from other human groups that once lived outside Africa. These genetic fragments serve as markers of past encounters that were invisible in the fossil record alone.

The presence of such inherited genetic material indicates that the meeting of different human populations was not always hostile or rare. At times, these encounters led to interbreeding. The descendants of those unions eventually became part of expanding human populations whose genes continue into the present. In this sense, the disappearance of certain ancient groups does not necessarily mean they vanished entirely. Portions of their biological heritage may persist within the genomes of modern people.

These discoveries encourage a different perspective on the human past. Rather than imagining the rise of modern humans as a dramatic replacement event, it may be more accurate to view it as a complex demographic process. Populations moved, interacted, merged, and sometimes absorbed one another. Over thousands of generations, these processes produced the genetic mosaic that characterizes humanity today.

Climate played a crucial role in shaping this dynamic landscape. The Ice Age was not a single continuous period of cold. It consisted of repeated cycles in which glaciers advanced and retreated, sea levels rose and fell, and ecosystems shifted dramatically. Regions that were once fertile could become arid or frozen, forcing populations to migrate or adapt. Other areas opened new corridors for movement, allowing previously isolated groups to encounter one another.

These environmental changes created both challenges and opportunities. Populations that were well adapted to one set of conditions could struggle when those conditions changed. Others might expand into newly favorable territories. The shifting climate therefore acted as a powerful engine of demographic change, redistributing human groups across continents and occasionally bringing them into contact.

Within this environment, survival depended not only on biology but also on culture. The ability to transmit knowledge across generations—how to make tools, track animals, navigate landscapes, and cooperate within groups—became a critical factor. Cultural traditions could spread faster than genetic changes, allowing populations to adapt behaviorally even when biological evolution proceeded more slowly.

This cultural dimension complicates attempts to define exactly when "modern humanity" emerged. Anatomical features such as skull shape and skeletal structure can be traced through fossils, but behaviors such as language, symbolism, and ritual leave far subtler traces. Some of the earliest evidence of symbolic activity appears relatively late in the archaeological record, but it is difficult to know how long such behaviors existed before they left durable material evidence.

It is possible that cognitive capacities developed gradually within populations long before they became visible through art or complex artifacts. If so, the archaeological record may represent moments when those capacities became widespread enough—or expressed in durable enough forms—to survive across millennia. In this sense, the emergence of modern culture might resemble a slow ignition rather than a sudden spark.

Recognizing the interplay of biology, culture, and environment leads to a more nuanced understanding of human evolution. The story is no longer about a single lineage climbing toward modernity. Instead, it becomes a history of interconnected populations navigating a constantly changing world. Some lineages flourished for long periods before disappearing. Others merged with neighboring groups. A few expanded widely, carrying their genes and cultural traditions across continents.

Homo sapiens emerged from within this dynamic context. The ancestors of modern humans were part of the same evolutionary landscape that produced other human lineages. Their eventual global spread was shaped by demographic growth, technological innovation, and the ability to adapt to diverse environments. But their success also unfolded through interactions with other human populations whose histories intersected with their own.

Understanding this broader picture changes how we think about the past. It reminds us that humanity was once far more diverse than it is today. The world our ancestors inhabited was not empty of human relatives; it was populated by them. Some shared landscapes with our ancestors for thousands of years. Others contributed genetic or cultural elements that may still influence modern populations.

The disappearance of these other lineages remains one of the most intriguing questions in human history. Some groups may have been absorbed through interbreeding. Others may have dwindled as environments changed or as populations fragmented. Still others may have vanished through processes that left little trace in the archaeological record.

Whatever the mechanisms, their presence reshapes the narrative of human origins. Instead of a solitary species ascending toward dominance, we see a complex family of related populations whose paths intersected across deep time. The modern human story is therefore not just the story of one lineage. It is the outcome of a broader evolutionary drama in which multiple human groups played roles.

Exploring that drama requires moving beyond simplified models and acknowledging the complexity of the evidence. Fossils, genetics, archaeology, and environmental science each provide partial insights. When combined, they reveal a past that is richer and more interconnected than earlier generations could have imagined.

In the chapters ahead, we will examine how these different lines of evidence converge to reshape the story of humanity. From the tangled structure of the human family to the surprising revelations of genetic research, each discovery adds depth to the narrative of how our species emerged from a wider human world.

The deeper we look, the clearer it becomes that the origins of humanity cannot be reduced to a single path. They are the product of many paths crossing through time.

One of the most transformative developments in the study of human origins did not come from the ground but from the laboratory. For most of the history of paleoanthropology, the past could only be reconstructed through visible remains—bones, stone tools, and the geological layers in which they were buried. These remains were invaluable, but they offered only partial glimpses into ancient lives. The emergence of genetic analysis introduced an entirely new form of evidence. Instead of studying what ancient humans left behind, scientists could begin studying what ancient humans left within us.

Every living organism carries genetic material inherited from its ancestors. This genetic code is not simply a blueprint for biological development; it is also a historical record. Mutations accumulate gradually across generations, creating identifiable patterns that allow researchers to trace relationships between populations. By comparing the DNA of people living today, scientists can reconstruct aspects of population history stretching back tens of thousands of years.

In recent decades, this capability has expanded dramatically with the development of advanced sequencing technologies. It is now possible to examine entire genomes and identify subtle genetic variations shared across different populations. These variations reveal pathways of migration, population bottlenecks, and episodes of contact between groups that might otherwise remain invisible.

The implications of this genetic evidence have been profound. One of the earliest and most striking discoveries was that many modern human populations carry traces of DNA inherited from other ancient human groups. These genetic fragments are small, often representing only a tiny percentage of a person's genome. Yet their presence is unmistakable. They testify to encounters in the distant past between populations that had once been considered entirely separate.

Such discoveries reshaped the narrative of human evolution. The earlier model of clean replacement—where modern humans expanded outward and displaced all other human forms—began to give way to a more intricate picture. In this revised view, the expansion of modern humans involved interactions with existing populations rather than simple replacement. Some of these encounters led to the exchange of genetic material, leaving faint but measurable signatures that persist today.

Genetics also revealed something even more surprising: the existence of ancient human populations that were almost entirely unknown from fossils. Certain patterns in modern genomes indicated contributions from lineages that could not be explained by any known ancestral group. These signals pointed toward the presence of additional human populations—sometimes referred to as "ghost populations"—whose remains have either not been discovered or have not yet been recognized as belonging to distinct lineages.

The discovery of such genetic ghosts underscores the incompleteness of the physical record. Fossils provide direct evidence of anatomy, but they represent only those individuals who died under conditions favorable for preservation. Entire populations could exist for thousands of years without leaving identifiable skeletal remains. Genetics, by contrast, can preserve traces of these populations within the genomes of their descendants.

The result is a powerful new method for reconstructing the past. By examining the distribution of genetic markers across modern populations, researchers can infer events that occurred long before written history. These include ancient migrations across continents, population expansions following climatic changes, and episodes of interbreeding between human groups.

However, genetics does not offer a perfect window into the past. Like all scientific tools, it has limitations. Genetic signals become diluted over time as populations mix and recombine. Small segments of inherited DNA may break apart across generations, making them harder to detect. Interpreting these signals also requires complex statistical models that rely on assumptions about mutation rates, population sizes, and migration patterns. Different models can sometimes produce different estimates for the timing of ancient events.

Despite these challenges, the genetic revolution has fundamentally altered our understanding of human origins. It has revealed that humanity's past is far more interconnected than once believed. Populations did not simply replace one another in a clean sequence. They met, interacted, and sometimes merged. Their histories became intertwined in ways that only now can be reconstructed through the molecular record.

Another insight provided by genetics concerns adaptation. Certain genetic variants appear more frequently in populations that have lived for long periods under specific environmental conditions. For example, populations living at high altitudes often possess genetic adaptations that help them cope with reduced oxygen levels. Similar patterns appear in populations adapted to cold climates, tropical diseases, or specialized diets.

These adaptations illustrate how biological evolution and environmental pressures interact. When populations encounter new environments, natural selection may favor genetic variants that improve survival under those conditions. Over many generations, these variants can become widespread within the population.

In some cases, genetic research suggests that beneficial adaptations may have entered modern human populations through contact with other human groups. If two populations interbreed, advantageous genetic variants from one group can spread into another. This process introduces an additional dimension to the story of human evolution: adaptation not only through internal mutation and selection but also through exchange between populations.

Such exchanges highlight the cooperative aspect of evolutionary history. Survival in harsh environments often depends on the ability to integrate useful innovations, whether cultural or biological. Populations that maintained contact networks and exchanged knowledge or genetic traits may have been better equipped to respond to environmental challenges.

The genetic perspective also emphasizes the importance of demographic processes. Population size, migration patterns, and social structure all influence how genetic traits spread across generations. A small founding group moving into a new territory may carry only a fraction of the genetic diversity of the population it left behind. Over time, the descendants of that group can expand and shape the genetic makeup of entire regions.

Similarly, population bottlenecks—periods when a population's numbers decline sharply—can dramatically reshape genetic diversity. When a population recovers from such a bottleneck, the genetic traits carried by the survivors may become disproportionately represented in future generations. These demographic fluctuations leave detectable signatures in genetic data, allowing researchers to reconstruct episodes of expansion and contraction across deep time.

Another remarkable advance in recent years has been the recovery of ancient DNA directly from archaeological remains. Under favorable conditions, fragments of genetic material can survive within bones or teeth for tens of thousands of years. By sequencing this ancient DNA, scientists can compare the genomes of ancient individuals with those of modern populations.

This approach provides direct snapshots of the genetic landscape at specific moments in the past. It allows researchers to track how populations changed over time, how migrations reshaped regional genetic patterns, and how interactions between groups contributed to the emergence of modern humanity.

Yet ancient DNA is itself limited by preservation conditions. Genetic material degrades rapidly in warm and humid environments, meaning that some regions critical to human evolution yield little recoverable DNA. As a result, genetic reconstructions remain incomplete, particularly for areas where early human populations likely thrived.

Even with these limitations, genetics has become an indispensable partner to archaeology and paleoanthropology. Each discipline provides a different kind of insight. Fossils reveal anatomy and physical presence. Archaeological artifacts illuminate cultural behavior and technological innovation. Genetics traces relationships, migrations, and interactions that might otherwise remain hidden.

Together, these fields create a richer and more nuanced narrative of human origins. They reveal a species whose history is not defined by isolation but by connection. The ancestors of modern humans moved across continents, encountered other populations, exchanged genes and knowledge, and adapted to environments ranging from tropical forests to icy tundra.

In this broader perspective, humanity emerges not as a solitary line but as part of a dynamic evolutionary network. Our species carries the legacy of countless interactions across thousands of generations. Some of these interactions are visible through fossils and tools; others survive only in fragments of DNA passed down through time.

The study of human genetics therefore does more than reconstruct migration routes or identify ancient relatives. It reshapes our understanding of identity. Each genome contains echoes of journeys taken by distant ancestors—journeys through landscapes that no longer exist and through encounters that shaped the biological inheritance of modern populations.

The deeper researchers explore this genetic archive, the clearer it becomes that the human story is not a simple narrative of emergence and dominance. It is a complex tapestry woven from the lives of many populations, some known and others still hidden within the record.

The continuing exploration of this genetic history promises to reveal even more about the origins of our species. As sequencing technologies improve and new samples are recovered, previously unknown chapters of human evolution may come into view.

Each new discovery brings us closer to understanding how the many threads of ancient humanity came together to form the world's surviving human lineage.

The story of human origins is often told as if it has already been completed. Textbooks present tidy timelines. Museum exhibits guide visitors through ordered galleries of skulls and tools. Popular documentaries narrate the rise of our species as a sequence of discoveries that gradually solved the mystery of where we came from. Yet beneath this polished presentation lies a reality that researchers understand well: the story is still unfolding.

Scientific knowledge is not a static archive of facts. It is a living process in which ideas are continually tested against new evidence. As methods improve and discoveries accumulate, interpretations change. Some conclusions grow stronger with time, while others are revised or replaced. The study of human evolution is especially susceptible to these shifts because the evidence itself is fragmentary and widely dispersed across geography and time.

Every new fossil discovery, every improvement in dating techniques, and every breakthrough in genetic sequencing adds another piece to the puzzle. Occasionally these pieces confirm what researchers already suspected. At other times they challenge assumptions that once seemed secure. In recent decades, several such discoveries have forced scientists to rethink fundamental aspects of the human story.

One of the most important lessons emerging from these revisions is that the past rarely conforms to simple narratives. Evolutionary change does not proceed along a single path, and cultural development does not always follow a predictable progression. Instead, human history appears to be shaped by a combination of biological adaptation, environmental change, social interaction, and chance.

Climate, for example, has repeatedly altered the conditions under which human populations lived. During the Ice Age, glaciers advanced and retreated across large portions of the Northern Hemisphere. Sea levels dropped dramatically when water became locked in ice sheets, exposing land bridges that connected regions now separated by oceans. When the climate warmed again, those same land bridges vanished beneath rising seas.

These environmental shifts influenced migration patterns, resource availability, and population distribution. Groups that once lived in fertile landscapes sometimes found themselves facing harsher conditions. Others gained access to new territories that had previously been inaccessible. The movement of human populations across continents was therefore closely linked to the changing rhythms of the planet itself.

Within these shifting landscapes, survival depended not only on biological adaptation but also on cultural ingenuity. Humans developed technologies that allowed them to exploit diverse environments. Stone tools became more specialized, enabling more efficient hunting and processing of resources. Fire provided warmth, protection, and the ability to cook food. Over time, increasingly sophisticated social systems allowed groups to coordinate activities and transmit knowledge across generations.

These cultural innovations were as significant as biological evolution in shaping the human journey. Knowledge accumulated gradually within communities, forming traditions that could persist for thousands of years. Skills learned by one generation could be refined and passed to the next, allowing cultural change to occur more rapidly than genetic evolution alone.

However, cultural knowledge is also vulnerable. When populations decline or disperse, traditions can be lost. Languages disappear, skills vanish, and social structures collapse. Archaeological evidence suggests that human populations experienced several demographic bottlenecks—periods when numbers dropped sharply due to environmental stress or other factors. During such times, much of the accumulated cultural complexity of earlier generations may have been reduced or forgotten.

This pattern of cultural expansion and contraction adds another layer of complexity to the story of human origins. Instead of a steady march toward greater sophistication, the archaeological record sometimes reveals cycles of innovation, decline, and renewal. Technologies appear, disappear, and reappear in different regions. Cultural traditions spread across landscapes and then fade as populations move or vanish.

Understanding these cycles requires recognizing the fragile nature of human societies in deep time. For most of our species' existence, humans lived in small, mobile groups. Their survival depended on intimate knowledge of their environments and on cooperation within their communities. The collapse of even a few groups could mean the loss of entire knowledge systems.

In this context, the emergence of modern civilization appears less like an inevitable outcome and more like the product of numerous converging factors. Agriculture, permanent settlements, and complex social hierarchies developed only within the last ten thousand years—a brief moment compared to the hundreds of thousands of years during which humans lived as hunter-gatherers.

Yet even the rise of agriculture and cities did not erase humanity's deeper heritage. The biological and cultural foundations of modern societies were established long before written history began. The capacities for language, symbolic thought, cooperation, and innovation evolved gradually within earlier populations whose lives remain only partially visible in the archaeological record.

As researchers continue to explore this distant past, they encounter evidence that challenges the boundaries of traditional narratives. Genetic data reveals unexpected patterns of contact between populations. Archaeological discoveries push back the dates of cultural innovations once thought to be relatively recent. Fossils discovered in new regions suggest that human lineages may have been more geographically widespread than previously believed.

Each of these findings contributes to a growing recognition that the story of human origins cannot be confined to a single explanatory framework. Instead, it must integrate insights from multiple disciplines. Paleoanthropology provides anatomical evidence. Archaeology reconstructs cultural behavior. Genetics reveals patterns of ancestry and interaction. Climate science reconstructs the environmental context within which human populations evolved.

When these perspectives are combined, the result is a richer and more dynamic picture of the past. Humanity appears not as a solitary species progressing toward modernity but as part of a complex web of relationships extending across time and geography. Our ancestors interacted with diverse environments and with other human populations whose histories intersected with their own.

This realization invites a broader reflection on identity. Modern humanity often imagines itself as uniquely separate from the rest of nature. Yet the evidence of evolution reminds us that we are deeply connected to the biological world. Our species emerged from the same processes that produced all other life forms. The traits that define us—intelligence, creativity, cooperation—are the outcome of countless generations navigating the challenges of survival.

At the same time, the complexity of the human past underscores the limits of our current knowledge. Many questions remain unanswered. How many human lineages once existed? How frequently did populations encounter and interact with one another? What cultural traditions flourished in regions that left little archaeological trace? And how many chapters of human history remain hidden beneath sediments or encoded in genetic patterns we have not yet fully interpreted?

These uncertainties should not be seen as weaknesses in the scientific understanding of human origins. On the contrary, they reflect the vitality of a field that continues to evolve as new evidence emerges. The willingness to revise conclusions in light of new discoveries is one of the defining strengths of scientific inquiry.

The purpose of this book is to explore that evolving story in greater depth. Rather than presenting human evolution as a finished narrative, it approaches the subject as an ongoing investigation. By examining evidence from fossils, genetics, archaeology, and environmental research, the chapters ahead seek to illuminate the many threads that converge in the emergence of modern humanity.

Some of these threads confirm established knowledge. Others raise new questions about the complexity of our past. Together they reveal a species whose origins are both ancient and dynamic—a species shaped by adaptation, interaction, and resilience across vast stretches of time.

In exploring these themes, we return to the fundamental question that has fascinated thinkers for centuries: *What does it mean to be human?* The answer is not contained in a single discovery or a single theory. It emerges gradually from the combined efforts of researchers across disciplines who continue to uncover new evidence about our shared past.

The deeper we explore that past, the clearer it becomes that humanity is the product of an intricate evolutionary journey. Our species carries within it the legacy of many ancestors, many environments, and many encounters that shaped the course of history long before written records began.

Understanding this journey does more than satisfy curiosity. It offers perspective on the extraordinary resilience of our species and on the fragile conditions that allowed human societies to flourish. It reminds us that the story of humanity is still being written, not only in laboratories and archaeological sites but also in the choices we make about how to understand and preserve the world we inhabit.

In that sense, the exploration of human origins is also an exploration of our future. By studying how our ancestors adapted to changing environments and interacted with one another across vast distances, we gain insight into the forces that continue to shape humanity today.

The chapters that follow invite readers to look beyond simplified narratives and engage with the complexity of the human story. They present a past that is both more intricate and more fascinating than the familiar outlines often suggest—a past that continues to reveal new dimensions as science advances.

Our origins are not a closed chapter. They are a living inquiry into the deepest roots of what it means to be human.

CHAPTER 1 — EVOLUTION'S BLIND SPOTS

Most people first meet human evolution as a neat story with a satisfying arc. Somewhere in the distant past there are ape-like ancestors; then, as if stepping-stones were placed in a straight line across a river, a sequence of increasingly human forms appears—stooped to upright, crude to clever, animal to civilized. The story ends with us. This version feels intuitive because it matches a modern habit of explanation: take a vast, messy process and convert it into a single plot that can be told quickly, remembered easily, and repeated without confusion. We are trained to expect simplicity as a sign of understanding. Yet in the deep past, simplicity is often the opposite of accuracy.

The "too simple" human story is not usually presented that way because someone sat down and decided to deceive the public. It happens for more ordinary reasons, and those reasons are powerful enough to shape an entire civilization's memory. Education systems must compress. Documentaries must entertain. Museums must guide visitors along a corridor of clarity. Even scientists, when speaking to non-specialists, must choose what to emphasize and what to leave out. Complexity is expensive: it costs time, attention, and the willingness to sit with uncertainty. The deepest reason the story becomes simple is that simplicity reduces cognitive strain. It turns an overwhelming amount of data into something that feels like a picture rather than a blizzard of fragments.

But the moment you try to place real evidence into that picture, it buckles. Human evolution is not one tidy lineage marching forward. It is a shifting landscape of populations, migrations, bottlenecks, expansions, extinctions, and interbreeding—some of it visible in bones, some of it visible only in the molecular record, and some of it still invisible because the evidence did not survive. The simplified version persists because it is narratively satisfying, not because it is the only reasonable interpretation of what happened.

A simple story also serves a social purpose. Civilizations like origin myths that feel stable. Whether the myth is sacred or scientific, the structure is similar: a beginning, a development, a culmination. In older cultures, gods shaped humans from clay, breathed life into them, and set them in a world designed for meaning. In modern culture, a secular origin story often replaces the sacred one, but it inherits the same appetite for coherence. We want our beginnings to justify our present. We want the past to feel like a staircase that leads to us. The trouble is that evolutionary history is not obliged to provide moral comfort. It is not designed as a biography. It is an unfolding of chance, constraint, and adaptation across immense spans of time.

There is another pressure that forces simplification: the way human beings process categories. Our minds like clean boundaries. We are tempted to divide the past into "species," "stages," and "eras," and then treat those labels as if they were separate rooms in a museum, each with a clear doorway. But biology, especially in deep time, is not always so polite. Evolution often produces continua rather than boxes. Populations can look different while still exchanging genes. Groups can separate and then reconnect. Traits can appear in one place and spread through movement and interbreeding rather than emerging independently in a neat sequence. When the public hears "species," many imagine a wall. In reality, the wall is sometimes a fence, and sometimes it is just a line drawn on a map after the fact.

The simplified narrative also reflects a historical bias in the kinds of evidence we used to have. For much of the twentieth century, the fossil record dominated public understanding because it was tangible, dramatic, and photographable. A skull in a display case feels like direct access to prehistory. But fossils are rare, unevenly distributed, and brutally selective. They preserve what happened to die in the right place under the right conditions. They preserve almost nothing of behavior, language, symbolic thought, and social structure—the very things that make "human" feel like a meaningful category. And even when bones survive, they often survive as partial fragments that must be interpreted through comparative anatomy and inference. A fossil can be a window, but it is also a keyhole.

Because fossils are fragmentary, early narratives relied on a kind of stitching. A jawbone here, a cranium there, a tool assemblage nearby, a date estimate from surrounding sediments. From these pieces, researchers constructed a plausible sequence. The construction was often brilliant, and it pushed knowledge forward. But it also encouraged a linear presentation. If you only have a handful of snapshots from a thousand-mile journey, you will be tempted to arrange them into a route that looks continuous. You will assume the traveler walked from one snapshot to the next, even if, in truth, the traveler wandered, doubled back, split into groups, and met other travelers whose photos were never taken.

The modern era of genetics complicates that stitched story in ways that are both exhilarating and unsettling. DNA does not behave like a polite confirmation of the fossil record. It behaves like a witness who remembers things nobody wrote down. Genetics reveals relationships that bones cannot show and events that leave no obvious trace in stone. It exposes interbreeding between groups that were once presented as separate evolutionary branches. It suggests the existence of "ghost" populations—ancient groups that contributed genes to modern humans but are not yet clearly identified in the fossil record. It forces us to admit that what we call “modern humans” did not arise in isolation, sealed off from other human types, but in contact, exchange, and sometimes intimate blending.

This is where the simplified story becomes not merely incomplete but actively misleading. When people are taught that Homo sapiens replaced other hominins as if sweeping them aside in a clean wave, they inherit an expectation: a single victor, a single line, a single ascent. Yet the genetic record implies a world in which different human groups overlapped for long periods, met, competed, shared landscapes, and in some cases merged. The reality is less like a ladder and more like a braided river system, where streams separate and reunite, carrying sediments from different sources into a single downstream flow.

Still, even with genetic evidence available, the public story often remains simple because complexity threatens a comforting identity. Many people unconsciously want “fully human” to mean “pure,” as if our lineage must be unmixed to be legitimate. But nature does not share that preference. Hybridization is common across life. It can be a source of resilience and novelty. If our species carries traces of other human types, that does not make us less human. It makes us more historically real. It means we are not a solitary miracle perched above the rest of the family tree. We are a surviving, changing population that carries the marks of contact with neighbors who were also human in their own ways.

There is also an institutional reason the story is presented too simply: science must be cautious, and caution often looks like minimalism to outsiders. Researchers are trained to avoid claims that run ahead of evidence. They tend to focus on what can be measured, replicated, and defended. In public communication, this caution turns into a narrow corridor of “safe statements,” and everything outside the corridor is either left out or treated as fringe. That is not always bad. It protects against fantasy and overreach. But it can also produce a strange effect: the public receives a clean narrative that reflects what is least controversial, not what is most true in full detail. The result is a story with the edges trimmed off. And the edges are often where the most interesting questions live.

Anomalies are especially vulnerable to this trimming. If a fossil does not fit the expected pattern, there is a temptation—sometimes conscious, often unconscious—to treat it as an error of dating, an odd pathology, a misinterpretation, or an outlier that can be safely ignored until later. If a genetic signal implies a population we cannot name, it becomes an abstract statistic rather than a living group that once existed. If evidence is ambiguous, it is postponed. None of this requires malice. It is simply how a disciplined field protects itself from collapsing into speculation. Yet the discipline has a side effect: the public's version of the story becomes smooth, while the real story remains jagged.

This jaggedness matters because human evolution is not merely a scientific topic. It is a background theory of identity. People carry it as a silent assumption about who they are and what they are capable of. A simplified narrative can subtly flatten the range of possibilities. It can make the emergence of cognition feel like a slow, inevitable climb rather than a complicated interplay of biology, environment, culture, and contact. It can make human uniqueness feel obvious, as if it were always present and simply waited to be expressed. In reality, the traits we consider most "human" may have arrived in pulses, spread unevenly, and been amplified by social transmission as much as by gene frequencies.

Here we approach one of the most sensitive points: the question of missing chapters. If human history includes admixture events, lost populations, and abrupt transitions, then the story is not just "simple but true." It becomes "simple because we have been narrating around the difficult parts." That framing can slide too easily into paranoia—the idea that "they" are hiding the truth. A more grounded interpretation is that complexity threatens consensus, and consensus is what gets printed in textbooks. When you must teach millions of students, you teach what is broadly agreed upon and easy to test, not what is unresolved and hard to picture. That is how institutional knowledge works. It is not a conspiracy; it is a filter.

And yet filters can distort reality. When the filter becomes the story, the public ends up believing that the past itself was simple. That belief then feeds back into what people are willing to consider plausible. A complex, hybridized origin begins to sound like a radical claim, even when it is increasingly supported by genetic research. The mind resists it because it violates the narrative it was given at the beginning. Once a simple story is internalized, anything more nuanced feels like an attack on certainty.

This volume is built on the premise that the jagged edges deserve attention—not because every anomaly is a revelation, and not because every myth is literal history, but because human origins appear to be more intricate than a single-file march. There may be chapters we misunderstand because the evidence is subtle, dispersed, or difficult to interpret. There may be chapters we have only glimpsed because the fossils are absent and the genetic signals are partial. There may be chapters that survive indirectly—in patterns of adaptation, in traces of admixture, in the distribution of traits, and perhaps even in the persistent human tendency to tell stories of "others" who were not quite human, who taught, mingled, vanished, or were absorbed.

To say "the story is presented as too simple" is not to say science is worthless. It is to say science is alive, and living things have unfinished limbs. The honest posture is not to discard the mainstream model, but to recognize its current shape: a map drawn with incomplete terrain, updated in patches, revised as new instruments reveal what could not be seen before. The point is not to replace one myth with another, but to admit that our origin story is still under construction—and that the most consequential pieces may lie precisely where the old narrative was most eager to smooth, simplify, and move on.

The first step, then, is to become comfortable with a different kind of clarity: not the clarity of a straight line, but the clarity of a complex system. When we look again at human evolution with that readiness, we begin to see why the "too simple" story took hold—and why it can no longer hold the full weight of what we are learning about our past.

The simplest evolutionary story does more than compress evidence. It adds a direction. It quietly suggests that time itself is an escalator carrying life upward, from lesser forms to greater ones, until it arrives—almost inevitably—at modern humans. This is the myth of linear progress, and it is one of the most persistent habits of thought in the modern world. It doesn't always announce itself with grand language. Sometimes it hides in phrases that sound harmless: "more advanced," "primitive," "higher," "lower," "missing link," "the next step." But the structure is always the same. The past is imagined as a straight road, and every earlier species is treated as a rough draft leading toward us.

In biology, however, evolution has no built-in arrow pointing at improvement. It doesn't work toward a goal. It doesn't know what will count as "better" a million years from now. It is simply the changing of populations across generations under pressures that fluctuate, collide, and sometimes vanish. Traits that are advantageous in one environment can become liabilities when the environment changes. Solutions that work brilliantly in one niche can fail in another. Survival is not a moral reward for superiority; it is a temporary alignment between a living system and the conditions surrounding it. To call that alignment “progress” is to smuggle in a human value—our love of upward narratives—where nature offers only adaptation.

The myth persists because it feels emotionally right. In most human societies, “progress” is a core civic religion, whether acknowledged or not. We tell ourselves that knowledge accumulates, technology improves, and the future becomes brighter, more capable, more refined. Even when we fear collapse, we imagine collapse as a deviation from an expected upward path. That expectation then spills backward into how we interpret the deep past. We imagine earlier humans as less complete, less intelligent, less aware, because that matches the story we like to tell about history itself. But evolutionary time is not the same as civilizational time. Culture can accelerate in ways that biology cannot, and biology can change in ways that do not map neatly onto “better.”

Human evolution is especially vulnerable to the illusion of progress because we judge everything by our own traits. We look for bigger brains, more complex tools, more symbolic artifacts, more evidence of planning, and then arrange the record as if it were a curriculum. The earliest chapter becomes an introduction, the next a midterm, the next a graduation ceremony, and finally the modern world is the diploma. Yet the evidence does not obligingly line up in that order. There are long periods of stability in which technology changes little. There are bursts of innovation followed by reversals. There are regions where tool traditions remain consistent for vast spans, while other regions show rapid shifts. Sometimes the most striking changes are not biological at all but social—networks, cooperation, teaching, imitation, and the accumulation of shared skills across generations.

Linear progress also distorts our view of other human types. If you assume a ladder, then anyone not on the top rung becomes a failed attempt. Neanderthals become "brutish cousins," Denisovans become shadowy intermediates, archaic humans become stepping-stones. But in their own environments, these were not half-formed. They were complete populations with their own adaptations, their own survivals, and likely their own cultural richness. The fact that they are gone does not prove they were inferior. Extinction often has less to do with weakness than with circumstance: changing climates, shrinking habitats, fragmented populations, diseases, competition, and the unpredictable consequences of contact. A small population can be wiped out by a run of bad years, even if it is well adapted. A well-adapted population can be absorbed into another, leaving little fossil trace of its final centuries. And sometimes a group disappears not because it failed, but because its genes and traits persist inside the survivors.

This last point is where linear progress becomes almost deceptive. If you imagine that one species replaces another, you imagine clean borders: they vanish, we appear. But the more we learn about human prehistory, the more it resembles a complex demographic drama rather than a heroic replacement. Populations meet and exchange genes. They sometimes merge. They sometimes carry each other's advantageous traits forward. The ladder image struggles to accommodate this because the ladder demands separation. A web, or a braided river, can accommodate it easily. The myth of progress is not just about direction; it is about purity of sequence. It wants ancestry to be clean.

Even our language betrays the myth. Consider how easily we slip into describing evolution as if it were a craftsman upgrading a design: "selection favored," "traits evolved for," "this feature developed in order to." These phrases can be useful shorthand, but they create the impression of intention. In truth, traits arise through variation, and selection filters what works under current conditions. No one is designing the feature. No one is guiding it toward a final version. And crucially, what "works" can be something as simple as leaving slightly more descendants. A trait doesn't need to be "better" in any grand sense; it only needs to be good enough in a specific place, at a specific time, under specific constraints.

Once you loosen your grip on the progress myth, you start to see the human story differently. You stop asking, "Which species was closest to us?" and begin asking, "How did different populations adapt to their worlds, and how did their worlds reshape them?" You stop assuming that a larger brain always means a smarter life, and begin paying attention to what brains actually do: regulate energy, manage social complexity, coordinate movement, store learned skills, navigate landscapes, interpret signals, and build internal models of reality. Bigger is sometimes useful, but brains are expensive organs. They demand calories, impose risks during childbirth, and require long childhoods. They do not grow merely because time passes; they grow because the costs are outweighed by benefits under specific conditions.

And then there is the problem of the "modern." The progress myth quietly tells us that modern humans are the finished product. Yet the very word "modern" is a cultural label that we project onto biology. Anatomical modernity does not guarantee behavioral modernity, and behavioral modernity is itself a contested concept because it depends on what you count as “fully human.” Is it art? Burial? Symbolic ornaments? Complex language? Long-distance trade? A theory of mind? Fire control? Cooperative hunting? The record suggests these traits are not a single switch that flips on once. They appear at different times and in different places, sometimes in partial form, sometimes in richer form, sometimes leaving no trace because much of culture is made of perishable materials—wood, fiber, hide, song, spoken story. If we insist on a straight ladder, we will keep arguing about which rung counts as “true humanity.” If we allow a mosaic, we can accept that humanity may have emerged as a patchwork of capacities distributed unevenly and assembled through interaction.

Linear progress is also reinforced by the way we picture time. We imagine the past as emptier than the present, as if the world began simple and gradually filled with complexity. But deep time was full. It was full of ecosystems, full of predators and prey, full of changing climates and landscapes, full of hazards and opportunities. The human niche did not emerge in a vacuum. Early humans, and the other human groups alongside them, were already living in worlds shaped by powerful constraints. They had to survive ice-age oscillations, shifting rainfall patterns, advancing and retreating deserts, changes in sea level, and the movements of large animals across vast ranges. They faced competition not only from other humans but from predators and from the environment itself. In such contexts, “progress” is not the right word. The right word is responsiveness.

The myth also sets a trap for interpreting archaeological silence. If you expect a steady upward trajectory, then gaps in evidence feel like missing steps. The mind immediately fills them. It assumes that what came before must have been cruder, and what came after must have been refined, so the gap must contain intermediate stages. Sometimes that is true. But sometimes gaps exist because evidence is absent, not because behavior was absent. Coastal settlements drowned under rising seas. Open-air sites eroded. Organic artifacts rotted away. Entire regions remain underexplored or politically inaccessible. The ladder model pressures us to imagine what must have happened, instead of admitting what we do not know. And once imagined, those "must have" steps can harden into assumed reality.

There is, however, a deeper reason the myth of linear progress clings to human evolution: it protects a particular idea of human exceptionalism. If we are the endpoint of an upward climb, then our dominance feels like destiny. It becomes easier to believe that our intelligence is a unique flame that suddenly ignited, rather than a set of traits that emerged gradually and unevenly, perhaps with contributions from multiple human groups. It becomes easier to imagine that other hominins were simply waiting to be replaced, rather than living complex lives in parallel. It becomes easier to explain why we are here and they are not, without having to confront contingency—the unsettling fact that if conditions had shifted differently, another human type might have persisted, or we might not have.

This is where the myth begins to collide with the core theme of missing chapters. If human history includes admixture, sudden expansions, bottlenecks, and lost populations, then the ladder model is not merely simplistic. It becomes an obstacle. It prevents us from asking the right questions. It tells us to look for a single line, a single progression, a single narrative of improvement. But the evidence increasingly suggests a more complicated picture: multiple human groups interacting across time, with traits moving through contact, with population histories shaped by climate shocks and migrations, and with cultural leaps that may not align neatly with anatomical changes.

None of this requires a leap into fantasy. It simply requires adjusting our mental image of what evolution is. Evolution is not a staircase. It is a branching process with dead ends, mergers, and long plateaus. It produces specialists as often as it produces generalists. It can shrink brains as well as enlarge them. It can favor strength, endurance, social bonding, or metabolic efficiency depending on conditions. The story of humans, in particular, may be a story of adaptability itself—of surviving by switching strategies, by learning from one another, and by carrying forward whatever worked, regardless of which population first developed it.

When you let go of linear progress, you also become more sensitive to the possibility that ancient people—and ancient human types—cannot be neatly ranked. Some may have been superbly adapted to cold climates, to rugged terrain, to low-oxygen altitudes, to particular diets, or to certain social structures. Some may have carried genetic variants that later became valuable to us. Some may have had cultural traditions that left few durable traces. The absence of an artifact does not equal the absence of a mind.

The progress myth is seductive precisely because it turns all of this into a single, reassuring curve. But the real curve is not smooth. It is jagged, and its shape depends on where you stand and what you measure. The human story becomes more interesting—more human, in fact—when we accept that our origins were not a straight ascent but a complicated survival, carried through shifting worlds by beings who were never simply "primitive," only different, and often far more capable than the ladder model allows.

A fossil looks like certainty. It is weight, mineral, contour—something you can hold in your hands and photograph under museum lights. Compared to ideas about vanished peoples and invisible genes, a fossil feels like the past made solid. That feeling is why the fossil record became the backbone of the public story. It offers the drama of discovery and the authority of the tangible. But the fossil record, for all its importance, is also an archive of absence. It is not a full biography of our origins. It is what remained after a brutal filtering process, and it cannot show many of the very things we most want to know about the humans who came before us.

The first limitation is simple and devastating: most living things do not fossilize. Most bodies return to the soil, are consumed, scattered, dissolved, or destroyed by weather and time. For fossilization to occur, death must meet rare conditions—rapid burial, low oxygen, the right chemistry, minimal disturbance, and then an even rarer chain of luck that preserves the specimen long enough to be discovered. The result is that the fossil record is not a census. It is a collection of accidents. Entire populations can exist for thousands of years and leave nothing we will ever find. Entire regions can be full of life and leave almost no fossils because their soils, climates, and erosion patterns are hostile to preservation.

Even when fossils do preserve, they preserve unevenly. Hard tissues—teeth, dense bone—survive far more often than fragile structures. This creates a distorted portrait. Teeth tell us about diet, development, and sometimes ancestry, but they cannot tell us about social bonds, beliefs, language, or music. A skull can suggest brain size and facial anatomy, but it cannot reveal the inner world that skull once contained. The fossil record is biased toward what is durable, not toward what is meaningful. It rewards the parts of a body that mineralize well, not the parts of a life that mattered most to the one living it.

Then there is the problem of time resolution. Fossils do not come with timestamps. They are dated through surrounding sediments, associated volcanic layers, chemical signatures, and comparative methods—tools that can be powerful but often come with ranges and uncertainties. A fossil might be placed within a broad time window, and that window may overlap with others in ways that make cause-and-effect stories tempting but fragile. When the public hears "this species lived from X to Y," it often imagines a clear timeline. In reality, those boundaries are fuzzy. A species does not begin at midnight on a certain date and vanish at sunrise on another. Populations expand and contract. Traits spread gradually. Local extinctions happen while other pockets persist. The fossil record captures scattered points along that shifting front.

Geography makes this even more complicated. Fossil sites cluster where conditions favor preservation and where researchers have access. Some landscapes expose ancient strata easily; others bury them under vegetation, cities, or modern sediments. Political stability, funding, and research traditions shape what gets explored. If a region is understudied, its past can appear emptier than it truly was. If a region happens to preserve remains well, it can dominate our narratives. The fossil record is partly a record of the Earth's geology and partly a record of human research history. That means what we "know" about human evolution has always been shaped by where we have looked and where the ground has been willing to give us its secrets.

Even in well-studied regions, fossils rarely come with the context we crave. A skeleton in articulation—bones still arranged as they were in life—is rare. More often, remains are scattered, broken, and mixed. A jaw may be found without a skull; a skull without a body. A single tooth can become a major data point because that is all that survived. From these fragments we infer species, relationships, and migrations. The inferences may be reasonable, but the public often imagines a stronger foundation than exists. A lineage built on a few key specimens can look like a smooth sequence in a textbook illustration, even when the underlying evidence is more like isolated islands in a vast sea.

And then we reach what the fossil record cannot show at all: behavior, cognition, and culture in their living richness. Bones do not fossilize conversations. They do not fossilize the tone of a mother teaching a child to shape stone, the rituals performed under a winter sky, the strategies used to negotiate conflict in a small group, or the sense of awe that might have arisen at an eclipse. The fossil record can hint at behavior indirectly—through injuries, muscle attachments, tooth wear, the presence of tools or hearths—but these are shadows on a wall. They tell us something happened, not what it meant.

Meaning is the missing dimension. Most of what makes humans recognizable as humans is not anatomy alone but symbolic life: the ability to represent, to imagine, to bind communities through shared stories and practices. That symbolic life is usually made of fragile materials—song, gesture, memory, spoken language, perishable artifacts, and rituals performed in places that leave little trace. A sophisticated social world can exist almost invisibly from an archaeological perspective. A group can have deep knowledge of landscapes, seasons, plants, animals, and stars, and leave behind nothing but a few scattered hearths. If you expect the fossil record to confirm intelligence by leaving durable monuments, you will repeatedly underestimate ancient minds.

This is why "absence of evidence" becomes such a dangerous phrase in paleoanthropology. It can be a sober reminder not to overreach. But it can also become an excuse to dismiss possibilities that are simply hard to preserve. For most of our species' existence, humans lived without cities, without writing, without stone architecture. If you judged our modern minds solely by the physical traces left by small mobile groups, you might conclude that "full humanity" arrived late. Yet the capacity for complex language and thought may have existed long before it left durable signatures. The fossil record cannot easily tell you when stories began, when metaphors were first spoken, or when music first braided emotion into communal rhythm.

The same limitation applies to social structure. The fossil record cannot show kinship systems, alliances, leadership styles, norms of sharing, or the moral codes that hold groups together. But these social systems are not decorative. They shape survival. Cooperation can determine whether a group thrives in harsh climates. Teaching and imitation can accelerate the spread of skills. Shared belief can coordinate action. These are evolutionary forces in their own right, yet their traces are mostly intangible. When we imagine human evolution as mostly anatomical change, it is partly because anatomy is what the fossil record provides most readily. We are looking under the streetlight because that is where the light is.

The fossil record also struggles with admixture—the mixing of populations. Bones can sometimes suggest mixture through intermediate traits, but anatomy is not a straightforward genetic ledger. Traits are influenced by environment, development, and complex inheritance patterns. Two populations can interbreed and still look distinct in many features. Conversely, unrelated populations can converge on similar traits under similar pressures. Without genetics, many interbreeding events are invisible. This means the fossil record can give the impression of clean separations even when gene flow occurred. In other words, the fossil record can support the illusion of a ladder because it often cannot show the braided river beneath.

It also cannot show most of what happened during population bottlenecks and expansions—those periods when small groups become founders of later populations, carrying only a subset of genetic diversity. Fossils do not come in sufficient abundance to track demographic waves with precision. You can sometimes infer changes in settlement patterns from archaeology, but the demographic details—how many people, how quickly they moved, how often they met others—often remain beyond reach. Yet these demographic patterns shape evolution profoundly. A small founder group can amplify certain traits simply by chance. A population crash can erase lineages. A rapid expansion can spread adaptations widely. The fossil record is not built to capture these dynamics in fine detail.

Even when fossils are abundant, they often fail to capture the full range of variation within a population. Modern humans vary widely in skull shape, stature, and facial features, and much of that variation exists within a single species. Now imagine trying to reconstruct our own species from a handful of fragmentary fossils scattered across continents, separated by tens of thousands of years. We might mistakenly split ourselves into multiple species based on normal variation. Or we might mistakenly combine distinct groups because we lack enough specimens to see differences. This is not a criticism of paleoanthropology; it is the nature of the data. The fossil record is sparse enough that classification is sometimes less like reading a clear script and more like interpreting a torn manuscript with missing pages.

And because classification is difficult, the fossil record cannot always answer the most emotionally charged question: who counted as "us"? The moment we begin dividing ancient humans into named categories, we risk treating those categories as essential realities rather than provisional tools. In deep time, boundaries blur. A group can be distinct enough to look unique yet close enough to exchange genes. A population can change gradually across a landscape, creating clines rather than clean breaks. The fossil record struggles to represent that fluidity because it is sampled at points, not continuously. It is a set of dots, and our minds are eager to draw lines between them.

This does not mean the fossil record is unreliable. It is indispensable. It is the only direct evidence of anatomy and physical presence across deep time, and without it we would be guessing in the dark. But it does mean we must treat it as one instrument, not the entire orchestra. Fossils tell us what bodies looked like, where some individuals lived, and sometimes how they died. They can indicate locomotion, diet, health, growth patterns, and in some cases social care when injuries suggest survival with support. That is already astonishing. Yet fossils cannot tell us how knowledge moved between groups, how myths formed, how cognition leapt forward, or how contact and interbreeding shaped the human mosaic.

If this volume is about missing chapters, then the fossil record is the first place to recognize missing pages—not as a failure of science, but as a feature of the kind of archive the Earth provides. The deep past is not a library with complete shelves. It is a landscape where a few pages survived the fire. Those pages matter, but they cannot tell the whole story.

To recover more, we need other forms of evidence—especially genetics, which can preserve relationships and contact events even when bones vanish. And we need a willingness to think in systems rather than ladders: to accept that the true narrative may be complex, hybridized, and partly hidden, not because it was deliberately concealed, but because the past itself is a brutal editor.

If fossils are the stones of the past, genetics is its echo. Bone and tooth can show you what something looked like, but DNA can show you where it came from, who it met, and how many unseen ancestors stand behind it. That is why genetics has become the most disruptive instrument in the study of human origins. It does not replace paleoanthropology, but it changes the rules of the game. It turns the human body into an archive that traveled through time, carrying traces of ancient events even when the physical record is missing. In that sense, genetics behaves like a time machine—not because it takes us back to a specific moment with perfect clarity, but because it allows us to infer deep histories from what still lives inside us now.

The power of genetics begins with a simple idea: inheritance is a record. Every generation passes on a shuffled version of what came before, and that shuffling is not random in the way people imagine. It follows statistical patterns shaped by recombination, mutation, drift, selection, and population movement. Over thousands of generations, these patterns accumulate like signatures. Some become faint, others remain bold. But the overall structure carries information about ancient bottlenecks, expansions, migrations, and interbreeding. Even without a single fossil, you can sometimes see that a population once shrank dramatically. You can see that two groups once separated and later reunited. You can see that a population received genetic input from another group that is no longer around—or no longer identifiable in the archaeological record.

This is one of the most profound shifts in how we can approach prehistory. For much of modern science, the past could only be reconstructed from external remains: bones, stones, sediments. Genetics offers something different: internal evidence, carried forward by the descendants themselves. It is as if each living human carries a palimpsest—an overwritten manuscript—where older lines have faded but not fully vanished. The challenge is learning how to read it without imagining it says more than it does.

At the most basic level, genetics helps map relationships. By comparing DNA sequences across individuals and populations, researchers can infer patterns of shared ancestry. Some parts of the genome change quickly, others slowly. Some regions are heavily shaped by selection, others more neutral. When a mutation arises, it can spread through a population or remain rare. Over time, lineages accumulate distinctive sets of variants. If two populations share many variants, they likely share recent ancestry or recent gene flow. If they differ, they may have been separated for longer. This logic—applied carefully and with appropriate models—turns genetic variation into a kind of historical signal.

There are special genetic systems that act like particularly readable timekeepers. Mitochondrial DNA, inherited mostly through maternal lines, can track female lineage histories with relative clarity because it does not recombine in the same way nuclear DNA does. The Y chromosome, passed through paternal lines, offers a paternal counterpart. These lineages can reveal branching patterns and estimate the timing of common ancestors. But they are narrow windows. They represent only two threads in an immense tapestry. Relying on them alone can mislead, because human history is not a single maternal or paternal line. It is the combined flow of countless ancestors. The real power lies in whole-genome analysis, where millions of markers can be used to detect subtle signals: admixture, population structure, and shifts in effective population size.

Admixture is where genetics becomes almost scandalous to older, simpler narratives. The public story long assumed that modern humans emerged and then replaced other hominins in clean succession. Genetics revealed something messier: modern humans carry inherited DNA from other human groups. Interbreeding was not an impossible rarity; it happened often enough to leave detectable traces. That alone forces a recalibration of how we imagine the ancient world. It suggests contact zones, overlap, and intimacy where older models imagined separation. It suggests that "humanity" in deep time was not a single isolated island but an archipelago with bridges.

Even more disruptive is the discovery of ghost populations—genetic contributions from groups that are not yet clearly known from fossils. These signals appear when patterns in modern genomes cannot be explained by known ancestral populations alone. Researchers infer that there must have been additional lineages that mixed with ancestors of modern humans. Sometimes these ghost populations are later linked to new fossil finds or to ancient DNA extracted from remains. Sometimes they remain theoretical, inferred from statistical mismatches in genetic data. In either case, the implication is unsettling: the past contained human groups we have not properly met yet. They are present in us as faint signatures without names, without faces, without stories—except the story their genes imply.

This is why genetics feels like time travel. It brings into focus events that left no obvious physical trace. A population might have existed in a region with poor fossil preservation. It might have been absorbed by another group, leaving few distinctive skeletal traits. It might have been small, scattered, and later wiped out by climate change. Fossils may never surface. Yet gene flow from that population can persist if it conferred advantages or if it became part of a later expansion. In that case, the population is gone but not erased. Genetics keeps a remnant.

Genetics also helps illuminate adaptation in ways fossils struggle to show. Certain traits are shaped by selection—meaning they spread because they improved survival or reproduction in a given environment. When researchers identify variants that are unusually common in particular populations and linked to clear physiological effects, they can infer something about historical pressures. High-altitude adaptation is one of the clearest examples: populations living for long periods in low-oxygen environments can carry variants that help their bodies cope. The presence of such variants can reveal not only adaptation but also potential sources of that adaptation, including borrowing through interbreeding. When an advantageous genetic package appears to have been introduced from an archaic human group, it reframes the story: interbreeding was not only contact; it was sometimes an exchange of survival tools.

This is important for the core theme of missing chapters because it suggests that some of what made Homo sapiens successful did not arise solely within one isolated population. Parts of our resilience may be inherited from encounters—genetic gifts, so to speak—picked up along the way. That does not diminish human ingenuity; it adds a layer of realism. In a harsh and changing world, survival often comes from combining strengths, not from purity of lineage.

But a time machine is only as good as its calibration. Genetics does not provide a cinematic replay of the past. It provides probabilistic reconstructions. There are limits, and those limits matter. One major constraint is that genetic signals degrade over time. Recombination breaks up ancestral segments in the genome, chopping long blocks of inherited DNA into smaller fragments. The further back an admixture event occurred, the more fragmented its signal becomes, and the harder it is to detect and time precisely. Mutations accumulate, but many become indistinguishable from background noise. Population movements and subsequent mixing can blur earlier structures. In this sense, the genome is not a perfect archive. It is a living, evolving system that overwrites its own history as it goes.

Another limitation is that genetic interpretation depends on models and assumptions. Researchers must infer ancient population sizes, migration rates, divergence times, and admixture events from present-day data and from ancient DNA when available. These inferences rely on statistical frameworks that can be sensitive to assumptions about mutation rates, generation times, and population structure. Different models can sometimes produce different timelines. The broad outlines may be robust, but the fine details can shift. That is why careful genetic science is cautious: it distinguishes between strong signals and plausible scenarios. It reports confidence intervals. It revises when new data appears.

Ancient DNA has transformed this landscape because it allows direct sampling of the past rather than inference from the present alone. When DNA can be extracted from human remains thousands or tens of thousands of years old, it becomes possible to compare ancient genomes directly to modern ones and to each other. This can confirm admixture events, clarify relationships, and reveal migrations that would otherwise be ambiguous. But ancient DNA has its own constraints. It survives best in cold, stable environments. Hot and humid regions—many of which were central to human prehistory—destroy DNA quickly. That means ancient DNA is geographically biased, leaving some regions and periods still largely reconstructed through indirect inference. The time machine works better in some climates than others, and that unevenness shapes our knowledge.

Despite these limitations, genetics offers something almost unprecedented: a way to detect invisible history. It can reveal demographic crashes that match climatic upheavals. It can show that populations once thought stable were replaced or reshaped. It can uncover ancient connections between regions that appear separate archaeologically. It can test hypotheses about contact. And most dramatically, it can force recognition that modern humans carry within them a complex ancestry, including contributions from other human types. In a single stroke, it undermines the fantasy of a clean line.

This is where genetics becomes not only a tool but a philosophical challenge. If the human story is a tapestry of admixture, then identity itself is not a singular origin. It is a composite. We are the descendants of multiple lines, some of which left bones and tools, some of which left only molecular traces. We are not the endpoint of a ladder; we are the current expression of an ongoing, hybridized history. That does not make the story less meaningful. It makes it more honest, and in many ways more mysterious, because it implies that our origins include interactions and inheritances we have barely begun to map.

And yet genetics also tempts overreach. Because it can detect what fossils cannot, it can invite the imagination to run ahead of evidence. A genetic anomaly can become a mythic narrative too quickly. A statistical ghost can be treated as a fully formed lost civilization. Responsible interpretation keeps the boundary clear: genetics can show that something happened—some contact, some mixing, some divergence—but it cannot always tell you the cultural circumstances. It cannot tell you whether the meeting was hostile or peaceful, whether it involved small exchanges over centuries or dramatic contact events. It cannot tell you what stories were told around the fire. It cannot tell you what people believed about themselves.

Still, as a time machine, genetics is unmatched in one crucial way: it makes forgetting harder. The Earth can erase bones. The sea can swallow coastlines. The desert can grind artifacts to dust. But if a population's genes live on in descendants, the past remains present in a literal sense. The archive walks around inside living bodies. That is why this book must lean on genetics alongside fossils, archaeology, and myth. Not because genes are destiny, but because genes are evidence—evidence that the human story is layered, hybrid, and far less tidy than the myth of linear progress ever allowed.

Every field that deals with complex evidence develops a kind of immune system. It has to, or it collapses under noise. In human origins research, that immune system is skepticism—methodical, disciplined, and usually healthy. It keeps the field from being hijacked by wishful thinking, sensationalism, and the endless human temptation to make the past say what we want it to say. But the same immune system can also misfire. When evidence does not fit prevailing expectations, the first reflex is often to treat it as an error: a misdated layer, a contaminated sample, a pathological specimen, a misidentified tool, an outlier that will dissolve once the methods are refined. Sometimes that reflex is exactly right. Yet over time, the reflex can become a habit, and the habit can become a filter strong enough to remove real signals along with the false ones.

To understand why anomalies are so often treated as errors, it helps to begin with the nature of the evidence itself. Paleoanthropology and archaeology operate at the edge of what can be known. The record is fragmentary, uneven, and full of gaps. Dating methods have uncertainties. Sites are disturbed by geology and time. Bones can be warped or broken. Artifacts can be displaced by water, animals, or human activity long after their original deposition. Even genetics—so powerful and so precise in the laboratory—must contend with contamination, degradation, and model dependence when it is used to infer deep history. In such an environment, mistakes are not rare; they are expected. The field has learned, sometimes painfully, that dramatic claims based on thin evidence can quickly collapse. So the culture of research becomes cautious, and caution expresses itself as suspicion of the unusual.

There is also a statistical reason anomalies are treated as errors: in messy data, outliers are often noise. A single odd measurement can come from a broken instrument, a mislabeled sample, or an unexpected local factor. If you build your understanding around outliers too quickly, you risk creating a story that cannot generalize. Scientists are trained to look for patterns that replicate across multiple lines of evidence. A fossil that appears too early, a tool that seems too advanced for its context, a genetic signal that does not align with existing models—these are not ignored out of arrogance. They are questioned because the cost of being wrong is high. Once a claim enters textbooks and media, it is difficult to retract cleanly. The incentive, therefore, is to demand extraordinary rigor from extraordinary findings.

But that is not the whole story. Anomalies are not only a methodological problem; they are also a psychological and social problem. The human mind is a pattern-making engine that prefers coherence. Within a scientific field, coherence takes the form of paradigms—frameworks that organize facts into a workable picture. Paradigms are necessary. Without them, evidence would be a pile of disconnected observations. Yet paradigms also exert pressure. They shape what counts as a meaningful question and what counts as a plausible answer. When an anomaly appears, it threatens coherence. It demands either an adjustment of the paradigm or a containment of the anomaly. Containment is easier. It preserves the overall structure. And so there is a subtle incentive, even among sincere and careful researchers, to interpret anomalies in ways that minimize disruption.

This incentive is amplified by the social reality of science. Careers are built on credibility. Funding is scarce. Peer review can be unforgiving. If you become associated with claims that later fail, the stain can linger. The safest route is to work within the broad consensus, to publish results that align with established models, and to treat anomalies as tentative until they are supported by overwhelming evidence. In principle, this is a rational strategy. In practice, it can create a conservative environment where certain kinds of questions are discouraged, not because they are logically impossible, but because they are professionally risky.

The result is a paradox. Science is celebrated as a system that corrects itself, and it does. But self-correction is not automatic at every moment. It often requires time, replication, and generational shifts in what researchers consider acceptable. In the meantime, anomalies can be pushed to the margins. They become footnotes, unresolved debates, or quiet controversies discussed in specialist circles but rarely translated into the public narrative. The public then receives a smoother story than the field actually contains.

Another reason anomalies are treated as errors is that they can be genuinely difficult to interpret without inventing. Take a fossil with unusual morphology. Is it a new species? An unusual individual within a known species? A hybrid? A case of pathology or developmental variation? Without a broader sample, classification becomes speculative. The responsible response is to avoid over-claiming. But the responsible response can also be framed as dismissal, especially by outsiders who expect a definitive answer. In reality, much of the field's caution is simply a refusal to pretend certainty.

This tension becomes even sharper when anomalies intersect with popular fascination. Human evolution is a magnet for mythic thinking. People want lost civilizations, hidden ancestors, and dramatic revelations. Media often amplifies the most sensational interpretation because it sells. Researchers, aware of this, can become even more defensive. They fear that an anomaly will be hijacked by exaggerated narratives. So they emphasize error-checking, alternative explanations, and methodological skepticism. Again, this is often justified. But it can create an environment where unusual data is treated almost as a threat—something to be neutralized rather than explored.

Genetic anomalies offer a modern example. When unusual signals appear in genomic data—patterns that imply unknown admixture or unexpected divergence times—the first response is typically to test whether the signal could be an artifact: contamination, sampling bias, model misspecification, or statistical noise. This is necessary. Genomics is powerful, but it is also complex, and false positives can occur. Yet as methods improve, some anomalies persist. Ghost populations remain. Unexpected admixture events become confirmed. The anomaly, once treated as error, becomes a new chapter. This is how the time machine of genetics has repeatedly expanded the human story: by turning what first looked like a glitch into a recognized reality.

There is also a deeper philosophical bias at work: the desire for neat causality. If you believe evolution proceeds mainly through gradual change, then sudden leaps look suspicious. If you believe populations remain isolated, then evidence of interbreeding looks like contamination. If you believe the timeline is well established, then early dates look like mismeasurement. In other words, anomalies are often defined not only by their unusual nature but by their contradiction of an expected rhythm. The rhythm is what we think nature "should" do. When evidence breaks rhythm, we assume the evidence is wrong rather than the rhythm.

That expectation of rhythm has a history. For decades, the public narrative emphasized gradualism and clean replacement. Those models were not invented out of thin air; they were built from the best available evidence and the theoretical tools of the time. But once they became cultural common sense, they acquired an emotional authority. They became "how evolution works." When new evidence suggests a more punctuated, networked, and hybridized story, it feels disruptive not only to science but to identity. It challenges the sense that humans are the endpoint of a clean ascent. It suggests we are composites. It suggests that other human types were not just stepping-stones but contributors. The anomaly becomes existentially uncomfortable, and discomfort can masquerade as skepticism.

This is where the language of error becomes especially seductive. If something threatens the story, it is comforting to label it a mistake. A mistake can be fixed without changing the narrative. A real anomaly, by contrast, demands rethinking. It demands admitting that our current map is incomplete. In any field, that admission is hard; in a field tied to human origins, it can feel like moving the foundation stones beneath a whole worldview.

Yet anomalies are not automatically revelations. Many truly are errors. Fossils can be misdated. Samples can be contaminated. Artifacts can be intrusive. Claims can be exaggerated. It is important to say this plainly because the alternative—treating every anomaly as suppressed truth—leads straight into fantasy. The discipline of skepticism exists for a reason. The goal is not to overthrow skepticism but to use it wisely: to distinguish between anomalies that vanish under scrutiny and anomalies that persist, replicate, and connect with other lines of evidence.

When anomalies persist, they do not always announce their meaning clearly. Sometimes they require new methods to interpret. Sometimes they require larger datasets. Sometimes they require a conceptual shift. The history of science is full of cases where anomalies were mocked or dismissed until instruments improved or a better theory emerged. In the context of human evolution, this pattern is especially relevant because the record is so incomplete and the methods are evolving rapidly. Genomics, improved dating techniques, remote sensing, and new site discoveries are constantly rewriting what is plausible. What looked like an impossibility twenty years ago may now be mundane. What looks like an outlier today may be the first glimpse of a hidden pattern.

The crucial question, then, is not whether anomalies exist—they do—but how we treat them. If we treat them only as errors, we risk missing the very chapters that could make the human story more accurate and more complete. If we treat them only as revelations, we abandon the standards that make knowledge reliable. The mature stance is to hold anomalies in suspension: to test them ruthlessly, to consider mundane explanations first, but also to remain open to the possibility that the anomaly is pointing to something real.

This volume's theme—suppressed, misunderstood, anomalous chapters—must be approached in exactly that way. "Suppressed" does not need to mean a hidden hand at work. It can mean suppressed by the practical constraints of evidence, by the conservatism of paradigms, by the social risks of reputation, and by the public's appetite for simple stories. An anomaly can be suppressed simply by being inconvenient. It can be pushed into the margins because it doesn't yet fit. And then, slowly, as more evidence arrives, the margin becomes the center. The story grows.

In the end, anomalies are treated as errors because science is trying to protect itself from error. The tragedy is that sometimes the protection comes at the cost of curiosity. The discipline that guards against fantasy can also guard against discovery if it becomes too rigid. The task is not to dismantle the immune system, but to ensure it does not attack the very signals that could reveal who we really are.

A civilization's origin story is never only about the past. It is a mirror held up to the present. It tells people what kind of beings they believe themselves to be, what they think they deserve, what they fear, and what they imagine as possible. That is why the question of human origins carries such emotional weight. It is not a neutral puzzle like the orbit of a comet. It touches identity, meaning, and the deep, often unspoken assumptions that shape how people interpret everything from morality to destiny. When the official story is presented as settled, it offers comfort. When the story becomes complex again, it can feel like the ground shifting under our feet. Yet complexity may not be a threat. It may be a truer reflection of reality. To reopen the question of who we are is not to abandon science, but to take it seriously enough to follow where it leads—even when it leads away from tidy narratives.

The first thing reopening requires is admitting that "who we are" has layers. There is the biological question: what lineages contributed to modern humans, what adaptations shaped our bodies, and what genetic events left signatures we can still detect. There is the behavioral question: how and when capacities like language, symbolic thought, long-term planning, and complex cooperation emerged and spread. There is the cultural question: how stories, rituals, and knowledge systems formed, survived, and sometimes vanished. And there is the mythic question: why human beings across the world so often remember the past in the form of beings who were not quite human—giants, teachers, sky people, gods, and hybrids. These are not the same question, but they overlap. The mistake of the simplified narrative is that it collapses them into one linear path and then declares the matter closed.

Reopening the question begins with a different posture: humility toward the archive. The deep past is not an open book. It is a partial record, preserved unevenly, interpreted through tools that change over time. Every generation of researchers inherits not only new methods but new blind spots. That is not a scandal; it is how knowledge evolves. What matters is whether the field remains flexible enough to revise its models when evidence demands revision. Over the past few decades, evidence has demanded revision repeatedly. Genetics has revealed interbreeding where older models imagined replacement. It has revealed ghost populations where older models imagined completeness. Archaeology has revealed bursts of innovation that complicate gradualist expectations. Climate science has revealed instability—rapid shifts, harsh bottlenecks, and ecological pressures that could have reshaped humanity more dramatically than older, steadier climates would allow. Even without any speculative framing, the mainstream picture is already far more intricate than the public story usually admits.

This matters because once the public story becomes "settled," it can create an illusion that real inquiry has ended. People assume we know the shape of the family tree, the timing of key transitions, and the boundaries between human types. They assume the past is mapped and labeled. In reality, we are in the middle of a revolution in methods, and revolutions do not produce final answers quickly. They produce expanding questions. They reveal how much was missing. They force new frameworks. If we take that seriously, we must treat our current narrative not as a finished sculpture but as a working model—useful, but provisional.

Reopening also requires a more honest understanding of how scientific consensus functions. Consensus is often framed as truth, but it is better understood as an agreement about what is most defensible with current evidence. It is the center of gravity, not the edge of possibility. It tends to lag behind new data because it must be stable enough to teach, publish, and build upon. That stability is valuable. But stability can also create a gap between what specialists are actively debating and what the public believes. Reopening the question of who we are means narrowing that gap. It means acknowledging that the center of gravity has been moving and will likely continue to move.

At the biological level, the reopened question begins with a simple statement that would have sounded provocative not long ago: modern humans are not a pure line. We are a composite species in the sense that our genomes carry inherited material from other human groups. Some of these groups are known; some may still be unknown or poorly defined. The implications are profound. It means we did not rise in splendid isolation. It means our survival and expansion involved contact. It means traits that helped us endure may have entered our lineage through admixture rather than arising within one "sapiens-only" pipeline. It means the very definition of "species," in deep time and in the human case, is less absolute than people assume. This is not speculation. It is an increasingly foundational part of the modern scientific picture, even if the details remain actively researched.

At the behavioral level, reopening the question forces a confrontation with how little the fossil record can tell us about mind. Anatomy is not consciousness. Brain size is not thought itself. Tool traditions are not the full range of intelligence. A people can be sophisticated in ways that leave few durable traces. If we are honest, we cannot point to a single moment when "humanity" switched on. We can point to markers—artifacts, symbolic objects, complex hunting strategies, long-distance exchange—but these markers likely represent visible peaks of capacities that existed in less visible form earlier. This means the emergence of modern cognition may not fit cleanly into the popular timeline. It may have been distributed, uneven, and dependent on social transmission as much as genetic change. In other words, it may have been as much about culture as biology.

At the cultural level, reopening means recognizing that humanity's most powerful tool is not stone or fire but shared knowledge. Knowledge can persist across generations with astonishing fidelity when embedded in ritual, story, and daily practice. It can also be lost abruptly when populations collapse, languages die, or ecological shocks scatter communities. If the Ice Age world produced repeated demographic resets—bottlenecks, migrations, extinctions—then culture may have been repeatedly interrupted. This would leave a strange pattern in the record: bursts of complexity, gaps, apparent "sudden" appearances of sophisticated behavior, and regions where knowledge seems to reemerge rather than evolve smoothly. Such a pattern would not require magic. It would require fragility—human brilliance repeatedly pushed through narrow survival funnels.

This brings us to the mythic layer, which cannot be treated as a substitute for evidence but also cannot be dismissed as irrelevant noise. Humans do not merely live through events; they remember them. They compress them into stories. They translate trauma, awe, and contact into symbolic language. When cultures across the world carry persistent themes—non-human teachers, divine unions, giants, sky beings, hybrid heroes—it is worth asking why the human imagination returns to these motifs so reliably. The cautious stance is to say these are mythic archetypes, psychological patterns, metaphors for power and difference. That may be true, and it is often the safest interpretation. But reopening the question of who we are means allowing another possibility to sit at the edge of the discussion without taking over: that myth may sometimes preserve distorted memory of real encounters and real differences—between populations, between human types, between elites and outsiders, between newcomers and established groups. In such a case, "non-human" in myth might not mean extraterrestrial; it might mean unfamiliar, exceptional, or simply other in a way the culture struggled to name.

The outline of this volume is designed to move across these layers in a disciplined sequence. It begins with blind spots and the false comfort of linear progress. It moves into the reality of a web-like human family history. It examines the Denisovan puzzle as a case where genetics forced recognition of what bones barely revealed. It looks at high-altitude adaptation as a window into borrowed resilience. It turns to the apparent abruptness of cognitive leaps and the archaeological explosion of meaning. It then addresses the charged domain of giants, megalith builders, biological outliers, and hybridization myths—territory where evidence, rumor, speculation, and cultural fascination often tangle. The task is not to collapse into sensationalism but to keep boundaries clear: where evidence is strong, we treat it as strong; where it is suggestive, we say so; where it is mythic, we explore it as myth while remaining open to what myth might be doing.

Reopening the question also demands intellectual honesty about what we are trying to achieve. The goal is not to overthrow mainstream science as if it were an enemy. The goal is to push beyond the simplified public story and take seriously the complexities that mainstream research itself has uncovered. Where the outline includes speculative framing—intervention, hybrids, sky teachers—the responsible approach is to treat such ideas as interpretation rather than established fact, and to ask what real-world phenomena could generate those narratives. Did interbreeding and the presence of other hominins inspire myths of “others” and “mighty ones”? Did sudden cultural innovations create stories of knowledge being given rather than discovered? Did surviving elites after catastrophes become remembered as semi-divine founders? These questions can be asked without claiming certainty. They can be explored as hypotheses that connect biology, culture, and myth.

In practical terms, reopening the question of who we are means restoring permission to wonder while keeping the discipline to test. It means refusing the false choice between sterile certainty and uncontrolled speculation. It means admitting that the human story is not a single clean line, and that the evidence already points to a more tangled ancestry and a more fragile cultural pathway than older narratives allowed. It means treating anomalies not as embarrassing errors to be swept away, but as signals that sometimes deserve patient attention. And it means acknowledging that identity is not diminished by complexity. If anything, complexity makes us more astonishing: not a solitary ascent, but a lineage forged through contact, survival, and inheritance across many branches of a wider human world.

If we reopen the question properly, we do not lose ourselves. We find a deeper version of ourselves—one that is less mythically pure but more historically real. We find a species shaped by mixing, by movement, by catastrophe and adaptation, by cultural memory and cultural loss. We find that the mystery is not an embarrassment to be removed from the story, but a rightful part of it. The missing chapters are not merely curiosities. They are the hidden scaffolding of what we became, and perhaps the clearest reminder that the human story is still unfolding, still being discovered, and still capable of surprising us.

CHAPTER 2 — THE HUMAN FAMILY TREE THAT WASN'T A TREE

If you could step into the late Pleistocene world and see it as it truly was, the most startling discovery would not be mammoths or ice sheets or unfamiliar coastlines. It would be the company. The popular imagination pictures Homo sapiens emerging into an empty stage—alone, unique, destined to inherit the Earth. But the deeper you look, the more that stage fills with other actors: multiple human groups, closely related yet distinct, sharing continents and climates, overlapping in time, sometimes meeting, sometimes avoiding one another, sometimes competing, sometimes exchanging genes and ideas. The "human family tree" was never a solitary trunk with a single crown. It was a forest of lineages, and Homo sapiens was one branch among many.

The idea that we were alone is a modern illusion born from the fact that, today, we are alone. All other human species are gone. Their absence makes our uniqueness feel inevitable, as if nature always intended one human type to remain. Yet the archaeological and genetic record suggests the opposite. For much of the last several hundred thousand years, the human genus was diverse. Different populations evolved in different regions under different pressures, producing different combinations of anatomy and adaptation. Some had robust builds suited to cold climates. Some appear to have lived in environments that demanded unusual physiological resilience. Some may have been highly regional, occupying particular ecological niches and never expanding far. The world that produced us also produced others, and the simple fact that we survived does not mean we were the only form that ever mattered.

To grasp what it means for Homo sapiens to be "one branch," it helps to strip away the unconscious hierarchy in the word human. In everyday speech, "human" means us—our faces, our minds, our cultures.

In evolutionary terms, however, human is a category of related beings, some of whom are not alive today but were once as real as we are. They walked, hunted, sheltered, raised children, and navigated the dangers of their worlds. They were not half-formed rehearsals. They were living populations with their own histories, and in some cases, their own long endurance. When we treat them as mere steps leading to us, we commit a quiet error: we turn coexistence into a ladder.

Homo sapiens did not arise as a single, uniform entity either. Even within our own lineage, the story is not a clean birth of "modern humans" who instantly looked and behaved like people in the present. Early Homo sapiens existed in a landscape of variation. Populations spread, isolated, reconnected, and diversified. Some groups carried traits we now associate with later humanity; others retained features that look more archaic to modern eyes. The label sapiens can suggest a sharp boundary, but the reality was likely more gradual and geographically complex. In that complexity, sapiens was not a singular essence but a population history—a set of lineages that became more connected over time, shaped by movement and mixing.

This is the point where the metaphor of a tree begins to fail. A tree implies clean splits: one branch becomes two, which become four, each diverging in isolation. But humans and other hominins did not always evolve in strict isolation. Populations can split and later reconnect. Gene flow can continue across boundaries that look distinct in fossils. Interbreeding can stitch branches together. The result is not a tree with clean separations but a braided structure, a dynamic system of lineages that sometimes diverge and sometimes merge. Even if you keep the word "branch," the image must change: this was a branch in a tangled canopy, not a branch on a lonely trunk.

Why does this matter so much? Because the old picture—sapiens as the solitary peak—encourages a myth of inevitability. It makes our existence feel like the culmination of a long upward drive. But if sapiens was one branch among many, then our story becomes contingent. We begin to see that survival is not guaranteed by superiority. It depends on population size, geography, climate stability, contact, disease, and luck. A species can be highly adapted and still vanish if its populations become too fragmented or if its environment shifts beyond its tolerance. Another can persist simply because its demographic network is more resilient. In deep time, the winners are not always "better." They are often simply the ones who endured the right sequence of conditions.

The multi-branch reality also reframes what we mean by "the human revolution." When we imagine sapiens alone, we tend to credit all innovation to a single lineage's internal genius. But if multiple human groups existed at the same time, then the environment of innovation changes. Ideas and techniques can spread through contact, imitation, and borrowing, not just through isolated invention. Cultural complexity can emerge through networks rather than through solitary leaps. Even biological traits can move through interbreeding. This does not deny the brilliance of sapiens; it embeds that brilliance in a world of interactions where other human groups may have had their own capacities and may have contributed, directly or indirectly, to the story that led to us.

Consider the human range as a kind of living experiment conducted across diverse environments. Ice-age Eurasia imposed different constraints than tropical Africa. Mountain plateaus demanded different physiology than coastal plains. Forests shaped movement and hunting differently than open steppe. In such a world, it is not surprising that different human lineages emerged with different toolkits—anatomical, physiological, and behavioral. Some lineages may have specialized. Others may have remained flexible. Homo sapiens is often described as uniquely adaptable, but adaptability is not a mystical trait. It is a combination of biology and culture: metabolism, social learning, cooperation, communication, planning, and the ability to build and maintain networks. Those networks may have been as crucial as any anatomical feature, allowing sapiens to survive local disasters by receiving new members, exchanging mates, and spreading innovations across distance.

Here the focus sharpens: being one branch among many does not reduce sapiens; it clarifies what made sapiens distinctive. If multiple human types existed, then the question becomes comparative. What allowed sapiens to persist while others vanished or were absorbed? Was it simply numbers—larger, more interconnected populations? Was it broader ecological tolerance? Was it faster cultural accumulation and transmission? Was it a combination of traits that made sapiens more able to colonize varied environments? Or was it, in part, that sapiens could integrate with other groups through interbreeding, absorbing useful genetic variants and expanding the range of adaptive possibilities? These questions are not rhetorical. They are central to the modern scientific project of human origins.

But we must avoid turning the question into a new myth. It is tempting to say sapiens "won" because sapiens was "better." Yet the more we learn, the more it appears that other human groups were formidable. They survived harsh conditions for long spans. They built tools and controlled fire. They cared for injured individuals. They adapted to environments that were anything but gentle. In other words, the comparison does not support a simple story of crude cousins replaced by superior moderns. It supports a story of multiple capable human types inhabiting a world that repeatedly changed its rules.

From this perspective, “Homo sapiens” becomes a term that must be handled with care. It names a lineage, yes, but it also names a set of assumptions. In popular thought, sapiens equals fully human, and everything else is less. In reality, sapiens is one label among several applied to close relatives, and the boundary between “us” and “them” can blur in both biology and behavior. Some archaic humans likely had cognitive capacities closer to ours than we once assumed. Some aspects of symbolic behavior may have deep roots shared across lineages. Even if sapiens had particular advantages, it did not emerge in a vacuum. It emerged in a world already inhabited by other humans, in a landscape where contact was possible and sometimes actual.

The implications ripple outward. If sapiens is one branch, then the idea of a single “human nature” becomes more complex. Our nature may include inherited tendencies shaped not just by our direct sapiens ancestors but by admixture events with other lineages. It may include adaptations introduced from outside the sapiens branch that later became part of the sapiens inheritance. It may include cultural memories of difference—stories of “others” who were like us but not quite, remembered as giants, wild men, or teachers from beyond. These possibilities do not need to be treated as literal in every mythic detail, but they are no longer as easy to dismiss once we accept the multi-branch world.

There is also a psychological shift that comes with this chapter’s premise. People often want origins to confer dignity through singularity. If we are the sole heirs of evolution’s long climb, then our dominance feels justified. But if we are one branch among many, our dignity must come from something else: from the reality of survival, from the capacity to know, to imagine, to build meaning, and to carry forward the lives of those who came before us—including those whose names we do not know. The multi-branch view does not reduce human significance. It changes its foundation. It replaces a triumphalist ladder with a more sober inheritance: we are what remains of a wider human world.

This is why the "family tree that wasn't a tree" is not just a technical correction. It is a reorientation of identity. It asks us to picture prehistory not as a straight corridor leading to us but as a crowded landscape of relatives. It asks us to accept that many branches existed, and that their disappearance was not necessarily a verdict on their worth. Some may have been extinguished; some may have been absorbed. In either case, they are part of our story, not merely in the abstract but, in some cases, in our DNA and in the cultural memory of difference.

Reopening the human family tree in this way also sets the stage for everything that follows. Once you accept that multiple human lineages coexisted, the next questions become unavoidable. Who were the closest relatives we can name? What do we know about the ones we cannot yet name? How deep did interbreeding go, and how normal was it? At what point does the concept of "species" stop behaving like a clean category and start behaving like a rough tool? And what image should replace the ladder once we admit the world was a web?

The old story placed Homo sapiens at the center by default. The evidence invites a different posture: Homo sapiens as a participant in a broader human ecology, shaped by contact, movement, and contingency. We are not a single clean line. We are the survivors of a complex human landscape—one branch that endured, carrying within it the faint traces of other branches that once shared the Earth with us.

Once you accept that Homo sapiens was not alone, the late Ice Age world begins to look less like a stage waiting for its protagonist and more like a contested landscape filled with relatives. For most readers, the first of these relatives has a familiar name: the Neanderthals. The second has a name that still feels new, as if it belongs to a scientific rumor that hardened into fact: the Denisovans. And beyond these two—beyond the famous cousin and the genetic ghost that became real—there is a third category that matters just as much: unknown hominins, populations inferred, hinted at, and partially seen, but not yet fully defined. Together they transform the story from a simple replacement narrative into a world of overlapping human lineages.

Neanderthals carry an unusual burden in modern imagination. For more than a century they were treated as the symbol of primitive humanity: heavy-browed, blunt, dim, doomed. This caricature had little to do with the real beings who lived across Ice Age Eurasia. The more evidence accumulated, the more the caricature collapsed. Neanderthals were adapted to cold climates, built for strength and endurance in rugged environments. They were skilled toolmakers, capable hunters, and in many contexts, remarkably resourceful. Their brains were not small; in many cases cranial capacity overlaps with, and sometimes exceeds, modern human averages. The important question is not whether they were "smart enough" by modern standards, but what kind of intelligence their world required and rewarded—how their bodies, social lives, and cultures were tuned to the constraints they faced.

Those constraints were severe. Eurasia during the Pleistocene was a world of oscillations—shifting ice fronts, changing vegetation zones, prey moving across enormous ranges, winters that could starve a group if hunting failed. In such conditions, success depended on cooperation, planning, and deep practical knowledge. Neanderthals survived there for a long time. That alone should dismantle any assumption of incompetence. A lineage does not persist through harsh climates for tens of thousands of years by being a failed experiment. It persists because it works—because its adaptation is viable in a real ecological theater.

And yet Neanderthals disappeared as a distinct population. For a long time, that disappearance was explained as replacement: modern humans arrived, outcompeted them, and erased them. Genetics complicated that story by revealing that modern humans outside Africa carry inherited Neanderthal DNA. That single fact changes the nature of "disappearance." It suggests that at least some Neanderthals were not simply wiped out; they were absorbed. Their line did not end cleanly at a boundary in time. It ended unevenly, with gene flow into sapiens populations that expanded across Eurasia. Extinction, in other words, is not always a clean stop. Sometimes it is a merger.

Denisovans, by contrast, are defined less by fossils than by the genetic record. Their discovery arrived like a message from the future: a few fragmentary remains that would have meant little on their own, paired with DNA that revealed a distinct lineage. Denisovans remind us that we can live beside a human group for millennia and still fail to recognize them in bones alone. Their physical remains are sparse, but their genetic footprint is unmistakable in some modern populations, particularly in parts of Asia and Oceania. In the old fossil-centered worldview, that would have been almost impossible to imagine: an entire lineage, largely invisible in the traditional record, nevertheless leaving a detectable inheritance in living people.

This difference between Neanderthals and Denisovans is instructive. Neanderthals are visible and familiar because they left many fossils in regions where preservation and research have been intensive. Denisovans are elusive because the fossil record has not yet yielded an abundance of remains that can anchor them in geography and anatomy with the same confidence. Yet their existence is no longer optional. The genome forces recognition. It is as if the past is saying: you will not understand this world if you rely only on what fossilizes well and is easy to find. The human family was larger than the skeletons in museum drawers.

Between these two lineages, we can already see how complex the human landscape was. It was not a simple binary of sapiens versus Neanderthal. It included multiple archaic humans, some of whom likely overlapped geographically and temporally with both. Some likely met sapiens. Some likely met each other. When you introduce unknown hominins—those inferred from genetic signals and from fragmentary or enigmatic fossils—the landscape becomes even richer. The phrase "unknown hominin" can sound like a placeholder, but that is exactly what it is: a marker of our ignorance, and a clue that the record is incomplete in a systematic way.

Unknown hominins appear in two main forms. One form is genetic: signals in modern or ancient genomes that cannot be explained by known lineages alone. These signals suggest admixture with populations that diverged earlier or followed separate histories. They function like footprints without a body. The second form is fossil and archaeological: specimens or assemblages that do not fit comfortably into established categories, or that suggest a diversity of forms greater than the clean textbook sequence implies. Sometimes these finds later become associated with known groups through improved dating or new discoveries. Sometimes they remain ambiguous, hovering at the edge of classification.

The existence of unknown hominins should not be surprising. The human record is geographically biased. Vast regions of the ancient world are underrepresented in fossils, either because preservation is poor or because research access has been limited. Tropical climates degrade bone and DNA. Dense vegetation hides strata. Coastal sites were drowned by sea-level rise. Whole landscapes have been reshaped by erosion or buried under later sediments. If multiple human groups lived across these regions, their remains could be scarce or absent even if their populations were real and substantial. Genetics can sometimes detect their presence precisely because genes travel forward in living descendants even when bones do not.

This multi-lineage world raises a question that is both scientific and cultural: what does it mean to be human if there were multiple human types? The public is used to imagining humanity as a single species with a single essence. The Pleistocene reality suggests a spectrum of close relatives—different in anatomy, perhaps different in some aspects of cognition and culture, but not neatly divided into "human" and "not human." The boundary becomes less like a wall and more like a gradient. This is not a philosophical abstraction; it is a practical problem for classification. If lineages interbred, then "species" behaves differently than the public expects. If lineages overlapped for long periods, then the idea of a clean replacement becomes harder to defend. The story becomes less about a lone hero and more about a network of relatives shaped by contact and environment.

The presence of Neanderthal and Denisovan DNA in modern humans also forces a change in how we imagine sapiens' expansion. It suggests that when sapiens moved into Eurasia and beyond, they entered inhabited worlds. They did not simply colonize empty terrain. They encountered other humans who already knew those landscapes—who had survived there for long spans and had their own adaptations. Those encounters could have taken many forms: competition for resources, avoidance, occasional conflict, exchange, and interbreeding. The archaeological record may capture some of this indirectly in shifting tool traditions, site use, and the appearance of new behaviors. Genetics captures it more directly in admixture signatures. The combined picture suggests that human history is, in part, a story of meetings.

These meetings have been romanticized and weaponized in modern imagination. Some people want them to prove that Neanderthals were brutish and sapiens superior. Others want them to prove that ancient humans were one unified family without real difference. Both extremes are distortions. Neanderthals and Denisovans were distinct lineages with distinct histories. They were not simply “us” in a different costume. But neither were they monsters or dull-witted dead ends. The evidence points toward a more uncomfortable truth: they were close enough to us to interbreed and close enough to us to share many capacities, but different enough to have followed separate evolutionary paths for long periods.

Unknown hominins deepen this discomfort. They suggest that we are still missing pieces of our own story. There were likely groups with combinations of traits we have not yet mapped, living in places that were later erased by sea-level rise or climate shifts. There may have been populations that contributed to sapiens in certain regions and then vanished without leaving an easily identifiable fossil record. Some of these groups may have carried adaptations that were later absorbed. Some may have been isolated refugial populations. Some may have had cultural traditions that left only faint traces. The point is not to inflate this into fantasy, but to recognize that the human landscape was almost certainly more diverse than the names we currently have.

This realization changes the emotional tone of human evolution. It makes it less like a triumphal march and more like a survival story with many characters, many disappearances, and many unresolved threads. It also forces a new humility: the fact that Neanderthals and Denisovans were only fully appreciated when genetics entered the picture suggests that other lineages could still be waiting in the data, hidden until methods refine enough to reveal them. The family tree is not complete, and it may never be complete in the way people imagine completeness. But it can become more accurate, more nuanced, and less captive to the myth of a clean line.

In this chapter's framework, Neanderthals and Denisovans serve as proof of principle. They show that humanity's deep past included other human types who were capable, adaptive, and real. They show that interbreeding occurred. They show that fossils alone cannot map the full diversity of the human world. And they open the door to the unknown: to populations that are not yet named but may have left traces in the genetic record, in ambiguous fossils, or in the gaps where preservation erased what once existed.

As we move forward, the question becomes sharper and more technical. If we accept that there were unknown hominins, how exactly do we detect them? How do "ghost populations" appear in the genome, and what do they imply about ancient contact networks? The next subsection turns to that invisible evidence—the genetic signatures of people we cannot yet picture—and asks what it means when the genome remembers what the ground has forgotten.

A fossil gives you a face, or at least the outline of one. A genome can give you something stranger: a presence without a face, a footprint with no body attached. This is what researchers mean when they speak about "ghost populations." The phrase sounds supernatural, but it is a technical way of describing a very specific kind of evidence—genetic patterns that imply contact with, or ancestry from, a population we cannot yet identify directly in the fossil record, or cannot define cleanly as a named group. In other words, the genome sometimes remembers people the ground has not preserved.

The reason this is possible is that ancestry does not require bones to survive. It only requires descendants. If a population contributed genes to another population that later expanded, then fragments of that vanished group can persist for tens of thousands of years, carried forward in living bodies. Over time, those fragments are broken into smaller pieces by recombination, diluted by later mixing, and reshaped by selection. But they do not necessarily disappear. They become faint signals—subtle statistical patterns in DNA—detectable not by a single dramatic specimen, but by the collective structure of genomes across many individuals and populations.

This is a profound shift in what counts as evidence. Traditional reconstructions relied heavily on what fossilizes and what is found. Genomics allows inference from what is inherited. It turns the problem inside out: instead of asking only, "What did ancient humans leave behind?" it also asks, "What did they leave within us?" When those internal traces do not match any simple model of known populations, the logic points outward toward missing actors.

Ghost populations show up in multiple ways, and the most important thing to understand is that they are not a single phenomenon. Sometimes "ghost" means an archaic lineage that interbred with the ancestors of modern humans, leaving introgressed segments that look unusually divergent from typical sapiens variation. Sometimes it means a population that split from a known group, remained isolated long enough to develop a distinct genetic signature, and then later mixed back in. Sometimes it means an ancient population that contributed to a later population but left few or no direct remains in the regions we have excavated. In all cases, the ghost is not a mystical claim. It is a best-fit explanation for patterns in the data when known lineages alone cannot account for what we see.

To see why, consider how genetic history is reconstructed. Researchers compare genomes across populations and look for shared variants, differences in allele frequencies, patterns of linkage disequilibrium, and the distribution of genetic segments that appear older or more divergent than expected. They use statistical tools designed to detect whether a simple branching tree can explain the relationships, or whether the data demands admixture—gene flow across branches. When a model requires admixture from an unsampled source, that source is, by definition, a "ghost." It is the population you need to include to make the equations describe reality.

This is also why the ghost concept is inseparable from the limits of sampling. Even for well-known lineages like Neanderthals and Denisovans, our ancient DNA samples are sparse compared to the vastness of their former ranges. For other regions—especially many tropical and subtropical zones where DNA preservation is poor—ancient genomes are rare or absent. Yet modern populations from those regions carry deep genetic histories. If their ancestry includes contributions from ancient groups we cannot directly sample, then ghosts become not only possible but expected.

In some cases, the ghost is essentially a placeholder for "a population we haven't sequenced yet." This can happen even within Homo sapiens. Imagine a long-separated group that lived for tens of thousands of years in a region that has yielded few fossils and almost no ancient DNA. If that group later mixed with another, it could leave a signal that looks like admixture from an unknown source. The source might not be a different species at all. It might be a deeply divergent sapiens lineage. The ghost label does not automatically mean "non-human." It means "unsampled" or "unidentified."

In other cases, the ghost is more suggestive of archaic contribution. The logic here is straightforward. If a segment of DNA appears unusually divergent—meaning it looks like it separated from the lineage leading to modern humans much earlier than typical human variation would suggest—then one explanation is introgression from an archaic group. This is how the reality of Neanderthal and Denisovan ancestry became clear: modern genomes carried segments more similar to those archaic genomes than to typical sapiens variation. But even without having an archaic reference genome, researchers can sometimes detect unusually divergent haplotypes and infer that some introgression occurred from a lineage that split earlier than the main branches of modern human ancestry.

This is where the conversation becomes both exciting and delicate, because the farther back you go, the harder it is to separate true introgression from other processes that can mimic it. One of the main mimics is incomplete lineage sorting—a phenomenon where ancient genetic variation present in a common ancestor persists across descendant lineages in a patchy way. Incomplete lineage sorting can make some segments look unusually old or divergent even without later admixture. Another mimic is structure within ancestral populations: if ancient humans were already subdivided into semi-isolated groups, then deep divergence can appear without requiring interbreeding with a separate "species." Genetics is powerful, but it is not a simple decoder ring. It demands careful modeling and constant vigilance against overinterpretation.

Even so, the broader implication remains: the human past contained more populations than the fossil record alone has revealed. Ghost populations are the genome's way of pointing toward missing pieces. They are also a reminder that human history was shaped by movement and mixture. When a model built as a clean tree fails, it is often because reality was not a tree.

There is another dimension to ghost populations that is often overlooked: they can reflect lost geography, not just lost species. During the Pleistocene, sea levels rose and fell dramatically. Coastlines shifted. Plains became seabeds. Islands connected and disconnected. Entire corridors of migration opened and closed. Many human populations likely lived in coastal zones because coastlines are rich in resources and support dense settlement. But those coastal zones are precisely the landscapes most likely to have been submerged after the last glacial period. If large parts of prehistoric habitation are now underwater, then many populations may be archaeologically invisible. Yet their genes could persist if they contributed to later inland populations. The genome, in that sense, can preserve a memory of drowned worlds.

Ghost populations also force us to rethink the idea of "replacement." If modern humans expanded into regions already occupied by other hominins and by other sapiens groups, then what happened was not always conquest and erasure. Sometimes it was absorption. Sometimes it was demographic swamping. Sometimes it was a long coexistence followed by gradual blending. In such processes, the original population can disappear as a distinct group while still contributing significantly to the descendants. The genome then becomes a ledger of mergers. It records not just victories but incorporations.

This leads to a more nuanced view of extinction itself. We tend to treat extinction as an absolute: a species exists, and then it does not. But in a world where interbreeding occurs, extinction can be partial. A group can be extinguished as a distinct lineage while leaving genetic fragments behind. Those fragments can even matter. They can contain adaptive variants—responses to cold, altitude, pathogens, or diet—that later become useful in the descendant populations. In that case, a ghost population is not merely a missing chapter. It is a contributor to the living present.

At the same time, the concept of ghost populations demands humility because it sits at the intersection of signal and interpretation. Genetic data is real, but how we explain it can change with new discoveries and better models. A ghost inferred today might be resolved tomorrow into a known group once a new ancient genome is sequenced or a new fossil site is found. Or it might remain unresolved because preservation never allows direct confirmation. Some ghosts may eventually be understood not as separate populations but as evidence of deep structure within early sapiens. Others may point to archaic lineages that have left almost no physical trace. The truth could be a mixture: multiple kinds of ghosts, emerging from multiple kinds of missing evidence.

This is why the language around ghost populations must be careful. It is tempting to treat them as proof of hidden civilizations or dramatic interventions. That leap is not justified by the data alone. A ghost population in genetics means: there was a population, it contributed ancestry, and we cannot fully identify it yet. It does not automatically tell you what that population's culture was like, what it believed, how it lived, or whether it possessed unusual knowledge. Genetics can suggest contact and timing; it cannot provide a full ethnography of the Ice Age. The discipline of this book must be to keep those boundaries clear while still recognizing how significant the implication is: there were more human histories than we have named.

And once you accept that, the wider theme of missing chapters becomes more concrete. The missing chapters are not only lost artifacts and eroded sites. They are also missing populations—whole branches of humanity that may have existed for long spans and then vanished into others, leaving their last testimony in the quiet mathematics of inheritance. The genome becomes a map with faint roads drawn where no ruins remain, pointing toward meetings and migrations we can sense but not yet fully reconstruct.

In the end, ghost populations are less about mystery for its own sake and more about a changed standard of realism. A simple story requires a simple cast. A realistic story requires acknowledging the cast we cannot fully see. When the genome keeps producing patterns that demand additional actors, the honest response is not to force the data back into a tidy tree, but to let the map expand. Human evolution becomes not a single line but a set of overlapping journeys, with some travelers remembered in bones and tools, and others remembered only as faint traces inside the living.

This sets up the next step naturally. If ghost populations exist, the obvious question is how often contact and blending occurred. Was interbreeding a rare exception, or was it a normal feature of the human world for long stretches of time? The next subsection turns to that question and treats it not as a shocking footnote, but as a central mechanism in the formation of our species.

When most people first hear that modern humans interbred with other human lineages, they treat it as a scandalous detail—an oddity at the margins of an otherwise clean story. The implication feels disruptive because the public narrative trained us to expect separation: species as sealed boxes, evolutionary branches as isolated paths, Homo sapiens as a distinct creature that rose and replaced its rivals. But once genetics forced the issue, and once researchers began mapping the extent and pattern of admixture, a different interpretation started to look less controversial and more realistic. Interbreeding was not a rare footnote. It may have been normal—an ordinary feature of the human world whenever related populations overlapped long enough to meet.

To grasp this, it helps to return to basic biology. Humans, like many mammals, are not designed to recognize and obey taxonomic categories. They respond to proximity, opportunity, social dynamics, and compatibility. If two populations are close enough genetically that their offspring are viable, and if they share landscapes, then mating across group boundaries is not surprising. It would be surprising if it never happened. In modern human history, where cultural boundaries can be intense, gene flow still occurs across groups whenever contact persists. In deep time, where cultural categories were different, smaller populations moved through shared corridors, and groups could meet in unpredictable contexts, gene flow becomes even more plausible.

The old assumption of strict separation was partly inherited from an older view of species. In popular imagination, a species is like a separate kind, an absolute boundary: horses cannot produce fertile offspring with cows, and so on. But life often exists in greyer zones. Many species interbreed under natural conditions, producing hybrids that can sometimes reproduce and sometimes not, depending on the genetic distance and ecological context. In deep time, the situation is even more fluid. Populations diverge, build differences, and yet may still remain compatible for long periods. In the human case, the evidence suggests that Neanderthals and modern humans, and Denisovans and modern humans, were compatible enough for interbreeding to produce descendants who contributed to later populations. That compatibility alone tells us the boundary was not absolute.

Once you accept that, interbreeding becomes a predictable outcome of overlap. Consider what overlap means in practice. Two groups occupy adjacent territories. Their ranges shift with climate. Hunting grounds expand and contract. Rivers and valleys act as corridors. Resources draw groups toward the same locations. Over years and generations, contact occurs—sometimes peaceful, sometimes tense, sometimes brief, sometimes sustained. In such a world, it would be unrealistic to assume that contact never involved mating. In small populations, even a few interbreeding events can leave lasting genetic traces, especially if the resulting offspring integrate into expanding groups.

Genetics does not merely confirm that interbreeding happened; it allows us to see that it happened repeatedly and in different places. That is the crucial shift. A single admixture event could be treated as an anomaly. Multiple events across time and geography begin to look like a pattern. The pattern suggests that when human groups met, mixing was among the possible outcomes often enough to matter. Sometimes that mixing may have been limited and localized. Sometimes it may have been substantial. But the repeated detection of archaic ancestry in different modern populations implies that contact was not a one-off accident. It was part of the fabric of Ice Age human history.

This also reframes what "modern human" means. If modern populations carry inherited segments from archaic humans, then modern humans are not purely the product of one lineage. They are a blend, even if the dominant ancestry is sapiens. The blend is not uniform either. Different populations carry different proportions and patterns of archaic ancestry depending on where their ancestors traveled and whom they met. This creates a world-history dimension inside the genome: the DNA you inherit reflects ancient geography. It carries traces of where your ancestors went and who lived there before them. The genome becomes a record not only of lineage but of encounter.

Interbreeding as a norm also changes the emotional interpretation of other human groups. In the older replacement narrative, Neanderthals and Denisovans were doomed rivals. In a more realistic encounter narrative, they become something else: not simply competitors, but neighbors, sometimes mates, sometimes contributors. This does not mean the relationships were always harmonious. They may have included conflict, coercion, and unequal power dynamics as well as cooperation. The genome cannot tell us the social texture. But it can tell us that the boundary was permeable. And permeable boundaries imply social complexity: contact zones, integration, and the gradual reshaping of populations through both movement and mixing.

Once interbreeding is treated as normal, it becomes possible to ask a deeper question: what role did it play in human adaptation? The answer must be handled carefully. It is tempting to imagine archaic ancestry as a package of upgrades, a dramatic boost that explains everything. The reality is subtler. Many introgressed genetic segments were likely neutral or even disadvantageous and were reduced over time by selection. Others persisted. Some persisted because they happened to be carried along by demographic expansion. Some persisted because they offered small advantages in particular environments—responses to pathogens, climate, altitude, or diet. In such cases, interbreeding becomes more than contact; it becomes a mechanism of resilience. It allows populations to acquire variants already tested in local conditions by groups who had lived there longer.

This is an important point because it challenges the idea that sapiens succeeded purely by internal superiority. Sapiens may have been adaptable and networked, but adaptability can include the ability to incorporate. If sapiens moved into environments where archaic humans had long histories, then interbreeding could allow sapiens to absorb local adaptations rather than reinventing them from scratch. This would not be a magical intervention. It would be an ordinary evolutionary mechanism: gene flow introducing variants that selection can then amplify if they are beneficial.

The implication is that interbreeding may have helped sapiens expand into diverse regions. Not as the sole reason, but as part of a complex toolkit. Sapiens carried cultural flexibility, cooperative networks, and technological innovation. Interbreeding could add genetic flexibility in specific contexts. Together, these factors could produce a lineage capable of spreading widely while maintaining enough biological resilience to survive new pathogens, climates, and ecologies.

Interbreeding also helps explain why the family tree metaphor fails so often. A tree suggests lineage divergence without reconnection. But in the human world, reconnection appears to have been real. Populations split, drift apart, and then meet again. The result is not a clean branching tree but a reticulated pattern—a network with cross-links. Interbreeding is the biological version of those cross-links. It turns separate branches into a braided structure.

This braided structure also complicates the concept of "species," which will become central in the next subsections. If Neanderthals and sapiens were separate species in a strict sense, their interbreeding and the viability of offspring challenge simplistic definitions. If they were subspecies or distinct populations within a broader species complex, then the old narrative of "species replacement" becomes less precise. In deep time, biological categories often blur because divergence and gene flow can coexist. Populations can be distinct enough to look different and follow separate histories, yet similar enough to exchange genes when they meet. This is not an exception in evolutionary biology; it is a feature of how divergence often works.

Interbreeding as the norm also has a cultural mirror. Humans are story-making beings, and stories often preserve a transformed memory of contact. When ancient myths speak of unions between gods and humans, of hybrid heroes, of mighty ones born from mixed lineages, those narratives do not need to be literal to be relevant. They can be symbolic expressions of real social experiences: outsiders entering a group, elites claiming special ancestry, newcomers bringing new knowledge, and the sense that some lineages were not entirely local. When we know that interbreeding occurred in deep prehistory, such myths begin to look less like pure fantasy and more like the human mind's way of encoding difference and mixture into memorable forms. This is not proof of any specific myth as history, but it does soften the old assumption that mixture is unnatural. In the human story, mixture appears to be ancient.

At the same time, calling interbreeding "the norm" does not mean it was constant everywhere. Human populations were often small and scattered. Many groups likely lived and died without encountering other lineages. Geography could isolate. Climate could separate. Some contact zones could be brief. Norm, here, means something more precise: whenever related human groups overlapped in time and space, gene flow was not improbable. It was one of the likely outcomes across enough encounters to leave an enduring genetic record.

This reframing changes the emotional tone of human evolution. It becomes less like a clean battle for supremacy and more like a complicated demographic process: movements, meetings, partial mergers, extinctions that are sometimes absorptions, and a living species that carries within itself traces of others. We are not the descendants of a single line marching alone. We are the descendants of a world in which humanity existed in plural forms.

And once you accept that, the next question becomes unavoidable. If interbreeding was normal, then the tidy categories we rely on—species, branches, replacement—start to wobble. What exactly is a "species" in deep time when lineages can diverge and still exchange genes? Where do we draw boundaries, and what do those boundaries mean? The next subsection turns to that problem directly, because without confronting it, we keep trying to force a web into the shape of a ladder.

The moment you begin to picture the Pleistocene world as crowded with human lineages that sometimes met and interbred, an uncomfortable question rises to the surface: what, exactly, is a species?

The word feels solid in everyday life. It suggests a clean boundary. A lion is not a tiger. A horse is not a donkey. Humans are humans. But the further back you go into deep time—and the more you study populations that diverged and then reconnected—the more that solidity dissolves. "Species" is not a law written into nature. It is a tool humans use to describe patterns in nature. In deep time, especially in the human case, the tool starts to blunt.

This does not mean species are fictional. In many contexts they are real and useful categories that correspond to strong reproductive boundaries. But the popular idea of species as absolute, permanent divisions is a simplification. In evolutionary biology there are multiple species concepts, each emphasizing different criteria: reproductive isolation, morphological distinctiveness, ecological niche, genetic divergence, and more. These concepts often agree for living animals with clear boundaries. In deep time they can diverge dramatically because we cannot observe mating behavior, we cannot test fertility, and we often lack continuous samples to see how variation was distributed across space.

In the fossil record, species are frequently defined by morphology—shape and anatomy. That makes sense because bones are what we have. But morphology is an imperfect guide. Individuals within a single species can vary widely. Environment can shape development and body form. Convergent evolution can cause unrelated groups to look similar. Conversely, related groups can look different due to local adaptation. When you have limited fossil material, often scattered and fragmentary, it becomes very easy to over-split or over-lump. Over-splitting means declaring multiple species where there may have been one variable population. Over-lumping means merging distinct lineages because the sample is too small to see meaningful differences. Either error reshapes the story.

The human genus is particularly vulnerable to this problem because it spans hundreds of thousands of years and multiple continents. Populations could be isolated for long periods and then reconnect. They could spread across vast ranges with gradual clines of variation rather than sharp boundaries. They could experience repeated bottlenecks that change genetic makeup without necessarily changing skeletal form in obvious ways. A fossil species category, in such a context, can become a snapshot label placed on an evolving, moving target.

Genetics might seem like the solution—an objective measure of divergence. But genetics introduces its own complexities. Populations can exchange genes while still remaining distinct in many traits. In such cases, divergence is real, but it is not absolute. The genome becomes a mosaic: most ancestry follows a certain lineage history, while some segments come from another lineage through introgression. Over time, selection can remove or retain those segments. The result is that two populations can be clearly distinct yet not completely isolated. If your species definition requires absolute reproductive isolation, then many such populations would not qualify as separate species. Yet they may still have behaved as separate groups for long spans, with limited gene flow and distinct adaptations. The boundary becomes not a wall but a gradient of permeability.

This is where deep time breaks the everyday intuition. Species, in practice, often form through a process rather than an event. Divergence begins with partial isolation—geographic, ecological, or behavioral. Over time, differences accumulate. Eventually reproductive incompatibilities may arise. But this process can take long spans, and during much of it, populations exist in an intermediate state: distinct but still capable of interbreeding. The human story likely contains many such intermediate states. When two lineages like sapiens and Neanderthals diverged, they did not instantly become biologically incapable of mixing. They remained compatible enough for interbreeding to produce descendants who contributed to later populations. That does not erase their distinctness. It reveals that distinctness does not always equal absolute separation.

The problem becomes even more complicated when you add space. Imagine a lineage spread across a vast range with gradual variation from west to east. At one end, individuals look distinct; at the other end, they look distinct in a different way. In between, there is continuity. Where, in that continuum, do you draw a species boundary? In a modern context, you might avoid drawing a hard boundary at all and instead speak of populations or subspecies. In deep time, with scattered fossils, researchers may name separate species based on differences that, in reality, could have been part of a continuum. Or they may do the opposite, merging fossils that were separated by real barriers. Either way, the category is being imposed on a complex distribution.

There is also the problem of time depth itself. A species label applied to a group that lived for hundreds of thousands of years can hide internal evolution. Modern humans over just tens of thousands of years show changes in body form and adaptation in response to climate and diet. Over longer spans, ancient lineages surely changed as well. A fossil labeled "Neanderthal," for example, can represent a range of populations across time and space, not a single uniform type. The same is true for sapiens. When we treat these labels as fixed entities, we miss the reality that they were shifting populations, not static categories.

Why, then, do we cling so tightly to species labels in the human story? Partly because we need language. Without categories, we cannot talk about patterns at all. But there is also a deeper cultural reason: species categories support the myth of clean identity. If sapiens is a separate species that replaced others, then our ancestry feels pure and singular. If the boundaries blur, then we must admit we are composites, shaped by contact. That admission can be unsettling because it challenges the old ladder story. Species labels can become a psychological comfort: they make the past feel organized and the present feel uniquely justified.

In the scientific realm, the stakes are more technical but still significant. Species labels influence how we interpret fossils, how we model population history, and how we infer behavior. If Neanderthals are a separate species, then interbreeding becomes extraordinary and requires special explanation. If they are a subspecies or a closely related population within a broader species complex, interbreeding becomes less surprising. The label shapes the story's tone. But the underlying reality—distinct lineages with some gene flow—remains. The question becomes which label best captures that reality without distorting it.

This is why many researchers increasingly emphasize population-level thinking. Instead of asking, "Was this a separate species?" they ask, "How distinct was this population, and how much gene flow occurred?" They treat divergence and interbreeding as variables rather than absolutes. This approach fits the evidence better because it can accommodate a world of partial separation and reconnection. It also aligns with what genetics keeps revealing: human history is full of admixture events that do not fit a clean branching tree.

Deep time also complicates species because selection can create sharp trait differences even with gene flow. Two populations can interbreed occasionally but still remain distinct if they occupy different niches and selection reinforces different traits. Conversely, two populations can be genetically divergent but look similar if selection pushes them toward similar solutions. This means morphology and genetics do not always align neatly. A fossil might look "different enough" to name a new species, while genetically it might have been part of a network with gene flow. Or a fossil might look "similar enough" to merge, while genetically it might have been distinct. Without direct genetic data for every fossil, classification becomes a best-available approximation.

In the human case, the breakdown of species categories is not a failure of science but a revelation about how humanity evolved. It suggests that the human genus was not composed of neatly separated boxes. It was composed of lineages that diverged and sometimes reconnected, that maintained distinct adaptations while still occasionally exchanging genes. It suggests that humanity, in deep time, functioned more like a species complex—a cluster of closely related populations with varying degrees of separation.

This has implications beyond classification. It changes how we interpret the emergence of modern humans as a global species. It implies that what we call "Homo sapiens" may have been shaped not only by internal evolution but by exchange with other lineages. It suggests that the boundaries of "us" were permeable for long spans. And it hints that some traits we consider uniquely modern may have deep roots shared across multiple human types, or may have been influenced by contact rather than arising in isolation.

If "species" breaks down, then the story needs a different metaphor. Not the ladder of progress, not even the tidy branching tree, but a web—an interconnected structure where divergence, contact, and merging can all occur. That web is not chaos. It has patterns. It has corridors of movement, zones of isolation, pulses of expansion, and moments of connection. But it cannot be reduced to a single clean line.

This is why the chapter's conclusion must point toward a new image. If the human family tree was not a tree, then our ancestry is better pictured as a network of streams—separating, rejoining, carrying sediments from multiple sources into the river that became modern humanity. The next subsection turns to that image directly, because once you understand why species categories blur, the web is no longer a poetic metaphor. It becomes the most realistic way to describe what the evidence is actually telling us.

The ladder is a comforting shape. It has rungs you can count, a direction you can trust, and a clear top where you can place yourself. It turns deep time into a moral story: life climbs, becomes better, becomes us. But once you look closely at the evidence—once you accept multiple human lineages, ghost populations, repeated admixture, and the blurred boundaries of "species"—the ladder collapses. It collapses not because the past becomes unknowable, but because it becomes more realistic. The best image is not a ladder. It is a web: lineages splitting, drifting, reconnecting, exchanging, and sometimes merging so thoroughly that the old boundaries dissolve into a composite inheritance.

A web is not an invitation to chaos. It is a different kind of order. It recognizes that populations are not isolated points but dynamic groups moving through landscapes, meeting and separating according to geography, climate, and demography. In a web model, the fundamental unit is not a rung on a ladder or even a neat branch of a tree. It is a population—people who share mating networks, cultural transmission, and local adaptations. Populations can expand, contract, fragment, and reconnect. When you scale that across continents and millennia, the result is not a simple line but a living mesh of relationships.

One reason the web image fits so well is that human evolution was shaped by repeated environmental pulses. The Pleistocene was not a stable backdrop. It was a world of oscillations—ice advances and retreats, shifting rainfall patterns, expanding deserts, rising and falling sea levels, changing coastlines, and migrating prey. These changes reshaped where humans could live and how they could move. When conditions were favorable, populations expanded and met. When conditions were harsh, populations fragmented into refuges—mountain valleys, forest pockets, coastal corridors, river basins—where they could survive while other regions emptied. These cycles naturally create a web. Expansion creates contact. Contraction creates isolation. Over repeated cycles, lineages diverge and then reconnect.

The ladder cannot represent this because it assumes continuity of direction. It assumes that one form replaces another in a clean sequence. The web can represent replacement, but it does not require replacement as the default. In a web, disappearance can be extinction, but it can also be absorption. A population can vanish as a distinct group while leaving genetic and cultural traces inside another population. The result is a history where loss and survival intertwine. The past does not simply end; it is incorporated.

This incorporation is not speculative. Genetics makes it difficult to deny. Modern humans carry inherited segments from archaic lineages. That means some connections in the web were not merely geographic. They were biological. And where biology connects, culture often follows. Not always in a direct, provable way, but plausibly enough to change how we imagine the encounter zones. When groups meet, they do not exchange only genes. They exchange information: routes, resources, toolmaking techniques, hunting strategies, and ways of reading the landscape. Even a brief contact can transmit knowledge. Over time, repeated contacts can reshape cultural traditions. The web is therefore not only genetic. It is also informational.

A web model also forces a new honesty about the limits of our categories. The ladder encourages us to label: this species, then that species, then us. The web encourages us to describe: how distinct were these populations, and how much did they exchange? Instead of treating "species" as a definitive wall, we treat it as a variable boundary—sometimes strong, sometimes weak, sometimes porous. That does not mean anything goes. It means the relevant question becomes quantitative rather than absolute: what was the degree of separation, and what were the conditions under which contact occurred?

In this framework, Homo sapiens becomes less like a solitary apex and more like a successful node—an expanding network that absorbed, displaced, and incorporated other networks across time. This is a subtle but profound shift. It does not reduce sapiens to mere opportunism; it highlights what may have been our key strengths. A networked species is resilient. It can recover from local disasters because its wider web remains intact. It can maintain cultural complexity because innovations spread through multiple groups rather than dying with one small band. It can incorporate newcomers, increasing diversity and adaptability. In a world of constant environmental shocks, these network properties can matter more than any single anatomical feature.

The web image also clarifies why the fossil record can mislead. Fossils come to us as points—isolated finds that can be arranged into sequences. But those points are samples from a continuous reality. When you have only scattered points, it is tempting to connect them into a line. The web model warns against that temptation. It asks you to treat each point as part of a wider field of populations, most of which left no fossils. Instead of drawing a straight line from one skull to another, you imagine regions of overlap, corridors of movement, and zones of isolation. You imagine missing nodes and missing strands because you know, in a living web, many connections remain unseen.

This matters because the web model is the only one that can comfortably accommodate ghost populations. A ghost population is, by definition, a missing node. It is inferred because the relationships between known nodes cannot be explained without it. In a ladder or a tidy tree, missing nodes are embarrassing—they threaten the story's completeness. In a web, missing nodes are expected. A web can survive with hidden strands. In fact, most real-world networks are partly hidden. You can infer the presence of unseen connections from the way the visible system behaves. That is exactly what population genetics does.

The web model also challenges one of the deepest assumptions people carry about human uniqueness: that our story must be singular. In a web, singularity is a byproduct of survival, not a starting condition. Many human lineages existed. One network remained large and connected enough to persist globally. Others contracted, fragmented, and disappeared as distinct clusters. Some were extinguished. Some were absorbed. The outcome is one surviving species, but the path is plural. Modern humanity becomes the downstream river formed by multiple upstream streams, not a single spring.

This is where the web begins to resonate with the volume's broader theme: missing chapters. A web implies that many strands were cut, many nodes vanished, and many connections were never recorded. The human story becomes less like a finished family portrait and more like a partially burned tapestry. Some threads are intact. Others are missing. The remaining pattern suggests what the full design might have been, but the full design cannot be perfectly recovered. That does not mean we stop trying. It means we adjust our expectations. We stop demanding a single clean narrative and instead seek the most plausible structure that fits the evidence we have.

Importantly, a web model also helps separate mainstream complexity from speculative interpretation. You do not need to invoke extraordinary interventions to explain why human evolution looks tangled. Climate oscillations, population movement, and interbreeding are enough to create a network. Cultural transmission is enough to produce sudden-looking bursts of innovation if knowledge spreads through contact and if populations repeatedly bottleneck and expand. A web is the natural outcome of ordinary processes operating over long spans. Where this volume later explores more mythic framing—giants, hybrids, sky teachers—the web model provides a grounded base. It gives a realistic demographic and genetic architecture that can generate powerful stories without requiring the stories to be literal. It also leaves room for hypotheses where evidence is ambiguous, while keeping the distinction between what is supported and what is interpretive.

The ladder model makes us look for a single turning point: when "we" became "us." The web model suggests many turning points, distributed across regions and times, with different lineages contributing to the eventual composite. It suggests that modern humanity is not an isolated miracle but a historical outcome shaped by connection. It suggests that what we call "human" emerged through interactions—between genes, environments, and cultures—rather than through a solitary ascent.

In practical terms, adopting the web model changes the kinds of questions we ask as we move into the next chapters. Instead of asking only, “Which species came before us?” we ask, “Which lineages overlapped with us, and what did we exchange?” Instead of asking, “Why did they fail?” we ask, “What demographic and ecological pressures reshaped the network?” Instead of treating archaic ancestry as a scandal, we treat it as evidence of a normal feature of the system: permeable boundaries in a world of mobile populations.

And once that is accepted, the Denisovan mystery becomes not a strange side story but a central case study. Denisovans are, in many ways, the web made visible: a lineage revealed by genetics, connected to modern populations through admixture, and still only partially understood in anatomy and geography. They represent the kind of missing chapter that can only exist if the human story was a network. The next chapter will step fully into that case, because Denisovans are the clearest demonstration that the human past contained major lineages we would barely know without the genome’s time machine.

CHAPTER 3 — THE DENISOVAN MYSTERY

If Neanderthals forced science to revise its assumptions, Denisovans forced it to revise its imagination. Neanderthals were discovered first in bones, then slowly reinterpreted as the evidence expanded. Denisovans arrived in reverse. They were not introduced to the world as a recognizable face, a distinctive skeleton, or a long parade of fossil finds arranged in museum cases. They emerged out of the genome itself, as if the deep past had left a coded message that only a new kind of instrument could read. In the Denisovan story, DNA is not merely supporting evidence. It is the main witness. Fossils, so far, are little more than punctuation marks.

This inversion matters because it reveals how incomplete our traditional archive has been. For generations, human evolution was reconstructed largely from bones and stones—what fossilized, what endured, what was unearthed. That method is still essential, but it carries a bias: it privileges lineages that lived in regions with good preservation, accessible strata, and long research histories. Denisovans expose the consequence of that bias. They remind us that an entire major human lineage can exist, persist, spread, and influence the ancestry of living people, while leaving behind almost no anatomical record that we can confidently assemble into a picture. The Denisovan mystery is, at its core, a lesson in humility. The world was larger than the evidence we used to rely on.

What we call "Denisovans" are identified primarily by genetic distinctness. Their DNA does not match modern humans, and it does not match Neanderthals either, even though it is clearly related. The relationship is close enough to place Denisovans within the same broader human family, but different enough to mark them as their own branch—one with a deep history and, crucially, one that interacted with other branches. That is the heart of the revelation: Denisovans were not an abstract side lineage that lived and died unnoticed. They were part of the web. Their genetic signature is carried by some living human populations today, which means Denisovans were once close enough, in time and geography, to meet modern humans and leave descendants.

To say a species is known almost entirely from DNA is to say something radical about how knowledge now works. In older frameworks, you needed an anatomy to define a lineage. You needed a skull, a pelvis, a long bone—something you could compare, measure, and classify. Denisovans show that you can sometimes identify a lineage without having a clear anatomical type specimen in the traditional sense. You can infer its existence through patterns of divergence and inheritance. You can detect the echoes of ancient separation and later admixture. And you can do all this while still being unable to answer questions that feel basic to the public: What did they look like? How tall were they? How robust? How different were their faces? What kind of tools did they make? How did they live?

These unanswered questions are not signs that the Denisovans are imaginary. They are signs of how harshly the Earth edits the past. We have far more evidence for some lineages than others not because those lineages mattered more, but because preservation and discovery are uneven. In many environments—especially those that are warm, humid, or chemically aggressive—bone and DNA degrade quickly. Vast parts of Asia, where Denisovan ancestry appears most relevant, contain regions where ancient DNA preservation is challenging and where fossil recovery is difficult. Even when remains exist, they may be scattered, deeply buried, or destroyed by time. The Denisovan story therefore begins with a paradox: a lineage can be genetically undeniable and anatomically elusive.

This is why the Denisovans are often described as a "species known almost entirely from DNA." The phrase does not mean there are no physical remains at all. It means the physical remains are currently too limited to build a full-bodied portrait. A few fragments can anchor a genetic discovery, but they cannot yet define the full range of Denisovan anatomy across time and geography. And that range likely existed. Denisovans were probably not a single small cave population. Their genetic influence and inferred distribution suggest something larger—multiple populations, perhaps regional diversity, perhaps long occupancy in parts of Asia. In such a scenario, the Denisovans we glimpse genetically could represent a broad lineage with internal variation, not a single uniform type. Yet our fossil access to that diversity is still minimal.

DNA changes the logic of certainty because it can carry a lineage signature even when the skeleton record is absent. It does so in two main ways. First, ancient DNA extracted from archaic human remains can be compared directly to other genomes, revealing whether those remains belong to a known lineage or represent something distinct. Second, even without abundant archaic remains, modern human genomes can retain introgressed segments—pieces of DNA inherited from archaic populations through ancient interbreeding. When these segments consistently point to a distinct source that is not Neanderthal and not modern human, the inference becomes strong: there was another lineage, and it contributed to some modern populations. Denisovans were identified and then reinforced through exactly this kind of genetic triangulation. The past was inferred from the present, and then anchored by a small number of ancient remains.

The Denisovan case also forces a more nuanced understanding of what "species" means in deep time. If Denisovans interbred with modern humans and left descendants, then reproductive boundaries were not absolute. This does not dissolve Denisovans into modern humans, or modern humans into Denisovans. It means that evolutionary divergence can coexist with gene flow. The lineages can be distinct in their broader history while still exchanging genes when contact zones formed. Denisovans thus become a central example of the web model introduced in the previous chapter. They are not simply a branch that snapped off and disappeared; they are a branch whose strands still run through the modern human tapestry.

Yet we must be careful with the emotional pull of the word "mystery." The Denisovans are mysterious in a practical sense—unknown anatomy, uncertain range, limited archaeological association—but they are not mysterious in the sense of being beyond science. The mystery is the kind that emerges whenever evidence is uneven. We can see that Denisovans existed, that they were genetically distinct, that they interacted with other human groups, and that they left traces in living genomes. We cannot yet reconstruct their full physical form or cultural life with confidence. That gap between what we can infer and what we can describe is what makes the Denisovan story feel like a half-lit room.

In that half-lit room, imagination is always tempted to rush in and decorate the darkness. The discipline of this book, however, is to let the evidence set the boundaries. DNA can tell us certain things with unusual strength. It can show divergence relationships, approximate timing in broad strokes, and the presence of admixture. It can sometimes point toward adaptations if particular genetic variants appear to have been favored. But DNA cannot, on its own, tell us what a Denisovan community felt like from the inside. It cannot tell us whether Denisovans had complex symbolic traditions, what their languages sounded like, or what stories they told. Those questions belong to archaeology, anthropology, and—where the record is silent—careful hypothesis, not assertion.

Still, even within these boundaries, the Denisovan story carries a deeper implication that is central to this volume's theme. If one major lineage could remain largely invisible until the genomic era, then the human story almost certainly contains other lineages or sub-lineages that are still partly hidden. The Denisovans are not necessarily unique in their elusiveness; they may simply be the first to be revealed by a new method. Their discovery is proof that the old public narrative—one species replacing another in a clean sequence—was not merely simplified but structurally incomplete. It lacked actors. It lacked interactions. It lacked the sense that the Pleistocene world was populated by multiple human forms, some of which we may only ever know as genetic shadows.

The Denisovan mystery also reframes how we interpret human capabilities in deep time. If Denisovans were widespread enough to leave genetic traces across large regions, then they likely possessed robust ecological strategies. They survived in real landscapes under real pressures. They were not a fleeting anomaly. Their persistence suggests competence, adaptation, and social continuity. Even if we do not yet know their full archaeology, it is reasonable—within mainstream caution—to assume they were not biologically extraordinary in the sense of being outside nature. They were extraordinary in the sense that their discovery rewired our expectations about what could be missing from the record.

This is also why Denisovans become such a powerful bridge between genetics and the larger mythic themes that will appear later in the book. The mythic layer often speaks of "others": beings who are close to human but not fully human, remembered as powerful, strange, or exceptional. Without claiming that myth literally describes Denisovans, the Denisovan case does make one thing clear: the world genuinely contained "others" who were close enough to us to interbreed, different enough to be distinct, and real enough to leave traces in the living. In that light, it becomes easier to understand why human cultures might carry persistent narratives of near-human lineages, hybrid unions, and ancestral differences. The Denisovan story does not prove any myth. It does, however, make the existence of deep-time human diversity a matter of evidence rather than imagination.

As we reopen the Denisovan chapter, we are standing at the edge of a new kind of prehistory—one written not only in stone and bone, but in molecular inheritance. Denisovans are the clearest demonstration that the genome can reveal major lineages invisible to the older archive. They are the reminder that the human past was plural. And they are the invitation to treat DNA not as a footnote to the fossil record, but as a parallel record—a second library, written in a different language, capable of preserving what the Earth erased.

From here, the Denisovan mystery becomes sharper and more specific. If Denisovans were real and interacted with modern humans, what exactly did they contribute? In at least one striking case, genetic evidence suggests that a key adaptation for life at extreme altitude entered modern human populations through Denisovan ancestry. That case is not merely technical; it is a glimpse of how ancient admixture could shape survival in environments that would otherwise punish the human body. The next subsection turns to that high-altitude gene and what it reveals about Denisovan legacy.

Few places on Earth make the human body feel its limits as quickly and brutally as the high plateau of Tibet. The air is thin, the nights are cold, the sun burns through the atmosphere with a harsh clarity, and oxygen—a substance most people never think about—becomes the difference between functioning and failing. For a lowland visitor, the experience can be immediate: breathlessness after a short walk, headaches, nausea, insomnia, a strange sense of fragility in the chest. In severe cases, the body can spiral into life-threatening altitude sickness. And yet for thousands of years, Tibetan populations have lived, thrived, raised children, and built societies at elevations where the human physiology, in its baseline form, struggles to endure.

This is where the Denisovan mystery becomes tangible. It moves from the abstract world of genomes and divergence models into the lived reality of survival. Because one of the most famous examples of archaic genetic contribution involves high-altitude adaptation in Tibetans—an adaptation strongly associated with a gene region linked to oxygen response. The significance is not merely that Tibetans are adapted. Many populations adapt to their environments over time. The significance is that the genetic variant most strongly associated with Tibetan high-altitude physiology appears to have entered the modern human lineage through introgression from an archaic human group closely related to Denisovans. In other words, part of what makes Tibetan high-altitude life possible may have been borrowed from an older human lineage that had already faced those pressures.

To understand why this matters, you have to understand the problem altitude creates. At high elevation, the air pressure is lower, which means each breath delivers fewer oxygen molecules into the lungs. The body's immediate response is to compensate: breathe faster, increase heart rate, and produce more red blood cells to carry oxygen. In the short term, some of these responses help. But long-term, especially across pregnancy and childhood development, the standard lowland response can be dangerous. Excessive red blood cell production thickens the blood, increasing risks of clotting and cardiovascular strain. Severe hypoxia during pregnancy can lead to poor fetal growth, complications, and higher infant mortality. A population living permanently at altitude faces a relentless selective pressure: those who can manage oxygen efficiently without pathological side effects will leave more descendants.

Tibetan high-altitude adaptation is notable because it does not rely primarily on the same strategy many lowlanders use when acclimatizing. Instead of responding with an extreme rise in hemoglobin concentration, Tibetans often maintain relatively modest hemoglobin levels compared to what would be expected at their elevations, while still achieving effective oxygen delivery. Their physiology appears tuned to avoid the costs of chronic thickened blood. In evolutionary terms, this is a sophisticated solution: not simply more oxygen carriers, but a system that balances oxygen transport with cardiovascular stability.

A key player in this adaptation is a gene region involved in the body's response to hypoxia, commonly discussed through the lens of the EPAS1 region. The details can become technical, but the core idea is clear: certain variants in this region help regulate how the body responds to low oxygen conditions. In Tibetan populations, a particular variant is unusually common—far more common than would be expected by chance—suggesting strong selection for altitude fitness. What made this case so striking is that when researchers examined the genetic signature, the variant did not look like a typical modern human variant that slowly rose in frequency through selection. It looked like an introgressed fragment—an inherited segment coming from an archaic lineage. When compared against known archaic genomes, its closest match aligned with Denisovan-related ancestry.

This is the moment where admixture stops being a footnote and becomes a mechanism. Interbreeding is no longer simply evidence of contact. It becomes a way populations can acquire adaptive solutions that were already "tested" in other lineages. If an archaic group had lived at altitude for long periods—long enough for selection to favor variants that managed hypoxia effectively—then gene flow from that group could deliver those variants into sapiens populations entering or expanding into high-altitude regions. Once introduced, selection could rapidly increase the frequency of the advantageous variant in the new population. The result would look like a sudden evolutionary jump—not because biology violated its rules, but because the population did not have to wait for a beneficial mutation to arise from scratch. It acquired one through admixture.

This is why the Tibetan case is often described as an "archaic advantage." It shows how ancient encounters can shape modern survival. It also hints at the ecological range of Denisovan-related groups. If Denisovan ancestry is linked to a high-altitude adaptation, it implies that Denisovan or Denisovan-like populations were present in environments where hypoxia was a serious selective pressure—either living at altitude themselves or moving through such regions often enough for selection to matter. We do not need to picture a single Denisovan tribe perched permanently on a plateau. The broader point is that Denisovan-related populations occupied parts of Asia where altitude environments existed and where physiological adaptation could be advantageous. Their genetic legacy, preserved through introgression, suggests a history that is geographically and ecologically richer than the limited fossil record alone would indicate.

But we must be disciplined here. The Tibetan adaptation does not prove that Denisovans lived exactly where Tibetans live today, or that they were the first to inhabit the plateau. Genetic introgression could have occurred elsewhere and then spread through later population movements. The archaic variant could have entered a sapiens population in one region and later become highly selected when that population moved into high-altitude zones. Human history is full of migrations, replacements, and expansions. Genes travel. The genome preserves relationships but not always precise maps. What the evidence supports is the introgression of an archaic variant strongly associated with high-altitude adaptation. The story of exactly where and how that introgression entered the Tibetan lineage remains a matter of careful inference.

Still, even with this caution, the Tibetan case remains a remarkable proof of principle. It shows that archaic admixture could have functional consequences. It shows that the blending of lineages was not simply an evolutionary curiosity. It could shape the boundaries of what modern humans could tolerate. It also shows how the pace of adaptation can appear to accelerate when gene flow introduces variants that selection can immediately favor. In the simplified public narrative, evolution is imagined as slow, always waiting on new mutations. In reality, evolution can be faster when populations exchange genetic material. The human web does not merely mix; it can transmit adaptations across lineages.

This also reframes the meaning of "exceptional." Denisovans are often described as mysterious because they are genetically defined and anatomically elusive. The Tibetan gene demonstrates a different kind of mystery: the way an invisible lineage can reach forward through time and shape the survival of living people. It is not a mystical reach. It is inheritance. It is the hard logic of descent. Yet it can feel uncanny because it collapses distance. An archaic human group, long vanished, becomes part of the physiology of a modern population in one of the world's most extreme environments.

The Tibetan case also invites a wider question: how many other adaptations entered modern humans through similar pathways? The genome contains multiple introgressed fragments from archaic lineages. Most are neutral or were removed over time. But some persist, and persistence—especially when paired with signals of selection—suggests function. High-altitude adaptation is a clear example because the environment is extreme and the physiological challenge is well understood. In other domains, like immune response and pathogen resistance, the signals may be more complex but potentially just as significant. The Tibetan gene, therefore, is not merely a one-off curiosity. It is a lens through which to view archaic admixture as a creative force in human survival.

This is also where the Denisovan story begins to challenge the intuition that Homo sapiens was uniquely designed for every environment it conquered. Sapiens' success may have been partly due to cultural flexibility—clothing, fire, shelter, social organization—but biology matters too. If archaic admixture contributed to key adaptations, then sapiens' expansion becomes a story of both invention and inheritance. It becomes a story of cultural ingenuity combined with biological borrowing. The modern human is not just a self-made species. It is a species shaped by contact.

And yet, the Tibetan adaptation should not be mythologized into a narrative of "upgrades" in a simplistic sense. It is not evidence of intentional engineering. It is not proof of a master plan. It is evidence of normal evolutionary processes: selection, gene flow, and survival under pressure. The word "upgrade" may be useful as metaphor later in this volume's speculative framing, but here it must remain grounded. What happened was not magic. It was a genetic variant introduced through admixture that became advantageous under a severe environmental constraint.

The deeper lesson is that human evolution is not only about what happens within a lineage. It is also about what happens between lineages. The Tibetan high-altitude gene is a living example of the web at work: ancient divergence, later contact, genetic exchange, and then strong selection shaping a modern population. If Denisovan-related groups contributed one crucial adaptation that enabled long-term life at extreme elevation, then Denisovans are not merely a missing chapter in the sense of being unknown. They are a missing chapter in the sense of being foundational—part of the biological toolkit that modern humans carry into environments that would otherwise punish them.

From here, the Denisovan mystery opens outward. If one Denisovan-linked variant helped make life at altitude possible, then Denisovan-related populations may have been adapted to other harsh conditions as well. What kinds of environments did they endure? How did they persist in places humans "shouldn't" endure, at least according to lowland expectations? The next subsection broadens the focus from a single gene to a wider reality: survival in extreme environments, and what that survival implies about the capabilities and range of Denisovan-lineage humans.

There is a quiet arrogance in the way modern humans talk about habitability. We draw invisible boundaries and say, with confidence, where people "shouldn't" be able to live. We imagine that certain climates, elevations, latitudes, and ecologies are beyond the natural human range—too cold, too thin-aired, too rugged, too isolated. And in a narrow physiological sense, that arrogance has evidence behind it: bring an unadapted lowlander to extreme altitude, or place an unprepared community in an Ice Age winter, and survival becomes precarious. But the deeper history of our genus contradicts the arrogance. Humans—and human relatives—have repeatedly occupied places that, on paper, look hostile. They endured because evolution does not ask what seems reasonable. It asks what can persist across generations. The Denisovan chapter forces this lesson with unusual force, because Denisovan-related ancestry appears most strongly where environments become punishing and where adaptation becomes a matter of life and death.

The Tibetan high-altitude case is the clearest example because it ties a specific physiological challenge to a specific genetic region, and because the challenge is immediate and measurable: hypoxia, low oxygen. But the broader implication is larger than one plateau. Denisovan ancestry is not a single isolated curiosity. Its distribution and the kinds of adaptations associated with archaic introgression suggest that Denisovan-related populations were not fragile outliers living on the edge of extinction. They were capable human lineages that endured across a range of Asian environments, including those that demanded resilience, flexibility, and perhaps specialized physiological traits. The reason this matters is simple: we cannot understand the Denisovans as a minor footnote if their genetic legacy shows up most strongly where survival is hardest.

In the Pleistocene world, harsh environments were not exceptions. They were the rule. Climate oscillations repeatedly transformed landscapes. Regions that were lush became arid. Forests opened into steppe. Coastlines shifted by kilometers. Glaciers advanced and retreated, redrawing the map of viable habitats. For human groups, this meant living in a world where the "normal" environment could change within a few centuries—or less—forcing movement, innovation, and adaptation. In such a world, the ability to occupy marginal zones could become a decisive advantage. When the heartlands failed, refuges mattered. Mountain corridors, high plateaus, cold-temperate zones, and rugged interior landscapes could shelter populations when other areas became unstable.

This is one way to interpret Denisovan persistence without romanticizing it. Denisovans may have occupied, or at least traversed, a wide range of Asian habitats, including mountainous regions and ecological transition zones. In such regions, selection pressures can be intense and variable. Thin air, cold temperatures, seasonal resource scarcity, rugged terrain, and disease environments can all shape which traits become advantageous. A lineage that persists across such landscapes over long spans could accumulate a suite of adaptations—not because it was "superior" in an absolute sense, but because the environment repeatedly filtered out the less resilient strategies.

It is also important to remember that "endure" does not mean live comfortably by modern standards. It means survive long enough to reproduce, raise children, and maintain continuity. In the deep past, endurance often took the form of mobility and knowledge rather than permanent settlement. Populations could follow seasonal routes. They could exploit different altitudes at different times of year. They could develop toolkits and social strategies that reduced mortality: fire management, clothing, shelter construction, food storage, cooperative hunting, and flexible diets. These cultural strategies can expand a lineage's ecological range dramatically. Biology matters, but culture is often the bridge that allows biology time to adapt.

Still, biology remains central to the Denisovan puzzle because the strongest cases for Denisovan contribution are tied to physiology in extreme settings. High altitude is one. Another likely domain is immune response. Wherever humans entered new regions, they encountered new pathogens and parasites. In such contexts, introgressed genetic variants that improved immune function could be favored. The logic is the same as altitude adaptation: a lineage that had long lived in a region would have been exposed to its disease environment for thousands of years. If sapiens entered later, admixture could allow sapiens to acquire immune variants already shaped by local selection. This is not a claim of Denisovan "magic." It is an evolutionary common sense: local populations often carry local adaptations, including resistance to local pathogens.

Harsh environments also include cold and seasonal extremes, especially across northern Eurasia. The Pleistocene was a time of glacial cycles. Even outside ice sheets, climates could be severe. Living in cold zones requires not only technology but physiology: energy balance, fat metabolism, thermoregulation, circulation patterns, and strategies for dealing with seasonal food scarcity. It is plausible, within mainstream evolutionary reasoning, that archaic lineages occupying these zones would develop genetic variants that influenced metabolism and cold tolerance. Whether specific Denisovan variants did so is a matter for careful evidence, but the broader point stands: harsh climates exert strong selective pressures, and populations living there for long spans do not remain genetically unchanged.

Another kind of harshness is geographic isolation. Mountain systems fracture landscapes into valleys, ridges, plateaus, and microclimates.

Such fracturing can isolate populations and produce divergence. It can also create pockets of survival during broader climatic shocks. If Denisovan-related populations occupied mountainous Asia over long spans, they may have experienced a combination of isolation and intermittent contact—exactly the kind of demographic rhythm that produces both distinct genetic signatures and later admixture events. In such a scenario, Denisovans are not a single homogenous group; they are a lineage with regional branches, shaped by a geography that encourages fragmentation and refugial persistence.

This is one reason Denisovans remain so difficult to picture. A lineage spread across vast and varied Asia would not necessarily have left a single consistent anatomical "type" easy to identify. It could have contained multiple populations with different blends of traits, just as modern humans do. Fossil fragments from different regions might not look obviously like pieces of the same puzzle. Without genetic anchors, such fossils could be classified as ambiguous or assigned to other categories. Denisovans, in that sense, are not only missing because of preservation. They are missing because diversity itself can hide a lineage in plain sight.

The phrase "environments humans shouldn't endure" also invites reflection on how we define the human range. Modern humans today can live almost anywhere because we have technology and global supply chains. But that modern capability can make us underestimate ancient competence. Even without modern infrastructure, Paleolithic humans and their relatives could be astonishingly skilled at extracting survival from difficult landscapes. They knew how to read animal movement, how to track seasons, how to use plant resources, how to manage fire, how to craft tools suited to local conditions. In other words, part of what made harsh environments survivable was not only genetic adaptation but deep cultural knowledge. That knowledge could be transmitted across generations, but it could also be lost during bottlenecks. It could fragment with populations. It could survive in isolated pockets. The Denisovan story, built largely from genetics, may be pointing to lineages that were not only biologically adapted but culturally embedded in landscapes we still find forbidding.

This is where the Denisovan mystery begins to converge with the volume's broader theme of lost chapters and misunderstood histories. If Denisovan-related lineages occupied harsh and fragmented landscapes, then their archaeology may be scattered, hard to recover, and poorly preserved. Their cultural traces might not be dramatic monuments. They might be ephemeral tools, seasonal sites, caves used intermittently, and knowledge systems transmitted orally. Such traces are easy to miss, easy to misinterpret, and easy to erase. The genome becomes the loudest surviving voice not because genes matter more than culture, but because genes endure in descendants even when culture and material traces vanish.

At the same time, we must avoid the trap of turning harshness into mystique. Endurance does not imply superhuman capacities. It implies that selection favored certain traits and that culture provided adaptive strategies. Denisovans were not outside nature. They were inside the same evolutionary system as sapiens and Neanderthals—diverging, adapting, and sometimes mixing. Their endurance in difficult settings would have been the result of ordinary processes acting over long time scales. What makes it feel extraordinary is that we are only now becoming aware of how much of that endurance may still live inside us as inherited fragments.

The Denisovan story, therefore, becomes a story of boundary expansion. It expands the boundary of where humans could live in the deep past. It expands the boundary of how quickly adaptation can occur when gene flow supplies useful variants. It expands the boundary of what evidence can reveal when fossils are scarce. And it expands the boundary of our humility: we must accept that our understanding of ancient Asia's human ecology is still incomplete, and that the most important clues may not come from a perfectly preserved skeleton, but from statistical patterns written into living genomes.

From here, the natural question is why Denisovans appear to have left such a striking genetic legacy. If their remains are scarce, why does their DNA persist in living populations? Were Denisovans, in some genetic sense, unusually distinct or unusually relevant in their contact with sapiens? Did they carry variants that were repeatedly advantageous in certain environments, allowing those fragments to persist through selection? The next subsection turns to that possibility and examines what it means to say Denisovans were genetically exceptional—carefully, without exaggeration, but without ignoring the implications of what the genome is already showing.

Calling Denisovans "genetically exceptional" can sound like a dramatic flourish, as if we are suggesting they were a kind of evolutionary elite. That is not what the evidence requires, and it is not the right tone for serious inquiry. The safer, more precise meaning is this: Denisovans appear exceptional in the way they are revealed to us and in the way their legacy behaves in modern genomes. They are exceptional because we know them first through DNA rather than through abundant fossils. They are exceptional because their genetic imprint is unevenly distributed across living populations in a way that implies multiple encounters and complex geography. They are exceptional because some of the inherited fragments associated with Denisovan-related ancestry appear to have mattered—persisting, and in at least one famous case, contributing to adaptation under severe environmental pressure. In short, Denisovans are exceptional not as mythic beings, but as a lineage whose genetic story forces a new model of what human evolution looked like across Asia.

The first layer of Denisovan "exceptionality" is simply the scale of what we don't have. With Neanderthals, the fossil record gives us a physical anchor: bodies, variation, and a clearer sense of distribution across time and space. With Denisovans, the physical anchor is far thinner, yet the genetic signature refuses to be small. This imbalance creates a striking effect. It suggests either that Denisovans were widespread and diverse or that Denisovan-related populations were part of a broader complex of archaic humans across Asia whose remains we have not yet properly sorted. In either case, the genome is telling us that Denisovans were not a tiny, isolated cave population. They belonged to a larger human world in Asia, one whose breadth our current fossil record only faintly outlines.

A second layer lies in the internal complexity hinted at by Denisovan DNA itself. A lineage that contributed ancestry across large regions likely contained sub-branches—regional populations that diverged from one another over time. This is not speculation in a romantic sense; it is a demographic expectation. If Denisovan-related groups occupied vast areas with mountains, deserts, forests, and changing corridors, they would have experienced isolation and reconnection. Such patterns produce genetic structure: different Denisovan populations would not all be identical. Some might have been closer to certain sapiens contact zones. Some might have interacted with other archaic groups. Some might have persisted in refuges. The term "Denisovan" may therefore function, at least with our current evidence, as a broad label for a complex lineage rather than a single uniform population.

This matters because when modern humans carry "Denisovan ancestry," they may not all be carrying it from the same Denisovan population. The introgressed fragments could come from different Denisovan branches, acquired in different regions at different times. That would help explain why Denisovan ancestry is not uniform across Asia and Oceania, and why some inherited segments appear to reflect particular kinds of selection in particular contexts. In a web model, Denisovans are not a single node. They are a cluster of nodes—a sub-network within the broader human network.

A third layer of exceptionality comes from how Denisovan contributions appear where environmental pressure is severe. The high-altitude adaptation associated with the EPAS1 region in Tibetans is the most famous case. It stands out because it links archaic introgression with a clear physiological advantage in a clearly extreme environment. This case suggests not only that Denisovans interbred with modern humans, but that at least some Denisovan-derived genetic variants were beneficial enough to be strongly favored by selection in certain modern populations. That is unusual in its clarity. Much introgressed DNA tends to be neutral or selected against. The persistence and rise of certain segments implies that Denisovan-derived variation sometimes helped modern humans survive.

But why would Denisovans, in particular, have variants that proved useful? The simplest explanation is ecological history. If Denisovan-related populations lived in Asia for long spans, they would have been shaped by Asian environments—altitude, climate variability, local pathogens, diets, and ecological constraints that differ from those in

Africa where modern humans evolved. When modern humans expanded into these environments, they faced pressures they had not previously experienced. Admixture could provide shortcuts: pre-adapted variants already filtered by selection in local conditions. In this sense, Denisovan "exceptionality" may reflect the fact that their lineage history overlapped with environments that later became crucial for sapiens expansion. Their distinctness, combined with their long residence in those landscapes, could make their genetic variants more likely to matter when introduced into sapiens populations.

This is not to say Denisovans were uniquely advanced. It is to say they were locally adapted. And local adaptation becomes exceptionally valuable when a new population arrives. In modern human history, we see the same principle: populations living in particular disease environments often carry immune variants shaped by long-term exposure. Populations living with particular diets often carry metabolic variants favored by their local food ecology. Populations living in high UV exposure regions often show selection related to skin pigmentation. The Denisovan case extends that logic into deep time and across hominin lineages.

There is also a more technical kind of exceptionality that may be at play: the degree of divergence between Denisovans and modern humans. The more divergent two lineages are, the more "novel" the introgressed variants can be relative to the receiving population's existing variation. Novelty is a double-edged sword. Many novel variants can be incompatible with the receiving genetic background and thus selected against. But occasionally, novelty can introduce a beneficial solution that the receiving population did not have. In such cases, the divergent lineage becomes a reservoir of alternative variants—some harmful, some neutral, some useful. Denisovans, being distinct from both Neanderthals and modern humans, may have carried unique variant combinations that, under certain pressures, became valuable additions to modern human genomes.

However, it is crucial not to overstate how "exceptional" this makes them. The genome suggests that introgression from archaic humans was sometimes beneficial, but it also suggests that much archaic DNA was gradually purged over time. This is a normal evolutionary outcome: when two lineages that have been separated for long spans mix, many combinations are not optimal. Selection tends to remove deleterious introgressed segments. The segments that remain are those that are neutral or beneficial, or those that hitchhiked along with demographic expansion. When we look at the remaining Denisovan-derived fragments in modern populations, we are seeing a filtered legacy. It is not a raw imprint of Denisovan biology. It is the portion that survived the sieve of selection.

That filtering effect can create the impression of exceptionality because what persists is disproportionately the part that mattered. If a Denisovan-derived segment contributed to survival at altitude, it would rise in frequency. If another contributed to immune response, it might persist. The outcome is a Denisovan legacy that is not evenly distributed across the genome but concentrated in certain regions where selection favored retention. This can make Denisovan ancestry appear especially significant in those contexts, even if the broader contribution is modest. In other words, Denisovans may look exceptional because selection highlights the parts of their legacy that were useful.

There is another, more subtle dimension: Denisovans represent a form of knowledge about the past that cannot be easily replaced by fossils alone. They show how genomics can reveal entire chapters of human evolution that would otherwise remain invisible. In that sense, Denisovans are exceptional as a methodological turning point. They illustrate the power of DNA to uncover the hidden diversity of ancient hominins, especially in regions where preservation is poor and the fossil record is thin. The Denisovan story is a warning to our older confidence: if we could miss a lineage this significant, how much else might still be missing?

This methodological exceptionality also changes how we interpret other puzzling evidence. When archaeologists find tool traditions that do not fit cleanly into expected narratives, or fossils that resist easy classification, Denisovans become a reminder that our categories may be incomplete. An ambiguous fossil might not be "just an odd Neanderthal" or "just an early sapiens." It might belong to a lineage we have not yet properly mapped. Denisovans widen the space of plausible explanation without loosening standards of proof. They do not justify wild claims; they justify humility and openness to complexity.

In the broader structure of this chapter, Denisovan genetic exceptionality sets up a crucial expansion of scope. If Denisovans were diverse and widespread enough to contribute ancestry across broad regions, then their contact with modern humans was likely not limited to a single encounter. It may have occurred multiple times, in multiple places, across long spans. The legacy would therefore appear not as a single uniform signal but as a mosaic: different populations carrying different fragments, reflecting different contact histories. That mosaic is visible in the geographic pattern of Denisovan ancestry across Asia and Oceania, where some populations carry higher proportions and distinctive signatures. Understanding that mosaic is essential, because it tells us not only that Denisovans existed, but how they interacted with the expanding networks of Homo sapiens.

The next subsection turns directly to that distribution. It follows the Denisovan threads across the map, not as a simple percentage in a chart, but as evidence of real ancient encounters—contact zones, migrations, and admixture events that shaped the genomes of living people. In doing so, it will deepen the Denisovan mystery while also making it more concrete: Denisovans were not just a genetic shadow. They were an interacting lineage whose traces spread far beyond the small number of remains we have found.

If Denisovans were known only from a single cave and a single genetic surprise, they might remain a curiosity—an odd branch glimpsed once and then lost again in the fog. But the Denisovan story refuses to stay local. It sprawls across a map. It appears not as a uniform, evenly distributed trace in all people, but as a patterned legacy—strong in some regions, faint or absent in others, and shaped by geography in a way that implies real contact zones and real migrations. When you follow Denisovan admixture across Asia and Oceania, you are not simply tracking percentages in a database. You are tracing ancient meetings that reshaped populations.

The most striking feature of Denisovan ancestry is its unevenness. Modern human populations differ in how much Denisovan-related DNA they carry, and that difference is not random. It clusters strongly in parts of Oceania and in some populations of Island Southeast Asia, while being generally lower in most other regions. This pattern alone implies something important: Denisovan admixture was not a single event that affected all non-African modern humans equally. Instead, it suggests multiple admixture events or admixture that occurred in particular regions among particular populations, followed by migrations and expansions that carried those traces forward.

In a ladder narrative, this would be hard to explain because the ladder assumes clean succession. In a web narrative, it is expected. Populations move in waves. Different waves meet different neighbors. Some expand and dominate. Others remain localized. Some pass through corridors where admixture is possible; others do not. Denisovan ancestry, in this sense, becomes a genetic map of ancient pathways.

The strongest Denisovan signal in the world today is often associated with populations in Oceania, particularly in places like Papua New Guinea and in some groups linked to the broader Melanesian region. This is not a minor trace. It is a substantial component relative to the Denisovan signals seen in many other modern populations. The existence of this strong signal forces a conclusion that cannot be comfortably avoided: as modern humans moved into the eastern reaches of Asia and toward Oceania, they encountered Denisovan-related populations and interbred with them. This was not an isolated interbreeding event with no lasting consequence. The legacy is large enough to imply that the encounter mattered demographically.

That raises the next question: where did the encounter happen? The modern distribution suggests that some of the most consequential Denisovan admixture occurred somewhere in Asia before or during the movement of modern humans into the regions that would later connect to Oceania. But pinning down an exact location is difficult because genes move. Once admixture enters a population, later migrations can spread it, and later population replacements can erase or dilute it. The Pleistocene and early Holocene were times of enormous movement through Southeast Asia and the island chains leading toward Sahul—the Ice Age landmass that connected Australia and New Guinea when sea levels were lower. Coastlines were different. Islands were larger. Some straits were narrower. Human groups likely moved along coastlines and through island corridors, meeting other hominins who had long histories in Asia.

The key point is not to name a single precise place as if we had a pin on the map. The key point is the logic of the pattern. Strong Denisovan ancestry in Oceania suggests admixture occurred in the ancestry of populations that contributed heavily to the peopling of those regions. It also suggests that later population histories—bottlenecks, founder effects, isolation—could amplify that ancestry. If a small group with Denisovan admixture became the founding population of a region, the Denisovan signature could become proportionally stronger in descendants simply because of how inheritance works. Founder effects can magnify certain genetic components, making the legacy of an admixture event look larger in the modern population than it might have been across a broader ancient landscape.

At the same time, the distribution across Asia suggests complexity beyond a single "Oceania encounter." Some Asian populations carry Denisovan-related signals, though typically at lower levels than in many Oceanian groups. This implies either that admixture occurred in multiple places, or that Denisovan-related populations interacted with different waves of modern human expansion across Asia. Asia is vast. It contains mountain systems, deserts, river basins, forest belts, and island arcs. It contains refuges where archaic populations could persist while climates shifted elsewhere. It would be surprising if all Denisovan contact happened once, in one place, and then never again. A web model expects repeated opportunities for contact wherever ranges overlapped.

There is also reason to suspect that "Denisovan" admixture may not represent a single, uniform Denisovan source. If Denisovan-related populations were geographically widespread and internally structured, then modern human populations could have received ancestry from different Denisovan branches in different regions. In that case, two populations might both carry "Denisovan ancestry" but from slightly different Denisovan populations—different versions of the Denisovan lineage shaped by different local histories. This would help explain why the genetic signatures can vary: not only in quantity, but potentially in the specific segments retained.

This is an important nuance because it keeps us from treating Denisovans as a single tribe that met sapiens once. Denisovans, as far as the genome hints, may represent a lineage with regional diversity, spread across a large geographic space, with contact occurring in multiple zones. Their admixture into modern humans would then not be one story but many—overlapping episodes of contact embedded in larger waves of migration.

Oceania itself underscores how deeply geography shapes genetic legacy. The region's peopling involved movement through island chains, across water barriers, into environments that could isolate populations for long spans. Isolation tends to preserve genetic signatures because it limits later dilution from new incoming groups. Where later migrations did occur—especially in the Holocene—they could reshuffle ancestry patterns, but the deep Denisovan signal in some Oceanian populations suggests that at least some lineages remained relatively continuous and that Denisovan-derived segments persisted through long demographic histories. In the web model, Oceania can function as a kind of genetic refuge: a place where certain deep signatures remain visible because later mixing was limited or uneven.

But we have to be cautious about the word "refuge." It does not mean a pristine preservation of ancient ancestry. Human populations in

Oceania have complex histories, with multiple migrations and cultural transformations over time. Genetics can reveal some of these layers, but it can also be misread if one assumes modern populations are frozen snapshots of a single ancient moment. They are not. They are the product of ongoing history. Denisovan admixture is one layer in that history—an ancient layer that remained detectable.

The Asian side of the map also complicates simple narratives. Mainland Asia and Island Southeast Asia experienced multiple waves of modern human movement, likely including early dispersals along southern routes and later expansions. Different waves could have encountered different archaic populations. Some admixture events could have occurred early and then been diluted by later incoming populations with less archaic ancestry. Other events could have occurred later and remained localized. The resulting genetic landscape would look patchy—and that is exactly what we see when Denisovan signals are compared across populations. Patchiness is not a defect in the data. It is a signature of real demographic complexity.

This patchiness also helps explain why Denisovans remained hidden for so long. A lineage that contributed unevenly to modern humans, more strongly in some regions and less in others, would not be obvious if research focused heavily on certain populations and regions. Much early genetic work was biased toward European datasets, simply because of accessibility and historical priorities in research institutions. As sampling broadened across Asia and Oceania, the Denisovan story became clearer. In that sense, Denisovan admixture across the map is also a lesson about the sociology of science: what we discover depends in part on where we look.

If we step back, the Denisovan admixture pattern implies at least four broad realities. First, Denisovan-related populations were real and interacted with modern humans. Second, they likely had a wide geographic footprint, or at least a footprint wide enough to intersect with major modern human dispersals across Asia. Third, admixture was not a one-time anomaly but a process that may have occurred multiple times in different zones. Fourth, the modern distribution of Denisovan ancestry reflects not only ancient encounters but later demographic shaping—founder effects, isolation, expansions, and replacements that magnified or diluted the archaic legacy.

This is where the Denisovan mystery becomes both more grounded and more haunting. Grounded, because we are not dealing with myth; we are dealing with measurable inheritance patterns. Haunting, because those patterns imply a vast human landscape we can barely see—populations that were once living and local, now reduced to genetic fragments in descendants spread across continents and islands.

And this leads naturally to the chapter's closing movement. If Denisovan ancestry is distributed in a patterned way, and if some Denisovan-derived segments contributed to survival in extreme environments, then the Denisovan legacy is not merely historical. It is biological. It is functional in at least some cases. The next subsection turns directly to that idea with deliberate care. It asks what it means to speak of an "ancient upgrade" hidden in our cells—not as a claim of engineered intervention, but as a metaphor for how introgressed genetic variants can act like inherited solutions, quietly shaping what modern humans can endure.

The word upgrade is dangerous because it tempts the mind into the wrong genre. It can conjure images of engineers at a workbench, of intentional design, of a planned enhancement inserted into a species like a new component in a machine. That is not what the evidence demands, and it is not what disciplined inquiry should claim. But the word also captures something real, if it is used as metaphor rather than literal description. In a biological sense, an "upgrade" can simply mean an inherited solution—an adaptation that improves survival in a particular context, introduced into a population not by slow internal mutation alone, but by the meeting and mixing of lineages. When we speak carefully, that is what Denisovan legacy represents: ancient variants, shaped by selection in harsh environments, carried forward into modern humans through admixture, and still quietly active within our physiology.

To see why this metaphor has power, return to the problem of time. In the simplified narrative, evolution changes a population by waiting. It waits for random mutations to arise. It waits for selection to favor them. It waits for generations to pass. This does happen, and over long spans it can produce extraordinary transformation. But it is not the only mechanism available. When two populations that have been separated for a long time come back into contact, they carry different genetic experiments—different sets of variants shaped by different environments. If they interbreed, those experiments can be exchanged. A population does not need to invent every solution from scratch. It can inherit a solution that already exists.

That inheritance can look like an acceleration. A trait that would have taken thousands of years to evolve internally can appear, in functional form, much faster, because the variant responsible was already present in another lineage. This is not cheating. It is evolution operating through a different channel: gene flow. In the human story, Denisovan-related introgression offers some of the clearest evidence that this channel mattered. The Tibetan high-altitude adaptation is the most famous illustration because the environment is extreme and the physiological challenge is direct. But the broader point is that Denisovan-derived segments persisted in modern populations not simply because of chance, but in some cases because selection preserved them.

If you imagine the genome as a library, then introgression adds books your lineage did not write. You might not use most of them. Some might not fit with the rest of your collection and will be discarded over time. But a few contain knowledge that becomes essential under new conditions. Those few are kept, copied, and passed on. Over generations, the library becomes a composite—mostly authored by the dominant lineage, but containing crucial volumes inherited from others. The Denisovan "upgrade," in this metaphor, is the retention of those useful volumes.

The science behind this is not mystical. It is ordinary population genetics. Introgressed segments enter a population through interbreeding. Recombination breaks these segments into smaller pieces over time. Selection then acts on the variation. Deleterious segments tend to be reduced. Neutral segments drift. Beneficial segments can increase in frequency. The final result, thousands of years later, is a filtered legacy: a small subset of archaic DNA that persists because it either did no harm or did some good. When we look at modern genomes and find Denisovan-derived segments that are unusually common in certain populations, we are seeing the consequence of that filtering.

There is something philosophically striking about this process. It suggests that modern humans carry not only the inheritance of sapiens ancestors, but the residue of other human histories—experiments in adaptation that occurred in parallel and then were partially merged into our lineage. That residue is often invisible in daily life. We do not feel it as a distinct identity. It is not a second self inside us. But it can shape how our bodies respond to environments. It can influence physiology, immune responses, and perhaps other traits that affect survival and reproduction. In this sense, the Denisovan story is not merely about the past. It is about the living present as a composite of multiple pasts.

This composite reality also changes how we interpret the survival of Homo sapiens. In a ladder narrative, sapiens survives because it is the best. In a web narrative, sapiens survives because it is resilient—because it is mobile, networked, culturally flexible, and sometimes biologically permeable. Permeability here is not a moral category; it is a demographic fact. A lineage that can absorb others rather than remaining isolated can increase its genetic diversity and, in some cases, acquire adaptive variants that expand its ecological range. This does not mean admixture is always beneficial. It can introduce harmful variants too. But as a mechanism, it increases the repertoire of variation available to selection. In environments that change rapidly, variation is currency.

This is why it is reasonable, within careful bounds, to speak of ancient upgrades: not as engineered enhancements, but as inherited advantages. If Denisovan-related populations carried variants shaped by long residence in Asian environments, then modern humans entering those environments could, through admixture, gain access to that locally shaped repertoire. It is a kind of evolutionary borrowing. And like all borrowing, it comes with costs and complexities, but it can be decisive.

The metaphor becomes even more compelling when you consider the extreme environments that frame this chapter. High altitude is one. Another is pathogen exposure. When modern humans expanded beyond Africa, they encountered new disease landscapes. Archaic populations had long histories in those landscapes and may have carried immune variants that were advantageous. If some of those variants entered sapiens through introgression, then the "upgrade" would not be dramatic in outward appearance; it would be quiet, operating in immune signaling pathways and cellular responses. Yet it could have a large effect on survival, especially in children. In deep time, survival often hinges on these invisible battles.

Another domain is metabolism and adaptation to local diets and climates. Harsh winters, variable food supply, and different ecological resources could favor certain metabolic strategies. If archaic lineages carried variants suited to these conditions, introgression could introduce them into sapiens populations. Again, the effect would not look like a sudden invention; it would look like a population becoming more fit in a particular context.

But we must keep the boundary clear. Genetics can indicate that certain segments were introgressed and that selection may have acted on them. It does not always allow us to tell a complete functional story. Some retained segments may be beneficial in ways we have not yet identified. Others may persist because they are linked to beneficial regions, hitchhiking along. Some may persist because they are neutral and simply carried forward by demographic expansion. The temptation, especially in popular storytelling, is to turn every introgressed fragment into a purposeful upgrade. That is not warranted. The responsible way to tell the story is to acknowledge that introgression created a pool of archaic variants, that selection filtered that pool, and that some resulting segments appear to have contributed to adaptation under specific pressures.

This careful framing still leaves room for wonder. There is something genuinely awe-inspiring in the idea that a vanished human lineage can still shape the physiology of living people. The Denisovans, who remain mostly invisible in fossils, are nevertheless present in bodies. They are present in the way a highlander breathes and carries oxygen. They are present as faint genomic threads woven into modern populations across vast regions. This is not romance. It is inheritance. It is the deep past acting through biology. It also reframes the meaning of disappearance. Denisovans are gone as a distinct population, but their legacy is not zero. They did not simply vanish without trace. Parts of their genetic story continued because they were carried into sapiens networks that expanded and survived. In that sense, admixture is a kind of biological preservation. It preserves fragments of lineages that would otherwise be erased entirely. It is not a full preservation—most of what made Denisovans Denisovan is gone—but it is enough to matter. The web ensures that some strands persist even when nodes vanish.

This is where the Denisovan chapter completes its arc. It began with a lineage known almost entirely through DNA, a ghost made real by molecular evidence. It moved into a specific case where Denisovan-derived ancestry appears connected to a crucial adaptation for life at extreme altitude. It expanded outward into the broader distribution of Denisovan admixture across Asia and Oceania, showing that the legacy was not local or trivial. And it ends here, with the metaphor of an ancient upgrade—not as a claim of engineering, but as a disciplined way to describe what gene flow and selection can do: introduce solutions from one lineage into another, allowing a composite species to endure where it might otherwise fail. The Denisovan mystery, then, is not only a mystery of identity—what they looked like, where exactly they lived, how their cultures functioned. It is also a mystery of inheritance: how much of modern humanity is shaped by lineages we can barely see, and how many adaptive traits we carry that did not originate solely within our own sapiens line. Denisovans are the clearest demonstration so far that the missing chapters of human evolution are not merely missing pages in a book. They are missing peoples whose influence remains written into the biology of living descendants.

CHAPTER 4 — HIGH-ALTITUDE PEOPLES AND ANCIENT ADAPTATION

High altitude is one of nature's most unforgiving tests, not because it is visibly violent, but because it is quietly absolute. The mountains do not need to attack you with predators or storms to make their judgment felt. They simply reduce oxygen and let physiology do the rest. For most of human history, the thin air of the great highlands would have been an invisible wall. People could pass through, hunt seasonally, retreat when the body began to fail. But permanent life—birth, childhood growth, pregnancy, aging—requires more than courage. It requires a biological and cultural relationship with altitude deep enough to endure across generations. That is why the world's high-altitude populations matter so much in any serious discussion of human adaptation. Tibetans on the roof of Asia, Andean highlanders in South America, and Ethiopian highlanders in the Horn of Africa represent three of the most remarkable examples of human endurance in extreme hypoxic environments. They also represent something even more instructive: different solutions to the same problem, shaped by different histories.

The first temptation is to treat high-altitude adaptation as a single trait, as if "living at altitude" produces one universal human response. But the evidence points toward multiple adaptation strategies. The body can respond to thin air in several ways—through changes in breathing, blood oxygen transport, circulation, and developmental patterns. Populations living at altitude for long periods may converge on certain outcomes, such as maintaining adequate oxygen delivery, but they do not necessarily converge on identical mechanisms. This is exactly what makes the Tibetan, Andean, and Ethiopian cases so compelling. They are not three copies of the same adaptation. They are three distinct evolutionary pathways, three approaches to living in air that punishes the unprepared.

Tibetans are often discussed first because their adaptation has become a flagship example of archaic contribution and strong selection. Their physiology tends toward an efficient balance: many Tibetans can live at extreme elevation without the same degree of elevated hemoglobin concentration seen in many acclimatizing lowlanders. In practical terms, this means they can avoid some of the costs of chronic thickened blood while still maintaining functional oxygen transport. The body does not simply crank the "more red blood cells" dial to the maximum; it adjusts in ways that appear tuned to long-term health and reproductive success at altitude. This is not a guarantee of immunity from altitude stress, but it is a population-level pattern that points to deep adaptation.

Andean highlanders, by contrast, are often associated with a different physiological profile. Many Andean populations show elevated hemoglobin concentrations compared to lowlanders, a strategy that increases oxygen-carrying capacity by producing more red blood cells. This can be effective, and it has clearly supported long-term highland life in the Andes. Yet it also comes with trade-offs, particularly if hemoglobin levels become extremely high. The Andean solution, in broad strokes, seems to accept the increased red blood cell strategy more than the Tibetan pattern does. This difference is not a judgment about which is "better." It is evidence that there are multiple viable evolutionary responses to hypoxia, and that the path a population takes may depend on its genetic starting point, demographic history, and the specific altitude environment it inhabits.

Ethiopian highlanders add an even more intriguing variation to the story. Populations in the Ethiopian highlands have lived at significant elevations for long spans, yet some studies suggest they do not always show the same extreme hemoglobin elevation seen in many Andean groups. Their physiological patterns indicate that altitude adaptation can emerge in different configurations again—potentially involving cardiovascular function, oxygen saturation maintenance, and other regulatory pathways. Ethiopia's highlands are also geographically and historically distinct from Tibet and the Andes, with their own demographic and ecological histories. That distinctness matters. It means we are looking at three separate arenas where human populations confronted hypoxia and responded under different conditions.

What these three cases share is not a single mechanism but a single challenge: oxygen scarcity. At altitude, the body is forced to solve a basic supply problem. Every tissue depends on oxygen for metabolism. If oxygen is limited, the body must compensate through multiple interacting systems. It can increase ventilation. It can increase blood flow. It can alter how hemoglobin binds oxygen. It can modify developmental growth patterns, such as lung capacity and chest morphology, to improve lifelong oxygen uptake. It can fine-tune how cells sense hypoxia and respond at the molecular level. Over generations, selection can favor variants that make these responses more effective and less costly.

This is where the comparative approach becomes powerful. If Tibetans, Andeans, and Ethiopian highlanders have different adaptation profiles, then high-altitude life is not explained by a single "altitude gene" or a single evolutionary switch. Instead, altitude adaptation appears to be a multi-path problem. Different populations found different solutions because evolution works with what it has. It does not design from scratch; it modifies existing variation. A population's ancestral genetic background influences which variants are available for selection to amplify. Its demographic history influences how quickly selection can act and how strongly certain variants can spread. Its contact history—whether it interbred with other populations carrying useful variants—can introduce new material into the system. And its environment—exact elevation range, temperature, diet, disease exposures—can shape which trade-offs are tolerable.

The timelines for these adaptations also matter. High-altitude environments have not always been occupied continuously by the same populations, and the length of residence at altitude influences both genetic and cultural adaptation. The Tibetan Plateau and surrounding highlands have complex settlement histories, and not all human presence there has been permanent. Seasonal mobility can precede permanent settlement by long spans. In the Andes, highland habitation is tied to broader patterns of settlement and agriculture, and the highlands became centers of major civilizations that depended on long-term high-elevation living. In Ethiopia, the highlands have long supported dense populations and complex societies as well. Each region has its own story of when and how humans became highlanders, and those stories shape the biological pathways that became available.

Yet even without pinning down every archaeological and demographic detail, the basic evolutionary lesson is clear. When the same pressure is applied in different contexts, evolution can produce different solutions. That is why comparing these three populations is so valuable. It prevents us from treating the Tibetan case as the only model. It also prevents us from imagining that every highland population must share the same archaic inheritance. Tibetans may carry certain archaic-derived variants associated with hypoxia response, but Andeans and Ethiopian highlanders demonstrate that adaptation can occur through other routes as well.

In some cases, these routes may involve changes in gene regulation, developmental patterns, or physiological trade-offs that do not mirror the Tibetan profile.

At a deeper level, the comparison also reframes what "human limits" mean. Lowlanders often experience altitude as an assault because their bodies are not tuned for it. That experience can mislead us into thinking altitude is inherently incompatible with human life. The existence of long-term highland populations contradicts that assumption. Humans can inhabit hypoxic environments permanently, not through a single miracle trait, but through a spectrum of adaptations that integrate biology and culture. Culture matters because highland life requires specialized knowledge: how to manage cold, how to secure food in short growing seasons, how to build shelter, how to move through rugged terrain, how to raise children where oxygen is scarce. Biology matters because without genetic and developmental tuning, the costs of hypoxia can undermine reproductive success across generations. The highland populations demonstrate that the boundary of habitability is not fixed. It is negotiable, given time, selection, and cultural ingenuity.

This is also where the chapter's theme begins to widen beyond physiology. High mountains are not only biological challenges; they are cultural magnets. They become refuges, sacred zones, places associated with gods and ancestors, and in many traditions, places where knowledge descends from above. That cultural dimension will matter later in this chapter, but it is anchored here in the simple reality that highland populations often occupy landscapes that feel closer to the sky, harsher, and more separate from lowland life. Over time, that separation can shape identity and memory. It can create stories of distinct origins, of ancestral endurance, of people who are not quite like others.

For now, the key point is comparative clarity. Tibetans, Andeans, and Ethiopian highlanders show us three living examples of deep adaptation to hypoxia, each with its own physiological signature. Together they demonstrate that high-altitude adaptation is not a single trick but a set of strategies. They also set up the next question naturally. If these populations show adaptation patterns that appear strong and, in some cases, remarkably efficient, what does that imply about the pace of evolution? Do these adaptations fit comfortably within conventional timelines, or do they appear to have emerged with unusual speed, suggesting powerful selection or inherited variants that allowed rapid change? The next subsection turns to that tension—the feeling that high-altitude physiology, in some cases, seems to outpace what we expect—and asks what that means when we treat human evolution as a web rather than a ladder.

High-altitude adaptation unsettles people because it doesn't behave like the story we have been taught to expect. The popular model of evolution is slow and incremental, unfolding at a pace so gradual it becomes almost invisible. It is imagined as a gentle slope: tiny changes accumulating until, after immense time, a population looks different. But the physiology of long-term highland peoples—especially when compared to the immediate suffering of lowlanders at altitude—can feel like something sharper than a slope. It can feel like a leap. The body's ability to regulate oxygen response without spiraling into chronic pathology, to sustain healthy pregnancies, to raise children whose development is tuned to thin air, can appear too effective, too targeted, too complete for the casual timeline people carry in their heads. That sense of surprise is the starting point for this subsection: not to claim evolution was violated, but to show why the evidence forces a more realistic model of how fast selection can move when the pressure is intense.

Altitude creates an unusually clean selection environment. In many ecological settings, survival and reproductive success depend on dozens of variables that blur together: diet, pathogens, predators, social status, random accidents. At altitude, one variable dominates in a way that is brutal and measurable: oxygen scarcity. Hypoxia does not merely inconvenience the body; it attacks core functions. It strains the heart and lungs. It complicates pregnancy. It increases infant mortality risk if the physiology is not tuned. It reduces physical performance, which affects foraging, labor, defense, and social dynamics. This makes altitude a kind of evolutionary amplifier. If a variant improves oxygen management, even modestly, it can have outsize effects on survival and reproduction. Under such conditions, selection can be fierce. And fierce selection can produce rapid change.

Here is the first correction to the casual timeline myth: evolution is not always slow. It is often slow when selection is weak, when the environment is stable, or when the trait under selection has small effects. But when selection is strong—when the difference between functioning and failing is constant—allele frequencies can shift quickly. In populations living permanently at high altitude, the selective pressure is not occasional. It is continuous. Every pregnancy, every childhood, every winter, every illness occurs under hypoxic constraint. Over hundreds or a few thousand years, such pressure can reshape a population's genetic profile in meaningful ways, especially if the population remains at altitude long enough for selection to act repeatedly.

The second correction is demographic. The pace of visible adaptation depends not only on selection strength but on population size, structure, and gene flow. Small populations can change quickly in genetic makeup due to drift, but drift is random; it does not reliably produce adaptive solutions. However, small founder populations can amplify advantageous variants if those variants happen to be present early and selection favors them strongly. Large populations can maintain more genetic diversity, increasing the chances that useful variants already exist in the gene pool. And gene flow—movement and mixing—can introduce variants from outside, accelerating adaptation by providing pre-existing solutions. This is where the web model becomes essential. A population does not need to wait for a new mutation to appear if it can acquire an advantageous variant through contact.

High-altitude physiology often feels like it "defies timelines" because people imagine adaptation must start from zero. They picture lowland humans arriving at altitude with no relevant variation and then waiting for a perfect mutation to appear. In reality, variation is rarely zero. Human populations already contain many variants affecting oxygen sensing, hemoglobin regulation, vascular function, and metabolism. Even small differences in these systems can influence how well individuals tolerate hypoxia. If a population begins living at altitude, selection does not need to invent an entirely new physiological pathway. It can shift the balance among existing pathways. It can favor regulatory patterns that reduce harmful overreactions, like excessive hemoglobin elevation, while maintaining effective oxygen delivery. It can favor developmental patterns that expand lung capacity or alter circulation. Evolution often works by tuning systems rather than creating brand-new ones.

The Tibetan case intensifies this logic because it suggests a particularly powerful accelerator: introgression from an archaic lineage. If a Denisovan-related population had already experienced long exposure to hypoxic environments, it could have carried variants that were advantageous for altitude living. When modern humans encountered and interbred with that population, those variants could enter the modern human gene pool. Then selection could act on them rapidly in high-altitude contexts. The result would look dramatic—an effective adaptation emerging on a shorter timescale than expected—not because evolution became magical, but because the population acquired a head start. In the metaphor of "upgrade," the upgrade is simply a borrowed solution.

This is why high-altitude adaptation becomes a case study in how evolution can appear sudden. It is not truly sudden in the sense of happening in one generation. But it can be sudden relative to the intuitive timescales people hold in their minds. Thousands of years can be enough to produce major shifts when selection is intense and when the trait affects reproductive success. The same principle is visible in other well-documented human adaptations, such as lactase persistence in certain populations. The point is not to pile examples; it is to establish that human evolution can, under the right pressures, move faster than the popular story implies.

Andean and Ethiopian highland adaptations complicate the picture further by showing that there is no single timeline. Different populations may have occupied high elevations for different lengths of time. Their demographic histories differ. Their genetic starting points differ. Their contact histories differ. As a result, their physiological solutions differ. One population may rely more heavily on increased hemoglobin, which may be achievable through selection on variants that influence red blood cell production. Another may rely more on regulatory tuning that avoids extreme hemoglobin elevation. Another may show patterns suggesting different cardiovascular or developmental strategies. The presence of multiple solutions can make the timelines feel even stranger: if there are multiple pathways, why do they look so effective? The answer remains the same: strong selection acting on existing variation, shaped by demographic and contact history.

The “defies timelines” feeling also arises because high-altitude adaptation involves systems people assume are rigid. Oxygen regulation feels fundamental, like something that should not be easily altered. But biology routinely adjusts fundamental systems. Hemoglobin affinity varies across species adapted to altitude. Lung capacity, vascular networks, and cellular hypoxia pathways are all modifiable targets of selection. Humans are not exceptions. What feels rigid in one environment becomes plastic under another when survival demands it. The shock is not that adaptation exists; it is that the human body, which feels stable in daily life, can be tuned over generations into forms that make extreme environments livable.

There is another dimension to the timeline tension: cultural buffering. Culture can reduce mortality long enough for biological adaptation to take hold. A population can move into highlands and survive initially through technology—clothing, fire, shelter, diet—and social strategies—cooperation, seasonal mobility, trade with lowlanders. These cultural strategies do not eliminate hypoxia, but they can stabilize the population enough for selection to operate across generations. Over time, biological adaptation and cultural adaptation reinforce one another. This can create the impression that the population “arrived adapted,” when in fact the adaptation was a layered process: culture enabling persistence, selection shaping physiology, and both together producing a stable highland lifestyle.

The timeline tension also exposes a deeper misunderstanding about what counts as evidence for rapid change. People imagine that to prove rapid evolution, you need a fossil record showing dramatic anatomical transformation in a short time. But many crucial adaptations are not dramatic in skeletons. They are physiological and regulatory. They live in gene expression patterns, vascular function, and metabolic balance. These leave fewer obvious archaeological traces. They can evolve quickly without leaving dramatic bones. In that sense, high-altitude adaptation can feel like it "came out of nowhere" because the adaptation is largely invisible in the material record. The genome, once again, becomes the place where the story is most legible.

This is where the volume's theme of missing chapters becomes more than a metaphor. If a population acquires a crucial adaptation through archaic introgression, the most important event in that adaptation story is not a new tool or a new settlement pattern. It is an encounter—two lineages meeting, interbreeding, and passing an advantageous variant forward. That encounter might leave few archaeological signatures. It might leave no obvious bones. Yet it can reshape the physiology of thousands of descendants. The chapter is missing not because nothing happened, but because the crucial events occurred in the domain that fossilizes least: population contact, reproduction, and genetic transmission.

So does high-altitude physiology truly "defy" evolutionary timelines? Only if we cling to a simplified timeline myth. In reality, high-altitude adaptation shows that timelines are not single numbers. They are outcomes of interacting forces: selection strength, demographic structure, genetic variation, cultural buffering, and gene flow. Under strong pressure, with existing variation and occasional borrowed solutions, significant adaptation can occur on timescales that feel fast to human intuition. The highlands therefore do not violate evolution; they reveal evolution's real speed when the environment demands it.

This sets up the next question naturally. If some highland adaptations were accelerated by inherited variants, then where did those variants come from? Tibet provides one famous case where archaic introgression appears central, but the broader story likely involves multiple episodes of genetic borrowing—some from known archaic lineages, some perhaps from populations we have not yet identified. The next subsection turns to that possibility and examines how high-altitude adaptation may carry signatures of ancestry that is older, stranger, or simply less visible in the fossil record—genetic borrowing from unknown ancestors.

High-altitude adaptation is often presented as a clean, solved case: thin air produces selective pressure, selection favors helpful variants, and a population becomes more fit to live in hypoxic environments. That is true in broad outline, but it can also be misleading, because it implies that all the ingredients for adaptation must come from within the population itself. The deeper genetic picture suggests something more intricate. In some cases, the most powerful adaptations may not have been invented locally in the strict sense. They may have been inherited—borrowed from other populations through gene flow, including populations we can barely see in the fossil record or cannot yet name with confidence. The Tibetan case, with its archaic-linked altitude variant, is the most famous example, but it also serves as a gateway into a wider principle: human populations have repeatedly drawn on a reservoir of ancestral diversity, and some of that diversity may come from unknown ancestors whose existence is inferred more than documented.

To borrow genes is not to borrow in a cultural sense, with conscious intent. It is to exchange inherited variation through interbreeding. When two populations meet and produce offspring who become part of a community, the gene pools mix. Over generations, the mixture is recombined, filtered by selection, and shaped by drift. Some fragments vanish. Some persist. A few become crucial. In the context of extreme environments like high mountains, those crucial fragments can have outsized impact because the pressures are severe and continuous. A variant that improves oxygen regulation, reduces pregnancy complications, or stabilizes cardiovascular response can translate into more surviving children, and that advantage can spread quickly. The result is that a population's high-altitude success may contain hidden contributions from lineages that lived nearby long before, lineages whose bones may not be preserved, whose archaeological traces may be scattered, but whose genetic legacy remains.

The phrase "unknown ancestors" here does not require a leap into fantasy. It refers to something already implied by mainstream genetics: there were ancient human populations—sometimes sapiens, sometimes archaic, sometimes intermediate—that contributed ancestry to later groups but are not yet well represented by fossils or ancient genomes. The genome reveals their presence through patterns that cannot be explained by known sources alone. In some cases, these patterns look like deep divergence, suggesting admixture with a lineage separated for a long time. In other cases, they look like unsampled structure within early sapiens populations—deeply separated groups that later reconnected. Either way, the result is the same: the receiving population carries inherited variation that did not arise solely within its recent local line.

High-altitude environments are especially likely places for such borrowing to matter because mountains create contact zones and isolation zones at the same time. Mountain ranges fracture landscapes into valleys and plateaus, isolating groups for long periods. Isolation encourages genetic divergence. But mountains also create corridors—passes, ridgelines, river headwaters—that funnel movement. When climates shift, populations can be pushed upslope or downslope, bringing previously isolated groups into contact. The highlands can therefore function as both refuges and meeting grounds. In such regions, gene flow can occur between populations that have been separated long enough to accumulate distinctive variants. If any of those variants improve altitude tolerance, selection can elevate them rapidly.

This dynamic also explains why unknown ancestry can hide inside high-altitude adaptations. The highlands often preserve populations through periods of lowland instability. When climates become hotter, drier, or more chaotic, mountain zones can offer stable microclimates and reliable water sources. Populations can retreat into these refuges and persist while lowland groups contract or move. Over time, refuge populations can become genetically distinct. When conditions later allow expansion or renewed movement, these refuge lineages can contribute genes to broader populations. The genetic borrowing that results may leave a signature in traits under strong selection—like hypoxia response—because those are the traits most likely to be retained and amplified.

The Tibetan Plateau illustrates this principle with unusual clarity, but the broader argument does not require Tibet alone. The Andes and the Ethiopian highlands also represent long-term high-altitude occupation in landscapes shaped by complex migrations. In each case, a population's adaptation could involve both internal selection and inherited variants introduced through contact. The specific sources may differ. In Tibet, archaic introgression is strongly implicated in one key region. In the Andes and Ethiopia, the story may involve different genetic pathways and different demographic histories. The point is not to force every case into the same mold, but to recognize that the mechanism—borrowing from ancestral diversity—is available everywhere humans meet and mix.

This mechanism also helps resolve an apparent paradox. High-altitude adaptation sometimes looks too effective to have evolved quickly from scratch in a small, stressed population. But if the population acquired advantageous variants through admixture, then the adaptation could be both effective and relatively rapid. The borrowed variant would not be newly invented; it would be introduced. Selection would not have to wait for rare mutations; it would act on existing material. In such scenarios, the timeline feels compressed, but the process remains natural. The web accelerates evolution by expanding the pool of variation.

There is another layer to the "unknown ancestor" concept that deserves careful handling: not all unknown ancestry is archaic in the sense of being non-sapiens. Some unknown ancestry may be deeply divergent sapiens lineages that split early and remained separated, perhaps in different ecological zones, before later mixing. The public often interprets "unknown" as "non-human," but that is not necessarily what the data implies. Deep structure within Homo sapiens could produce genetic patterns that resemble admixture from an unknown source, especially if ancient population subdivision was strong. In a web model, early sapiens is not a single homogeneous wave; it is a set of populations with long internal histories. Some of the variants that later become important—especially under strong selection—could come from these early-divergent sapiens groups.

This is not an attempt to dilute the Denisovan story. It is an attempt to keep categories honest. The human past contained multiple lineages and multiple levels of difference. Some differences were between sapiens and archaic humans like Denisovans. Other differences were within sapiens itself. Both kinds of difference can contribute to modern adaptation through gene flow. When a population carries a variant that seems ancient and unusual, the correct question is not immediately "Which archaic species gave it to us?" The correct question is "What population history best explains this pattern?" Sometimes the answer will involve archaic introgression. Sometimes it will involve deep sapiens structure. Sometimes it will involve both, layered together.

Mountains make this layering especially plausible. They encourage the persistence of small, distinct populations. They also encourage recurrent contact because migration routes often pass through highland corridors. Over long spans, the same landscape can host multiple waves of human movement: early pioneers, later expansions, refuge populations, returning migrants. Each wave can bring new variation and receive old variation. Over time, the highlands become genetic palimpsests—records written over and over, with faint traces of earlier writing still visible beneath the new.

If that is true, then high-altitude peoples are not only examples of physiological adaptation. They are living archives of deep population history. Their genomes may contain signals of contact events that occurred thousands of years ago in landscapes where fossils are rare and archaeology is difficult. Those signals may be strongest precisely in genes tied to altitude fitness, because selection preserved them. In that sense, high-altitude adaptation can act like a spotlight, illuminating the inherited fragments that mattered most.

This spotlight effect also carries a warning. Because selection amplifies certain variants, it can make the ancestry story look simpler than it really was. If a borrowed variant becomes extremely common, it can dominate the narrative, making it seem like one introgression event explains everything. But the true history could involve multiple sources and multiple episodes of gene flow. The adaptation story might be a mosaic: one major variant introduced from one source, other supporting variants arising internally or introduced from other sources, and cultural practices providing the stability needed for the biology to matter. The genome can highlight the major pieces, but it cannot always reconstruct the full social and demographic context of how those pieces arrived.

This is where we begin to feel the deeper theme of mountains in human history. Mountains are not simply hard places. They are places where time behaves differently. They preserve lineages. They isolate communities. They shelter refugees. They become corridors and boundaries at once. In such places, the human web becomes tight, complex, and layered. Genetic borrowing from unknown ancestors is therefore not an exotic claim. It is a realistic expectation in landscapes where populations have repeatedly fragmented and reconnected.

From here, the chapter pivots toward a broader historical lens. If mountains can preserve and concentrate deep ancestry, then they may also preserve populations during periods of widespread environmental shock. Highlands can function as refuges when lowlands collapse—whether through drought, flooding, disease, or broader climate instability. In that context, high-altitude peoples may represent not only adaptation but survival through catastrophe. The next subsection turns to that idea and explores mountains as refuge zones after catastrophe, where human lineages may have persisted, mixed, and later re-seeded broader landscapes.

Mountains have always done two things at once. They block movement, and they preserve life. They can be barriers that isolate, slow, and redirect migrations, but they can also be reservoirs—places where water remains when the lowlands dry, where cooler temperatures soften heat, where valleys hold forests and game, and where microclimates create pockets of stability in a world that has become unstable elsewhere. In the long arc of human history, these qualities make mountains more than dramatic scenery. They make them refuges. When catastrophe reshapes the lowlands—whether through climate swings, drought, flooding, or ecological collapse—high places can become the difference between extinction and continuity. That is why high-altitude peoples are not only physiological case studies. They may also be the descendants of survival zones: populations that endured through disruptions and later contributed to the repopulation of wider regions.

The word catastrophe can tempt us toward a single apocalyptic event, but in deep time the more common catastrophe is gradual in the human sense and sudden in the geological sense. Climate shifts can reorganize whole ecosystems within centuries or even decades. Rainfall patterns move. Rivers change course. Grasslands replace forests. Coastlines are swallowed or exposed. For small human groups dependent on local resources, such changes can be devastating. A stable hunting territory can become unproductive. A reliable water source can vanish. A coast that provided abundant food can retreat beyond reach. Catastrophe, in this framework, is not necessarily a meteor strike; it is the collapse of a habitat network that a population depended on.

Mountains can moderate such collapses because of their vertical diversity. In a lowland plain, climate change can sweep across the entire region with little variation—everything warms, everything dries, everything floods. In mountainous terrain, the same climatic shift can create a patchwork of outcomes. A valley shaded by ridges might retain moisture. A slope with a different exposure might preserve vegetation longer. Snowmelt can continue to feed streams even when rainfall becomes erratic. Elevation gradients allow people and animals to move short distances to find different ecological zones. This patchwork provides options. When the lowlands become hostile, options are survival.

This is not a romantic theory about mountains as paradise. Highland life can be harsh. Oxygen is scarce, temperatures drop quickly, and travel is difficult. But in times of lowland collapse, harshness can be preferable to impossibility. A population may accept the cost of thin air if the alternative is drought, disease, resource failure, or conflict in crowded lowland refuges. Over generations, a refuge strategy can become a permanent adaptation strategy. Culture changes first—routes, seasonal patterns, food sources—and then biology can follow as selection acts on the new environment.

This helps explain why high-altitude peoples are so significant in a broader story of human resilience. Their existence suggests not only that humans can adapt to hypoxia, but that humans repeatedly chose mountains as places to persist when conditions elsewhere deteriorated. Even if we cannot always pinpoint the exact catastrophe that drove each movement, the logic of refuge remains sound across many contexts. The highlands offer water sources, defensible terrain, and ecological diversity. They allow groups to fragment and hide. They allow lineages to survive in pockets when broad regions become unstable.

In the Ice Age world, this refuge function would have been magnified. Glacial cycles repeatedly altered lowland environments. In some regions, cold and dryness could reduce the productivity of plains. In others, changing monsoons could shift vegetation belts. Sea-level changes could drown coastal refuges. Mountains, with their microclimates and glacial-fed water sources, could remain habitable even as surrounding landscapes became stressed. Populations sheltered there could retain continuity while lowland populations contracted, moved, or merged. Over time, this creates a demographic pattern that fits the genetic evidence we often see: isolated highland groups becoming distinct, then later contributing ancestry to expanding populations when conditions improve.

This is also where the Denisovan story and high-altitude adaptation intersect with a wider idea about "lost populations." If archaic lineages like Denisovans persisted in parts of Asia for long spans, mountains could have been among the landscapes that helped them persist. Rugged terrain can shelter populations from competition and from rapid replacement because movement is difficult and ecological niches are diverse. A lineage could remain in a mountainous refuge even as other lineages expanded across lowlands. Contact zones would form at the edges—passes, river valleys, foothill corridors—where groups could meet, exchange, and sometimes interbreed. In such a scenario, mountains are not only refuges from climate; they are refuges from demographic takeover. They can preserve older lineages longer than open plains might.

This idea also applies within Homo sapiens. Even if we set archaic lineages aside, mountains can preserve deep-divergent sapiens populations during periods of lowland instability. A group isolated in a highland refuge might maintain unique genetic variants. When later expansion reconnects them with wider populations, gene flow can introduce those variants into the broader gene pool. If any of those variants are advantageous—especially for altitude tolerance—they can spread rapidly. This is one plausible pathway by which "unknown ancestry" can persist and later reappear in the genetic record. The refuge protects the lineage; the reconnection distributes its legacy.

Mountains also offer social refuge. During periods of ecological stress, conflict often increases. Resources become scarce. Groups compete. Migration corridors become crowded. In such times, rugged highlands can provide defensible terrain and isolation from lowland competition. Even today, many regions show a pattern of highland communities maintaining distinct languages and identities in part because the terrain resists assimilation. It is reasonable to extend that principle cautiously into deep time. A population that retreats into mountains may not only survive ecologically; it may also maintain cultural and genetic distinctness because fewer outsiders can easily penetrate the refuge. Over time, distinctness can become pronounced.

But refuge is not isolation forever. Eventually climates shift again. Lowlands recover or become productive in new ways. Populations expand out of refuges, following water, game, and opportunity. When this happens, the highlands can function as sources as well as shelters. They release people back into wider landscapes. In a web model of human evolution, this creates a rhythmic pattern: contraction into refuges during stress, expansion out of refuges during favorable phases, and repeated mixing in contact zones. Such rhythms are exactly the kind of demographic process that produces complex genetic signatures—mosaics of ancestry, introgressed fragments, and population structure that does not fit a simple tree.

This refuge rhythm also helps us reinterpret the pace of adaptation. If a population survives in a mountain refuge through a catastrophic lowland phase, it has time—generations under consistent highland pressures—for selection to act. When that population later expands, it carries its adaptations with it. To outsiders, those adaptations can look like sudden changes, as if a group "appeared" already suited to altitude or rugged terrain. In reality, the adaptation was incubated in a refuge. The catastrophe did not create the adaptation directly; it created the conditions that sustained selection long enough for the adaptation to take root.

The Tibetan Plateau, the Andes, and the Ethiopian highlands can each be seen through this lens. These are not only high places; they are long-term human landscapes with histories of expansion and contraction tied to climate, ecology, and social change. In each region, the highlands could have served as refuges during lowland disruptions, preserving populations and their adaptations. The details differ by region, but the structural principle remains: mountains preserve continuity.

This principle also adds depth to the emotional resonance of highlands in human memory. If mountains repeatedly served as refuges during times of hardship, then cultural narratives might remember them as places of survival, origin, and sacred protection. Stories of ancestors retreating to the mountains, of gods living on peaks, of wisdom descending from high places, can reflect not only metaphor but historical experience. People who survive often sanctify the place that saved them. Over many generations, survival landscapes become sacred landscapes.

That bridge takes us naturally into the next subsection. If mountains function as refuges and sources, then they are not only biological laboratories of adaptation. They are also cultural theaters—places where human groups develop strong identities, myths, and knowledge traditions tied to the sky, the peak, and the boundary between earth and heaven. The next subsection turns to that dimension and explores why sacred peaks recur across cultures and why high places are so often associated with sky knowledge—an association that can be read as metaphor, as social memory, and as a reflection of the way highland life forces people to understand weather, seasons, and the heavens with unusual intensity.

The moment a human being stands above the treeline, the world changes. The horizon expands. Weather becomes immediate and personal. The sun's path feels sharper, more commanding, as if it is closer than it is. Night arrives with a fierce clarity, stars cutting through thin air with a brightness that can make the sky feel inhabited. In the highlands, the heavens are not an abstraction. They are a daily authority. This is one reason sacred peaks appear again and again across cultures. Mountains do not only challenge the lungs; they transform perception. They make the sky feel near, and they make survival depend on reading cycles that lowlanders can ignore. Over time, that dependence can harden into knowledge—seasonal, astronomical, meteorological—and that knowledge can become myth, ritual, and identity.

It is tempting to treat "sacred peaks" as merely poetic: people loved the beauty of mountains, so they imagined gods living there. Beauty is part of it, but the deeper reasons are practical. Mountain environments are high-risk. Storms can arrive quickly. Temperatures drop abruptly. Snowmelt can mean life or death for valleys below. In such landscapes, human survival depends on patterns: the return of wet seasons, the timing of snow, the cycles of winds, the migration of animals, the moments when passes become navigable. These patterns are often tied, directly or indirectly, to celestial cycles—the sun's seasonal movement, the lengthening and shortening of days, the timing of moon phases, and the predictable rhythms of stars used for orientation. In many traditional societies, what we now separate as "astronomy," "calendar science," and "religion" were fused into a single lived system. The heavens were not studied for curiosity alone. They were studied because life depended on them.

This is why high places become associated with sky knowledge. A mountain community must track seasons with precision. When the window for planting is short, or when herds must be moved at the right moment to survive, timing becomes sacred. A small error in the calendar can cost lives. Lowland agriculture can sometimes absorb mistakes through redundancy and surplus. Highland life is less forgiving. The result is that observational skill becomes a kind of social power. Those who can read the sky—who can anticipate seasonal shifts, who can recognize the turning points of the year—become valuable. They become trusted. Over generations, the role can formalize into specialists: calendar-keepers, ritual leaders, weather-readers, navigators of seasonal law.

In Tibet and the greater Himalayan world, mountains are not simply elevated landforms; they are living presences in cultural imagination, tied to protective spirits, sacred geography, and the idea that certain places are thresholds between worlds. The high plateau, in many traditions, is not only a landscape but a cosmology made visible: peaks as pillars, valleys as channels, skies as realms with agencies and signs. This does not need to be literal to be meaningful. It reflects a real psychological and practical fact: in the highlands, the environment feels animated because it behaves with such authority. The mountain dictates when you can travel, when you can plant, when you must shelter. The sky dictates when storms will arrive, when cold will intensify, when light will return. Agency is a human way of describing powerful patterns. Sacred peaks often arise where patterns are strongest.

In the Andes, the sacredness of mountains is even more obvious in historical memory because highland civilizations placed peaks at the center of their religious and political worlds. The mountains were not distant backdrops; they were actors. They fed rivers. They held snow. They controlled fertility. They were addressed through offerings and ceremonies. The sky, likewise, was not a neutral dome. It was a calendar and a code. The sun's annual movement, the appearance of certain stars, and the behavior of weather systems were woven into ritual cycles that coordinated agriculture, state labor, and social unity. In such contexts, "sky knowledge" becomes state knowledge: the ability to predict and regulate the seasons becomes a form of authority.

In the Ethiopian highlands, mountains and high plateaus have long supported dense human settlement and distinctive cultural identities. Highlands can preserve languages, traditions, and religious practices because the terrain resists easy assimilation. Over time, this creates communities that see themselves as enduring, protected, and set apart—qualities easily translated into sacred geography. The sky, again, becomes a keeper of time. Even when formal calendars exist, traditional knowledge of seasonal signs often remains central in highland life. The point is not to claim identical myths across these regions. It is to recognize a shared structural logic: highlands encourage both distinct cultural identity and a deep attentiveness to the heavens because those attentions are adaptive.

This is where the theme of "sky teachers" can begin to emerge without forcing speculation. Across cultures, the beings associated with mountains are often not purely earthly. They are linked to the sky: gods on peaks, ancestors in the clouds, spirits who descend and ascend. Some traditions speak of beings who bring knowledge from above—knowledge of agriculture, of calendars, of metallurgy, of law. It is easy to treat this as pure mythology. But in a highland context, it can also be read as a social memory of knowledge specialists. People who live by the sky's cycles can seem to outsiders like they possess hidden instruction. They can seem like they have access to something beyond ordinary perception, because they do: they have long practice in reading patterns others have not had to master. In a world where survival depends on timing, knowledge becomes sacred, and sacredness becomes a language for authority.

The mountain also functions as a theater of initiation. Many highland traditions involve trials, pilgrimages, or rituals tied to peaks. The physical act of ascent becomes a symbol of transformation. That symbolism is not arbitrary. Climbing changes the body. It changes breath, heart rate, perception. It can induce altered states—fatigue, hypoxia, awe—that make experiences feel profound. These states can be integrated into ritual frameworks. Over time, the mountain becomes a place where people go to be changed, to receive visions, to meet the sacred. The sky, vivid and near, becomes part of that experience. A culture can build an entire epistemology—an entire way of knowing—around the idea that truth comes from high places.

This is also where navigation and orientation enter the story. Mountain terrain can be disorienting. Valleys twist. Weather obscures landmarks. Snow transforms familiar routes. In such conditions, the sky provides stable reference points. Stars can guide night travel. The sun's path can orient direction. Even the behavior of clouds can signal changes in wind and season. Over long spans, these practical observations can be ritualized. The star that signals a seasonal shift becomes a sacred marker.

The sunrise point on a solstice becomes a festival day. The appearance of certain constellations becomes a story about ancestors or spirits. Knowledge becomes memory, and memory becomes myth because myth is one of the most efficient ways to preserve complex information across generations.

This is not an argument that myths are literal records of astronomical events. It is an argument that myth often serves as a mnemonic structure: a way to compress and transmit knowledge in memorable forms. In highland settings, where knowledge of timing is survival-critical, mnemonic systems can become especially elaborate. Sacred peaks then become archives—not because they contain books, but because they anchor stories, rituals, and observations to a physical place. A mountain becomes a calendar in stone. Its profile becomes a horizon marker. Its snow line becomes a seasonal indicator. Its sacredness becomes a way of ensuring attention, ensuring transmission, ensuring continuity.

There is also a social dimension that should not be underestimated. Highlands often produce strong boundaries between communities. Those boundaries can create a sense of distinctness that invites mythic explanation. Outsiders may view highland peoples as strange, tough, and "other." Highlanders may view lowlanders as soft, dependent, and spiritually less close to the heavens. In such mutual perception, myths can form around difference. Highland elites, in particular, might claim special ancestry tied to peaks or to sky beings, reinforcing their authority. Over time, this can generate narratives of non-local origin, divine descent, or ancient instruction. In the context of this volume's broader themes, such narratives become important not because they are proofs, but because they show how human societies encode difference and authority in the language of the sacred and the celestial.

If mountains are refuges after catastrophe, then sacredness can also be a form of gratitude and protection. A refuge that preserved life becomes a place that must be respected, appeased, and ritually maintained. The peak becomes a guardian. The sky becomes a judge. The community's survival becomes tied to proper observance. In that way, sky knowledge and sacred peaks become intertwined: to keep the refuge, you must keep the calendar; to keep the calendar, you must watch the sky; to watch the sky, you must honor the place where the watching is clearest.

All of this sets up the final movement of the chapter. High-altitude peoples are physiological adaptors, yes, but they are also custodians of refuge landscapes and of sky-centered knowledge traditions. If mountains preserve lineages through lowland collapse, then highland populations may represent the survivors of older worlds—groups who endured when others failed and who carried both biological adaptations and cultural memory forward. The next subsection explores that idea directly, in the careful language of hypothesis rather than certainty: high places and the survivors of a lost world, where altitude is not only an environmental challenge but a historical shelter that may have protected certain lineages, certain knowledge systems, and certain ancestral threads that lowlands could not preserve.

To understand why mountains carry such gravitational pull in human memory, you have to imagine what the lowlands can become when the climate turns, when rivers fail, when coasts shift, when disease moves through crowded corridors, when the familiar resource map collapses. The lowlands are where most people live because they are generous: more oxygen, more arable land, easier travel, wider trade. But lowlands are also more exposed. They flood. They dry. They become crossroads of conflict when resources tighten. In deep time, the lowlands are where demographic storms pass most violently. The highlands, by contrast, can be hard, but they can also be stable in the way that matters most: they preserve pockets of continuity. That preservation invites a hypothesis that sits at the edge of mainstream narrative and the volume's larger theme of missing chapters—namely that some high-altitude peoples may carry not only adaptations to thin air, but the biological and cultural residue of survival through periods when entire lowland worlds were disrupted or erased.

This must be framed carefully. "Lost world" does not mean a vanished civilization in the modern sense, with cities and writing necessarily swallowed without trace. It can mean something simpler and more plausible: a lost ecological world—a network of habitats, routes, and populations that once existed but collapsed under climate shifts or catastrophic transitions. When that happens, the survivors are not necessarily those with the most advanced tools. They are those positioned where continuity is possible. In many cases, that position is the refuge: a valley with water, a mountain corridor with predictable seasons, a plateau with stable grazing, a highland region whose microclimates soften the extremes. Over many cycles, such refuges can preserve lineages that would otherwise vanish, creating modern populations that appear, genetically and culturally, to carry deep threads that are not easily explained by simple migration stories.

The physiological dimension of this survival is the most concrete. Highland adaptation is not an ornament; it is a survival interface. It shapes pregnancy outcomes, infant health, endurance, and long-term viability. In that sense, high-altitude peoples are living proof that certain lineages persisted long enough in high places for selection to sculpt them. That persistence already implies a kind of survival narrative: these populations did not merely visit the mountains; they remained. They built continuity where thin air made continuity difficult. Whether their ancestors went to the mountains by choice, by necessity, or by repeated cycles of retreat and expansion, the end result is the same: high-altitude peoples embody a long-term relationship between human biology and refuge landscapes.

But the idea of "survivors of a lost world" gains additional force when you bring together three realities we have already established in this volume's first chapters. First, human evolution was a web, not a ladder, with repeated contacts and admixture. Second, the record is fragmentary, and major lineages can remain invisible until genetics reveals them. Third, mountains are both isolators and corridors, preserving distinctness and enabling occasional reconnection. When you overlay these realities, a plausible picture emerges: highland populations could preserve deep-divergent ancestry—whether archaic, early sapiens, or a mixture—through lowland disruptions, and later contribute those ancestral fragments to wider populations during expansion phases. In this picture, highland peoples become not just adaptors, but carriers—carriers of biological variation that is older than the surface narrative suggests.

This does not require us to claim that all highland peoples are direct descendants of archaic humans or that their uniqueness is best explained by exotic ancestry. It requires only the recognition that refuges preserve diversity. In ecology, refuges preserve species through stress periods. In human history, refuges can preserve populations and cultural practices through demographic shocks. In genetics, refuges preserve variant pools that might be lost elsewhere due to replacement or bottlenecks. Mountains, with their isolating terrain and microclimates, are unusually effective refuges. Over time, they can act like vaults. They store lineages. They store languages. They store customs. They store knowledge systems linked to seasonality and the sky. And when conditions allow, those vaults can open—not by intention, but by migration and marriage—and their stored inheritance can flow outward into broader networks.

The cultural dimension of survival is subtler but equally important. If highlands preserve continuity through lowland collapse, they can also preserve memory in ways that lowlands often cannot. Lowland societies, especially when they become dense and interconnected, can change rapidly. They are vulnerable to sudden social reorganization, conquest, collapse, and assimilation. Highlands can preserve slower rhythms. They can maintain older ways of life because terrain makes total transformation harder. This is why mountainous regions across the world often host distinctive languages and traditions that persist long after lowlands have been homogenized by states and trade networks. In deep time, the same principle would apply in a different form: highlands could preserve cultural patterns through ecological resets.

This is where the phrase "lost world" can be read as cultural as well as ecological. A lowland world can vanish not only because the habitat changes, but because the people who carried its knowledge are absorbed into new populations or forced into new lifeways. In such cases, the survivors might not carry the full structure of the old world, but they may carry fragments—ritual patterns, myths of flood or fire, sacred geographies tied to peaks, stories of ancestors who retreated upward, and a sense that the mountains are places of origin and protection. These stories are not proof of a specific catastrophe. They are plausible cultural expressions of a recurring refuge reality: when things fall apart below, people remember the places that held.

The idea also intersects with the volume's exploration of myth as evolutionary memory. Many traditions include narratives of a "second beginning," a time when the world was remade after destruction, and survivors preserved life by retreating to high places. Sometimes this takes the form of flood stories where mountains are the first land to emerge and the place where life is renewed. Sometimes it is a story of fire, drought, or cosmic punishment that drives people into refuges. Taken literally, such myths are diverse and culturally specific. Taken structurally, they share a pattern that fits what mountains actually do: they become survival platforms when lowlands destabilize. In that sense, myth and ecology can converge without one being reduced to the other. The myth expresses the experience; the landscape provides the mechanism.

There is also a psychological reason highland survival becomes mythic. Living in extreme environments creates a strong group identity. When survival requires special knowledge, special endurance, and special timing, communities develop a sense of distinctness. That distinctness can be interpreted as ancestry—"we are not like others because we come from different origins." Over long spans, such narratives can harden into claims of sacred descent, of sky connections, of divine protection. In the context of this volume's later chapters, this becomes significant because it shows how biological difference and cultural identity can merge into stories of hybridization, of gods and humans, of teachers and chosen peoples. Even when such stories are not literal, they can reflect a real social process: survival in refuges producing distinct communities whose identity becomes mythologized.

But the most important contribution of the "survivors of a lost world" hypothesis is that it encourages a different way of reading evidence. It suggests we should pay attention to refuge zones when we look for deep ancestry and deep continuity. It suggests that highland populations are not only interesting because of their physiology. They are interesting because they occupy landscapes that preserve. If genetics shows unusual patterns in certain highland groups—signals of selection, traces of archaic introgression, deep divergence—those patterns may not be anomalies at all. They may be expected outcomes of refuge dynamics.

This also helps explain why the human story so often feels incomplete. If major events in our past were not city-building episodes but demographic contractions, refuge survival, and later expansions, then the physical record of those events could be thin. A small group surviving in a mountain valley might leave few artifacts compared to a later lowland agricultural society. Yet that small group could be the bridge through which a lineage persists. The genome would remember what the ground does not. In that way, high places become symbols not only in myth but in evidence: they are places where continuity can be hidden.

At the edge of this chapter, then, we arrive at a grounded conclusion without wrapping the whole book. High-altitude peoples demonstrate that human adaptation can be strong, diverse, and rapid under intense pressure. They also demonstrate that mountains are not peripheral to the human story. They may be central to it as refuges that preserved lineages and knowledge through repeated lowland instability. The survivors of lost worlds are not necessarily keepers of secret civilizations; they are, more plausibly, the descendants of refuge strategies—communities shaped by extreme environments, carrying biological adaptations and cultural memory that were forged under pressure and preserved by terrain.

With that foundation, the book is ready to pivot into a new arena: the question of sudden cognitive leaps—moments in the archaeological and symbolic record that seem to arrive abruptly, as if the human mind changed gears. The highland chapter has shown one mechanism by which evolution can move quickly when pressure is intense and when borrowed variants accelerate adaptation. The next chapter will ask whether something similar might be happening in the domain of cognition, symbolism, and meaning—where the evidence is more ambiguous, the narratives more contested, and the temptation to overreach even greater.

CHAPTER 5 — SUDDEN LEAPS IN BRAIN CAPACITY

If the human story has a single moment that feels like a rupture—an event horizon after which everything looks different—it is not the rise of a particular tool type or the spread of a particular population. It is the sudden density of mind. At some point in our deep past, humans begin to behave as if they are inhabited by a new interior world: an ability to plan across longer horizons, to imagine what cannot be seen, to coordinate in larger groups through shared symbols, to speak not only about what is present but about what is absent, remembered, feared, or hoped for. The record does not present this transformation as a perfectly smooth gradient. It presents it, at least in the way it is often experienced, as a problem: a cognitive expansion that seems too rapid, too concentrated, and too consequential to fit the simplest version of gradual evolution.

The phrase "rapid cognitive expansion" must be treated with care because it is easy to inflate. Human cognition did not appear out of nothing. Our lineage had already been evolving for millions of years, building brains, refining social intelligence, developing tool use, and expanding learning capacity long before the period most people associate with "behavioral modernity." Even chimpanzees display complex social strategies and rudimentary culture. Earlier hominins crafted tools and organized hunts. Neanderthals likely possessed sophisticated skills and social cohesion. So the problem is not that cognition begins. The problem is that, at a certain point, cognition seems to cross thresholds in ways that are visible archaeologically and culturally: a threshold of symbolic density, of technological diversity, of long-distance networking, and of imaginative expression. The world begins to carry human meaning in a new way.

One way to frame the problem is to distinguish brain size from cognitive complexity. "Brain capacity" can mean literal cranial volume, and it can also mean capacity in the metaphorical sense—the power of the mind to represent and manipulate information. These are related but not identical. The fossil record shows brain expansion over long spans, but brain size alone does not guarantee the kind of symbolic behavior that later becomes unmistakable. Some of the most striking shifts in the archaeological record involve not simply larger brains, but different patterns of behavior: ornaments, art, ritualized treatment of the dead, complex tool kits, and evidence of long-distance exchange. These behaviors suggest changes in how minds were used—how information was shared, stored, and transmitted—perhaps more than changes in raw volume.

This is why the problem persists even when we acknowledge gradual anatomical evolution. The question becomes: why do certain behavioral signatures appear clustered rather than evenly distributed? Why do we see long periods with relatively stable tool traditions, then bursts of innovation? Why does symbolic expression—marking objects, decorating bodies, creating representational art—become prominent in certain windows? Why does technological diversity widen, with specialized tools and new materials? Why do long-distance networks become more visible? The difficulty is not that evolution cannot, in principle, produce these outcomes. The difficulty is that the pattern looks punctuated. It looks like a system that hits tipping points.

In complex systems, tipping points are normal. A system can change slowly until a threshold is crossed, and then a small shift produces a large effect. Human cognition, embedded in culture, may behave like such a system. A small biological change—perhaps in neural development, vocal control, memory capacity, or social cognition—could produce a large cultural effect if it improves learning, communication, or cooperation. Once culture becomes more complex, it can accelerate its own growth. Ideas accumulate. Tools improve. Social structures expand. Knowledge is transmitted more reliably. In such a scenario, the "rapid expansion" is not purely biological; it is bio-cultural. The mind changes, and culture amplifies the change.

But even this systems explanation leaves an open tension: what exactly crossed the threshold, and why did it happen when it did? If the ingredients for complex cognition existed earlier, why do we not see the same density of symbolic behavior earlier, at least not consistently? Some researchers argue that the record is biased—that symbolic behavior existed but did not preserve well, or that we have not yet found it. Others argue that behavior did change in a more significant way, and that the clustering reflects real evolutionary and demographic dynamics. The problem is not solved by choosing one side absolutely. It is sharpened by admitting both realities: the record is incomplete, and yet the pattern of clustering is strong enough to demand explanation.

Here, the themes of the previous chapters become relevant. We have established that human evolution is a web, shaped by population structure, interbreeding, bottlenecks, refuges, and expansions. This matters because cognition is not only a trait of individuals; it is a trait expressed through groups. A cognitive innovation that appears in one population might not spread if that population is isolated, small, or extinguished. Conversely, once a population becomes large and connected, innovations can spread rapidly. If climate shifts create repeated bottlenecks and then expansions, cultural complexity can be reset and then re-accelerated. In such a demographic landscape, the archaeological record can show bursts that are partly demographic: moments when networks became large enough for cultural accumulation to become visible and persistent.

In other words, "rapid cognitive expansion" may sometimes reflect changes in the scale and connectivity of human populations rather than a sudden mutation that rewired the brain overnight. When groups are small and scattered, innovations can be lost. When groups become larger and more connected, innovations can accumulate. The same basic cognitive potential can produce different archaeological footprints depending on demography. This is a powerful mainstream explanation because it does not require a miracle. It requires only that culture behaves like a cumulative process, sensitive to population size and network connectivity.

Yet even with this explanation, certain aspects remain stubbornly provocative. Symbolic behavior is not merely improved hunting efficiency; it is meaning-making. It is the ability to treat an object as representing something else, to mark the body with signs that communicate identity, to invest a place with ritual significance, to tell stories that bind a group. These behaviors imply shared mental models—collective imaginaries. The rapid expansion problem therefore touches not only on intelligence but on consciousness and social reality. It asks when humans became fully immersed in symbolic worlds, and why that immersion intensified.

The problem is made more complex by the fact that archaic humans were not cognitively empty. Neanderthals used sophisticated tools, controlled fire, and survived in challenging environments. There is evidence suggesting they engaged in behaviors that could be symbolic. If so, then the "expansion" is not a simple split between non-symbolic archaics and symbolic sapiens. It is a gradient with overlapping capabilities. That overlap pushes the problem into a more subtle form:

if multiple lineages possessed significant cognitive capacities, what changed in the late phases that made symbolic culture more widespread, more persistent, and more explosive? Was it a biological shift, a demographic shift, an ecological shift, or a combination?

The idea of borrowing—so central to the Denisovan and high-altitude story—also becomes relevant here, but carefully. Genetic admixture can introduce variants that influence brain development and cognitive traits, though linking specific archaic variants to complex cognition is far more difficult than linking variants to hypoxia response. The evidence is not as clean, and the risk of overclaim is high. Still, the web model suggests a plausible mechanism: if populations mixed, they exchanged not only genes but also cultural practices and knowledge. A cognitive "expansion" could therefore be partly cultural diffusion: different groups contributing innovations that, once networks expanded, combined into a richer cultural repertoire. The result might look like a sudden leap because multiple streams converged.

There is also a final layer to the problem: human children. Cognitive expansion is expressed most strongly through learning, language acquisition, imitation, and the internalization of cultural rules. The human brain is built not simply to think but to be shaped by others. This means the emergence of complex symbolic culture could be tightly linked to changes in social structure: longer childhoods, more cooperative breeding, more stable social networks, stronger teaching traditions, and the evolution of norms that protect knowledge transmission. A small shift in the social ecology of learning could produce a large shift in cultural complexity. Again, the leap could be real without being purely genetic.

So what, precisely, is the "problem" of rapid cognitive expansion? It is the tension between gradual biological evolution and the seemingly punctuated appearance of symbolic density and cultural complexity in the record. It is the challenge of explaining how brains that evolved over long spans generated cultural worlds that sometimes appear to bloom abruptly. It is the difficulty of separating true cognitive innovation from demographic visibility, preservation bias, and network effects. And it is the deeper question of why meaning—art, ritual, symbol, story—becomes so dominant, so quickly, that it redefines what it means to be human.

The next subsection turns to one of the clearest faces of this tension: symbolic thought. It asks how symbols emerge, why they matter, and why the evidence for them can appear without a neatly staged, gradual transition. In doing so, it keeps the line between evidence and interpretation sharp, because the more we move into the territory of mind, the easier it becomes to confuse what we want the past to mean with what the record can actually support.

Symbolic thought is one of the few human capacities that changes the texture of reality itself. It allows a mind to hold two worlds at once: the world as it is, and the world as it means. A shell is no longer just a shell; it becomes a marker of identity. A pigment is no longer just color; it becomes belonging, status, protection, transformation. A carved line is no longer a scratch; it becomes a sign. Once symbols take root, humans do not merely live in environments—they live in layered landscapes of significance, where objects, places, and gestures carry shared meanings that outlast any single moment. That is why symbolic thought sits at the center of the "rapid cognitive expansion" problem. When the archaeological record begins to show symbols clearly—ornamentation, engraved patterns, deliberate burials, ritualized spaces—it can look as if the human mind switched on a new mode. The transition, at least in the way the evidence is often experienced, does not always appear as a slow staircase. It appears as a threshold crossed.

To be careful, symbolic thought did not appear from absolute zero. Animals can communicate. Some animals use signals that function in ways that resemble symbols at a basic level—calls, gestures, displays tied to specific contexts. Humans, however, extend this into a full symbolic system: arbitrary signs tied to shared concepts, transferable across contexts, capable of building complex social realities. The transition from signal to symbol is not merely a matter of more intelligence. It is a matter of a new kind of collective cognition—one that depends on shared understanding, teaching, imitation, and stable social networks.

So why does symbolic evidence sometimes appear without what looks like a gradual transition? One answer is preservation. Symbolic behavior is often expressed in materials that do not survive well—body paint, textiles, wooden objects, spoken stories, ritual gestures, temporary structures. If early symbolic cultures relied on perishable mediums, they could have existed for long spans while leaving little direct trace. We might be seeing not the birth of symbolism, but the moment when symbolism becomes visible because it is increasingly expressed in durable forms: carved bone, engraved stone, beads, pigments stored and used in ways that leave chemical signatures, and burials that preserve intentional arrangement.

This explanation matters because it reminds us that archaeology is not a complete recording of behavior; it is a selective archive. The record favors what is durable and what happens to be preserved in conditions that protect it. If early symbolic thought was present but expressed mostly through speech and perishable performance, then the "sudden appearance" is partly an illusion created by the archive. We are not watching a mind wake up; we are watching the camera finally capture what was already happening off-screen.

Yet preservation bias cannot explain everything. Even allowing for missing evidence, there does seem to be a clustering of symbolic artifacts and practices in certain time windows and regions, and that clustering remains a genuine pattern that demands explanation. Why would symbolism become more common, more elaborate, or more widely distributed in some phases than others? Why do we see periods where ornaments and engravings become more frequent, where tool traditions diversify, where long-distance exchange becomes more visible? If symbolism existed earlier, why does it not leave similar density of trace? The answer may lie not only in minds but in networks.

Symbols thrive in networks because symbols are, at heart, social contracts. A symbol only works if other people share its meaning. A bead is not a badge if no one recognizes it. A ritual is not a ritual if it is not collectively understood. This means symbolism is sensitive to group size and connectivity. In small, isolated groups, symbol systems can exist, but they may remain local, subtle, and easily disrupted by demographic shocks. In larger, connected populations, symbol systems can become more complex and standardized. They can spread between groups. They can be taught more reliably. They can accumulate layers of meaning. In other words, symbolism becomes more visible when populations become more connected.

This is where the web model of human evolution again becomes central. If climate shifts and migrations repeatedly reshaped population networks, then there would be phases when humans were scattered and phases when they were more connected. During connected phases, symbols could spread and stabilize, leaving denser archaeological signatures. During scattered phases, symbols could remain local or be lost during bottlenecks. The archaeological record would then show bursts—not because the human mind evolved in bursts, but because the social system that supports symbolic culture expanded and contracted.

There is also an ecological dimension. Symbolic culture often intensifies when survival becomes uncertain or when social coordination becomes more important. In harsh environments or during volatile climate phases, humans may depend more on cooperation, trade, and group identity. Symbols help coordinate these dynamics. They mark who belongs. They signal alliances. They create trust within networks. They encode social rules. They also support psychological resilience by giving meaning to suffering and uncertainty. In this sense, symbolism is not just decoration; it is adaptive. It helps bind groups, stabilize cooperation, and transmit norms. If ecological volatility increases the need for these functions, symbolic culture may intensify—not because minds suddenly became capable, but because the pressure made symbolic systems more valuable.

Another explanation involves learning and childhood. Symbolic systems are not purely individual inventions; they are learned. A population that develops stronger teaching traditions, longer childhood learning windows, and more structured social transmission can support richer symbolic culture. Small changes in social learning—how often adults teach, how children are integrated into communal activity, how stories and rituals are repeated—can produce large changes in cultural complexity over time. Again, the leap can be social rather than genetic.

But the question remains: why does it feel like a threshold? Because symbols are not incremental in the same way tools can be. A slightly better stone flake is still a stone flake. A symbol, once established, reorganizes behavior around meaning. It creates identities and expectations. It produces social boundaries and shared narratives. The emergence of reliable symbolic systems changes what humans are doing at a fundamental level. Once a group consistently uses ornaments to signal identity, the social world becomes more layered. Once a group consistently marks objects with patterns that have shared meaning, memory can be externalized. Once a group creates rituals around death, the group's relationship to time changes. These are not small shifts. They are systemic transformations. That systemic quality makes symbolism feel sudden even if it emerged gradually, because the moment it becomes stable and widespread, it reconfigures human life.

There is also the issue of visibility in the record. A symbol system can exist for a long time without leaving many artifacts. Then, due to a change in materials, trade, or cultural emphasis, it begins to leave durable traces. The record then shows a steep rise in visible symbols. The steepness can be partly a measurement artifact, but it can also reflect a real intensification: a population choosing to invest more energy in symbolic display. That choice might occur when social complexity increases—larger groups, more interaction with outsiders, more competition, more need to signal identity and status. In such contexts, symbolic display becomes socially strategic. It becomes a way to navigate a crowded human landscape. The beads and pigments are not mere art; they are social technology.

A crucial point, especially for this book's overall thesis, is that symbolic thought blurs the line between "myth" and "history." Once humans live symbolically, they begin to encode experiences into narratives that can outlast any individual. They create myths, not as fantasies, but as structured memory systems that preserve lessons, origins, and identities.

This does not mean myths are literal records of events. It means myths can preserve deep truths about pressures, migrations, refuges, and contact. Symbolic culture makes long-term memory possible in a communal sense. It creates the conditions for stories of ancestors, of lost worlds, of non-human teachers, of catastrophes and renewals to persist and spread. The emergence of symbolic thought is therefore not only a cognitive milestone; it is the birth of cultural memory at scale.

This is where the appearance "without gradual transition" becomes especially provocative. If symbolic culture becomes widespread relatively rapidly, then the memory-making machinery of humanity turns on in a concentrated phase. The species begins to remember itself in new ways. It begins to store identity and knowledge in signs, rituals, and stories. That could help explain why, later, myths are so rich and so persistent. A cognitive and social threshold may have been crossed where symbolic systems became stable enough to carry memory across vast time spans. In that sense, the "rapid transition" is not only about art; it is about the emergence of a new mode of inheritance: cultural inheritance that can accumulate, transform, and survive demographic shocks in ways genes alone cannot.

Yet we must keep the distinction clear. The evidence for symbolism is not always clear-cut, and interpretation can outrun proof. A pattern etched into bone might be art, or it might be functional marking. A burial might be ritual, or it might be practical disposal. Pigment use might be symbolic, or it might be utilitarian. The responsible approach is to treat symbolism as a spectrum, not a binary. The problem is not solved by declaring a single "symbolic revolution." The problem is illuminated by recognizing multiple pathways by which symbolic practices could intensify:

preservation thresholds, network expansion, ecological stress, social learning shifts, and demographic cycles.

From here, the chapter moves into a cluster of capacities often associated with symbolic life: language, music, and mathematics. These capacities, too, can seem to appear abruptly—because they are hard to see in the archaeological record until they are already well developed, and because they depend on social transmission that can tip into rapid accumulation once networks reach a certain scale. The next subsection explores that sense of abruptness and asks why these capacities, which feel so foundational to human uniqueness, might look like sudden arrivals rather than slow steps.

Language, music, and mathematics are three of the most powerful signatures of human cognition, and yet they share a strange feature: they are among the hardest things to see directly in the deep past. A stone tool can be excavated. A hearth can be dated. A pigment smear can be analyzed. But language vanishes the moment it is spoken. Music disappears into air. Mathematics, unless written or embodied in durable structures, leaves only indirect traces—patterns, counts, alignments, repeated forms that might be practical or might be conceptual. This invisibility creates an illusion. It makes these capacities seem to appear abruptly, as if humanity lived in a long silence and then suddenly began to speak, sing, and measure the world with number.

The illusion is partly an artifact of evidence. We tend to recognize language when it leaves cultural footprints: complex social coordination, symbolic objects, long-distance trade, ritual life, and the ability to transmit technical knowledge reliably. But language itself likely predates many of these visible signatures. Likewise, we tend to recognize music when we find instruments—flutes, resonant objects, designed sound-makers.

But music almost certainly existed in voice and percussion long before instruments were carved. Mathematics is recognized when we find tallies, geometric patterns, structured architecture, or calendrical systems, but counting and spatial reasoning are older than any of those durable expressions. In each case, the capacity could exist long before the record captures it. The abruptness is often the moment of visibility, not the moment of origin.

Still, even if we account for invisibility, there remains a real question about clustering. There are phases in the record where symbolic artifacts become denser, where complex tool traditions spread, where art becomes more visible, where evidence for ritual and long-distance networks grows. These phases suggest that certain cognitive capacities were not merely present but became culturally dominant and socially amplified. Language, music, and mathematics could have existed earlier in simpler forms, but at some point they begin to shape the human world more intensely, leaving signatures that archaeologists can detect indirectly. The "abruptness" may therefore reflect not the invention of these capacities, but their escalation—the moment they become powerful enough in social life to reshape material culture.

Consider language first. Language is not just vocabulary. It is a system for sharing mental models: plans, counterfactuals, abstractions, histories, obligations, and invisible social rules. A population with richer language can coordinate more effectively, teach more efficiently, negotiate alliances, and transmit complex knowledge across generations. If language becomes more structured—if grammar becomes more powerful, if storytelling becomes more elaborate, if metaphor becomes routine—then culture can accelerate. This is where the tipping-point nature of language becomes crucial. Once a group reaches a certain level of linguistic complexity, it can store more information socially. It can coordinate larger groups with shared narratives. It can maintain norms across wider networks. In such a case, the apparent leap is a network effect: language unlocks cumulative culture, and cumulative culture expands language in return.

The archaeology of language is therefore indirect. We infer language from what it enables. A complex tool chain that requires precise teaching implies some form of structured communication. Long-distance trade implies negotiation, trust-building, and shared meanings. Ritual practices imply narratives and shared symbolism that are difficult to sustain without rich communication. None of this proves modern language in the strict sense, but it points toward increasingly sophisticated communication systems. The "abruptness" emerges when these indirect signs become common enough to suggest that language is no longer just a background faculty but a dominant social engine.

Music carries a similar logic, but with a different function. Music is often treated as entertainment, but in most traditional societies it is social technology. It coordinates group movement, strengthens identity, synchronizes emotion, and binds communities through shared rhythm and chant. It can mark rites of passage, create trance states, and embed memory in melody. In oral cultures, music and chant are also mnemonic devices—ways to preserve long narratives and complex instructions across generations without writing. In that sense, music is intertwined with the rise of symbolic culture because it helps store and transmit meaning.

Like language, music's early forms would be almost invisible archaeologically. The human voice is the first instrument. Clapping, stamping, and striking resonant surfaces require no preserved artifact.

The first durable evidence of music appears when instruments are crafted from bone, wood, or other materials that can survive. When such instruments appear in the record, they can make music seem "sudden," but the instrument is more likely a sign that a musical tradition already existed and became sufficiently elaborate to justify specialized objects. The instrument is evidence of intensity and cultural investment, not necessarily the birth of music itself.

Mathematics is the most difficult of the three because it is easy to mistake practical pattern for abstract concept. A series of marks might be counting, or it might be decoration. A repeated geometric motif might be aesthetic, or it might encode measurement. Yet mathematics, in its simplest forms, is older than writing and older than agriculture because it begins with the body: more versus less, one versus many, the counting of days, the tracking of seasons, the recognition of spatial relationships. Even animals have rudimentary numerical cognition. What makes human mathematics distinctive is not the ability to count, but the ability to externalize number into symbols and to build abstract structures—calendars, measures, proportions, and eventually formal systems.

In deep prehistory, the most likely arena for early mathematics is timekeeping. If a group must track lunar cycles, seasonal changes, migration patterns, and ritual calendars, then counting becomes social necessity. The sky becomes a numerical teacher: phases repeat, solstices return, stars rise and set in cycles. A group that encodes these cycles into ritual schedules is already practicing applied mathematics. This may be why mathematical thinking often appears intertwined with symbolic culture and with the sacred. The first mathematics may not have been written equations; it may have been calendrical knowledge preserved in myth, chant, and ritual timing.

So why do these capacities appear abruptly in our narratives? Partly because their early forms are invisible. Partly because once they become socially central, they amplify one another. Language makes it easier to teach complex music and to transmit numerical knowledge. Music makes it easier to preserve language and memory. Mathematics makes it easier to coordinate seasonal labor, navigation, and ritual cycles. When these systems intensify together, they can produce a cultural acceleration that leaves durable traces: ornaments that signal identity, tools that reflect complex instruction, ritual sites that reflect calendrical awareness, and networks that reflect expanded cooperation.

This mutual amplification also explains why the record can look like an "explosion" rather than a gradual incline. Culture is cumulative, but it is also fragile. In small populations, complexity can be lost. In larger, connected networks, complexity can persist and build. If climate shifts create phases of expansion and contraction, then the visibility of language-driven culture can rise sharply in expansion phases and dim in contraction phases. The underlying capacities may remain, but their expression changes depending on social scale. The result is punctuated visibility: peaks of cultural density that look like sudden arrivals.

This is where the broader thesis of this volume begins to tighten. If the record looks abrupt, it may be because critical ingredients converged: populations became sufficiently connected, symbolic systems became sufficiently stable, and perhaps—this is the careful hypothesis—the genetic and demographic web introduced variations that made certain groups especially capable of cumulative cultural growth. Not in a magical sense, but in a real evolutionary sense. A slight edge in learning, memory, social cognition, or vocal control could be amplified by culture into something that looks like a leap. And if archaic admixture contributed certain physiological advantages in harsh environments, it is at least plausible that the web of human contact also influenced cognitive traits—though the evidence there is far less direct and must be treated with more caution.

From here, the chapter turns toward the archaeological record as the main visible stage where these invisible capacities leave their footprints. If language, music, and mathematics are difficult to see until they are already mature, then what we can study is the moment when meaning begins to saturate the material world—when artifacts and sites become dense with signs of identity, ritual, exchange, and symbolism. The next subsection explores that saturation directly: the archaeological explosion of meaning, and why it feels like a sudden flowering of mind on the landscape.

At a certain point in the record, the human world begins to look inhabited not only by bodies but by ideas. This is difficult to describe without slipping into poetic language, because the change is not merely technological. It is semiotic. Tools are no longer only tools; they become signatures of style and tradition. Objects begin to appear that have little obvious survival function but immense social function—ornaments, pigments used deliberately, engravings that insist on being seen. Sites begin to show patterning that implies ritual sequence and shared rules. Burials begin to suggest that the dead are not merely disposed of but treated as participants in a social and symbolic order. The material world, in other words, begins to carry meaning in a way that feels intensified, sometimes concentrated, sometimes bursting into view with a density that makes earlier phases look comparatively quiet. This is what people often refer to—carefully or carelessly—as an archaeological "explosion."

The first thing to keep disciplined is the word explosion. It does not mean the human mind suddenly appeared. It does not mean earlier humans were mindless or cultureless. It means that the archaeological archive begins to preserve more unmistakable signals of symbolic and social complexity, and that those signals cluster in ways that are hard to ignore. The reasons for clustering may include real changes in behavior, changes in population size and connectivity, shifts in where people lived, and shifts in preservation conditions. The key is not to pick one cause and declare the problem solved. The key is to recognize that meaning becomes visible in the archaeological record at scale when several conditions converge: people are numerous enough to sustain complex traditions, connected enough to spread and stabilize them, and invested enough in durable symbolic display that their meanings survive as artifacts rather than disappearing as song and speech.

One way to understand the "explosion of meaning" is to look at what meaning does in human societies. Meaning binds. It creates identity. It marks belonging and difference. It stabilizes cooperation by embedding social rules in story and ritual. It channels emotion into shared forms—mourning, celebration, awe, fear. It also allows knowledge to accumulate because it provides frameworks for memory. When symbolic culture intensifies, the group becomes more than a collection of individuals. It becomes a narrative entity: a people with an origin, with signs, with obligations, with shared interpretations of the world. That transformation is difficult to capture in stone and bone, but when it becomes strong enough, it leaves traces.

Ornaments are a direct example. A bead is not a weapon. A pendant does not kill prey. Pigment applied to the body does not directly increase calories. Yet these objects persist, spread, and multiply in many contexts because they do something else: they signal. They tell others who you are, who you belong to, who you can marry, who you can trust, who you fear, who you honor. In small groups, such signaling can be informal. In larger networks, it becomes essential. As populations grow and interact more frequently with outsiders, identity becomes something that must be communicated quickly and repeatedly. Ornamentation becomes a kind of wearable language. When the archaeological record begins to show more ornaments, it may be telling us that social worlds expanded beyond intimate familiarity. People were meeting strangers more often. Meaning became portable because identity needed to be legible.

Engravings and deliberate markings can be read in a similar way. A mark is a claim: "this matters." It is the externalization of intention. Even if we cannot decode what an engraving meant, the act of engraving suggests that the maker believed meaning could be stored outside the mind. This is a profound step because it turns objects into memory devices. It creates a world where information can be placed into durable matter and retrieved later—by the maker or by others. In that sense, engraving is not merely art; it is cognition made material. It is the mind refusing to remain private, insisting on being shared through form.

Burial practices intensify this logic. Treating the dead in a patterned way implies that the dead occupy a social position that does not end at biological death. This does not require us to claim a fully developed theology. It requires only the recognition that the living are acting as if death is not merely disposal. They are investing energy in the dead body beyond immediate utility. They are arranging, marking, placing.

That behavior suggests a symbolic framework in which the dead remain within the group's moral world—honored, feared, remembered, guided. Such frameworks require shared narratives, because ritual is meaningless without shared interpretation. When burials become more visible, they hint at an interior world becoming collective: shared beliefs, shared fears, shared definitions of personhood.

Then there is the "meaning" embedded in tools themselves. Earlier tool traditions can persist for long spans with relatively slow change. Later periods, in some regions, show more diversification: specialized tools, composite technologies, and materials used in more varied ways. Diversity can be purely functional, but it can also be cultural. Different groups can develop distinctive styles, preferred forms, and signature techniques. Once meaning enters technology, a tool becomes part of identity. It becomes "how we do things," not merely "what works." This has evolutionary implications because cultural identity can preserve certain techniques even when alternatives exist, creating traditions that persist through learning and imitation. The record then begins to look richer, not only because people are more inventive, but because they are more traditional in a complex way—carrying styles across generations.

Long-distance exchange is another key indicator of meaning saturation. When materials move across large distances—shells, pigments, stone types not local to a site—it suggests networks of contact. Networks are not only economic; they are social. They require trust, negotiation, shared protocols, and often shared ritual. The more such networks exist, the more symbolic systems become necessary to manage them. A trade relationship is not just barter; it is a social bond. It often involves gift exchange, ritual hospitality, and obligations. Meaning becomes the glue that holds a network together. When long-distance materials appear more often, it implies that humans were building webs that extended beyond immediate territory, and that these webs required a richer symbolic vocabulary to function.

All of these elements—ornamentation, engraving, burial patterning, tool diversification, long-distance exchange—converge into something that feels like an explosion because they reinforce one another. A group with stronger symbolic identity is better at alliance-building. Alliances enable trade. Trade spreads ornaments and styles. Styles reinforce identity. Identity strengthens cohesion. Cohesion supports teaching and innovation. Innovation increases tool diversity. Tool diversity increases survival and expansion. Expansion increases contact. Contact increases the need for identity signals. The system becomes self-amplifying. Meaning creates networks, and networks create more meaning.

This is where demographic and ecological dynamics become crucial again. A symbolic system cannot persist if the population collapses repeatedly. Bottlenecks can erase cultural complexity, not because brains disappear, but because carriers of tradition die and social networks shrink. Conversely, when populations expand, cultural complexity can grow rapidly because there are more carriers, more opportunities for apprenticeship, more exchange between groups, and more memory stored in shared practice. This means the archaeological "explosion" may be partly the visible surface of a demographic transition: moments when human populations became dense and connected enough for meaning systems to become durable.

Climate could also play an indirect role. Volatile environments increase the value of cooperation and shared planning. They also increase uncertainty, which can intensify ritual and meaning-making as psychological stabilization. Under stress, humans often seek patterns, narratives, and shared ceremonies that transform fear into order. This is not a modern quirk; it is a social strategy that likely runs deep. If certain periods were marked by environmental unpredictability, symbolic culture might intensify because meaning becomes a tool for survival as much as food-gathering does.

Yet the “explosion of meaning” remains an interpretive challenge because meaning is not directly measurable. We can count beads, but we cannot directly measure the stories that made beads significant. We can excavate burials, but we cannot fully reconstruct the cosmologies that shaped them. We can map trade goods, but we cannot hear the agreements and rituals that made exchange possible. The record gives us objects and patterns; the mind gives us interpretation. The discipline is to keep those apart while still allowing the objects to speak in their own way: they speak of investment beyond immediate utility. They speak of shared attention. They speak of humans acting as if symbols matter.

This is where the chapter’s central tension sharpens. If meaning saturated the material world in certain phases, it suggests that cognitive and social capacities had crossed a threshold where symbolic systems became stable, cumulative, and widespread. The question becomes how that threshold was crossed. Was it primarily demographic—larger networks making symbols necessary and durable? Was it primarily cognitive—a shift in language, memory, or social cognition making symbolic systems more powerful? Was it ecological—stress and movement intensifying the need for cohesion and identity? Or was it, as the later speculative framing of this volume will cautiously explore, influenced by deeper factors like admixture-driven variation and the survival of specialized knowledge traditions through refuges?

The next subsection turns directly to the limits of explanation. It asks what standard evolutionary theory can account for comfortably in this story, and where the story becomes more ambiguous—where we can outline plausible mechanisms but cannot yet fully explain why the timing and clustering look the way they do. That is not a weakness of inquiry. It is the honest edge of what the evidence can support.

Evolutionary theory is not weak. It is one of the most powerful explanatory frameworks ever developed, and it has repeatedly absorbed discoveries that once seemed disruptive. It explains adaptation, divergence, extinction, and the shaping of traits under pressure. It can handle complexity. It can handle non-linearity. It can handle webs of ancestry and repeated interbreeding. The problem addressed in this chapter is not that evolution "fails." The problem is that when we try to apply our cleanest evolutionary narratives to the emergence of modern human cognition and symbolic culture, the fit is not always tight. The theory offers plausible mechanisms, but the timing, clustering, and intensity of certain shifts remain difficult to account for in a fully satisfying way—especially when we demand explanations that are both mechanistically grounded and consistent with the fragmentary nature of the record.

One limitation is simple and profound: cognition is not a single trait. It is a network of traits expressed through brain development, sensory integration, memory, social learning, language capacity, emotional regulation, and the ability to represent abstractions. Many of these traits are polygenic, meaning they are influenced by many genetic variants, each with small effects, and shaped heavily by environment and culture. Evolutionary theory can explain how such traits might change, but it struggles to produce clean narratives about when and why a complex suite of cognitive features would intensify together. In other words, the problem is not that evolution cannot produce complex minds. The problem is that complex minds do not leave clean fingerprints, and when their fingerprints appear as a cluster—language, symbolism, art, ritual, rapid innovation—we are left debating which changes were biological, which were cultural, and which were artifacts of preservation.

A second limitation is the visibility problem. The traits we are trying to explain—language richness, symbolic meaning, musical culture, abstract numerical thinking—are not directly preserved. Archaeology shows us durable artifacts, not internal states. Genetics can show us ancestry patterns, not the content of thought. Fossils can show us cranial capacity and some aspects of anatomy, but not the fine structure of neural circuitry and not the lived social worlds that shape cognition. This means evolutionary explanations often end up being forced to operate in the space between indirect proxies. We infer cognition from tools, art, and settlement patterns, but those inferences are always vulnerable to alternative interpretations. The record can therefore support multiple evolutionary narratives, and the theory cannot always adjudicate between them without more direct evidence.

A third limitation is that cultural evolution behaves differently from biological evolution. Culture can change quickly. It can spread horizontally between groups. It can accumulate and then collapse. It can leap across environments through imitation and teaching. A small biological advantage can be amplified dramatically by culture, and a cultural innovation can alter selection pressures, reshaping biology in return. Evolutionary theory recognizes gene-culture feedback, but modeling it with precision is notoriously difficult because it requires reconstructing population sizes, contact patterns, learning structures, and ecological contexts that are only partially known. When we try to explain why symbolic culture intensifies in certain windows, we can appeal to plausible gene–culture dynamics, but we cannot always demonstrate them cleanly. The theory provides a map of possible roads; it does not always tell us which road was taken.

This leads to one of the central tensions: the appearance of punctuated cultural change. We have already discussed that the "explosion" of symbolic artifacts could reflect thresholds in population size and connectivity. That is a strong mainstream explanation. Yet it does not fully dissolve the sense of timing. If demography and connectivity are the main drivers, why do we see certain bursts when we do, and why do they differ by region? Why do some innovations persist and spread, while others remain local or vanish? Why do symbolic traditions sometimes flourish and then appear to dim or fragment? Evolutionary theory can accommodate these patterns by invoking shifting climates, bottlenecks, and changing networks, but it often does so in a way that feels like patchwork: a cluster of plausible factors rather than a single coherent causal chain. That may be the truth, but it can also feel unsatisfying because humans crave a clean narrative.

A fourth limitation is that selection on cognition is hard to isolate. Natural selection works through differences in reproductive success. Cognitive traits can affect reproductive success, but often indirectly. Better planning, better social intelligence, better language could improve survival and mating opportunities. Yet the link is mediated by culture and ecology. A trait that is advantageous in one context might not matter in another. A highly innovative individual might be socially punished in a rigid group. A socially dominant individual might reproduce more regardless of cognitive nuance. Sexual selection may favor charisma and status rather than problem-solving accuracy. In such a landscape, selection on cognition can be strong, weak, or episodic depending on context. Evolutionary theory can model selection in principle, but in practice we often lack the data to quantify it in deep time. This makes explanations of "rapid cognitive expansion" vulnerable to speculation because we cannot easily measure the selective pressures that would have driven it.

A fifth limitation is the question of constraint. The human brain is metabolically expensive. It requires enormous energy investment, especially during childhood. It also comes with trade-offs: extended development, increased vulnerability, and the need for social support. Evolutionary theory can explain why brains can expand when the benefits outweigh the costs, but it also suggests that brain expansion is not an easy path. If the brain became rapidly more complex in a certain window, there must have been strong benefits or some change that reduced costs—perhaps improved diet, cooperative care, social structures that protected long childhoods, or ecological niches that rewarded complex planning. These are plausible, but again, they form a cluster of drivers rather than a single clean cause. We can outline the requirements, but we cannot always show exactly how those requirements were met in each region and time.

A sixth limitation—especially relevant to this volume's core thesis—is that evolutionary narratives have historically been tempted to simplify. For a long time, the dominant story treated modern humans as a clean lineage that replaced others with little mixing. That story is now known to be incomplete. Interbreeding occurred. The lineage is a web. This correction matters because it changes how we think about the evolution of traits. If modern humans absorbed genetic variants from other lineages, then some adaptations and perhaps some aspects of biology could have been influenced by introgression. Evolutionary theory can handle introgression, but the older narratives often did not emphasize it. We are therefore still catching up in our intuitive storytelling. The framework is capable; our cultural model of it is still adjusting.

The most difficult place where explanation remains incomplete is the relationship between cognitive potential and cultural expression. A population could have modern cognitive potential but not express it in visible archaeological ways due to social structure, ecology, or preservation bias. Conversely, a population could show a surge in symbolic artifacts because networks expanded, not because brains changed. In this ambiguity, evolutionary theory cannot always tell us which interpretation is correct. It can tell us what is possible; it cannot always tell us what happened. That gap is not a failure of the theory. It is a reflection of limited data.

And this brings us to the careful edge of the chapter. When a pattern remains underdetermined—when multiple plausible explanations exist and none can be proven decisively—human minds tend to reach for stronger stories. This is where the speculative framing of this book becomes tempting: the idea that something intervened, that an external intelligence accelerated cognition, that hybridization was not merely incidental but transformative, that myths of teachers and watchers preserve a memory of a real influence. Those ideas cannot be treated as established facts, and they must be separated from evidence. But they arise because the mainstream explanatory cluster still leaves a feeling of incompleteness, especially regarding timing and the sudden density of symbolic life.

In disciplined terms, the honest summary is this. Evolutionary theory can explain why complex cognition is advantageous and how it could evolve. It can explain thresholds, gene-culture feedback, and demographic effects that produce punctuated visibility. It can explain why symbolic culture might surge when networks expand. What it cannot fully explain—at least not yet, with the evidence we currently have—is a single satisfying, testable causal chain that accounts for the specific clustering, the regional variation, and the sense of rapid intensification in symbolic life without leaving major degrees of freedom. We can propose mechanisms, but we cannot always demonstrate their precise contribution. The record remains too incomplete, and cognition remains too complex.

The next subsection steps into that interpretive gap deliberately, not to declare an answer, but to examine why certain cultures have consistently framed cognitive and civilizational leaps as the result of intervention—gods, teachers, sky beings, civilizers—and how we should handle that framing when we are committed to keeping evidence and hypothesis distinct. It asks the forbidden question in a controlled way: did something intervene? And if we cannot prove it did, why does the idea recur so powerfully in myth, and what might those myths be encoding—literal events, social dynamics, or the human need to explain sudden change?

The question is older than science, older than writing, and in many ways older than the stories we use to answer it. When humans look back at their own origin—at the sudden power of language, the emergence of symbolic worlds, the rise of tools that feel more like thought than stone—they often arrive at the same intuition: this seems like too much, too quickly, to be explained by ordinary life alone. And so the mind reaches outward. It imagines a giver, a teacher, a shaper, an intervention. Across cultures, the leap into full humanity is frequently framed as a gift from beyond: gods who teach agriculture, spirits who reveal fire, sky beings who offer law, watchers who whisper knowledge, ancestors who return with secrets from other realms. The recurrence of this motif is not, by itself, evidence that an intervention occurred. But it is evidence that humans consistently interpret abrupt transformation as externally sourced. The question, then, must be handled with two hands at once: one hand holding evidence, the other holding meaning.

In the evidence-based frame, the most responsible answer is cautious. There is no direct scientific proof that an external non-human intelligence intervened in human evolution or cognition. The mainstream mechanisms—natural selection, demographic thresholds, gene–culture feedback, and the web of admixture—are sufficient in principle to explain complex cognition without requiring an outside agent. That is the baseline. It must be acknowledged clearly, especially because the temptation to convert mystery into certainty is strong. Yet the baseline does not eliminate the interpretive tension. It does not erase the fact that the cognitive and cultural record can feel punctuated, that the densest symbolic life appears clustered, and that the gap between physiology and meaning remains difficult to reconstruct. In that space, intervention narratives flourish.

The more interesting and disciplined way to approach the intervention question is not to treat it as a yes-or-no verdict. It is to treat it as a diagnostic. Why do humans repeatedly tell stories of external teachers? What does that reveal about the structure of cultural memory, the psychology of innovation, and the way knowledge moves between groups? If we can explain the recurrence of intervention motifs without invoking literal intervention, we learn something valuable about how humans encode discontinuities. And if there are aspects of the record that still resist explanation, we can articulate them precisely without turning them into proof.

Start with the simplest possibility: "intervention" as a metaphor for cultural transmission. In human history, most knowledge is not invented by the average individual. It is received—learned from parents, elders, specialists, strangers encountered through trade, migrants arriving with new techniques. For a group encountering unfamiliar technology or unfamiliar ritual expertise, the experience can feel like a revelation from outside. A new tool system can change daily life dramatically. A new calendrical tradition can reorganize time. A new agricultural package can transform survival. To those receiving it, the knowledge is given, not discovered. Over generations, the memory of being taught can crystallize into a myth of teachers: civilizers who arrive, instruct, and depart. The teacher becomes larger than human because the gift felt larger than life.

In this sense, intervention myths can encode real social dynamics. They can remember contact between groups of unequal power or unequal expertise. They can preserve the trauma of conquest and the awe of learning. They can compress long processes into narrative episodes because myth speaks in archetypes, not in demographic charts. The "sky" in the sky teacher does not have to mean outer space. It can mean the realm of authority—what is above, what is distant, what is beyond ordinary reach. A priesthood that controls calendrical knowledge can be "from the sky" because it reads the sky and claims legitimacy through it. A seafaring group arriving with new skills can be "from beyond" because they come from the horizon. A highland refuge population descending after a lowland collapse can be "from the mountains" and also "from the heavens" because peaks are where earth touches sky. Intervention, here, is the mythic form of cultural diffusion.

There is also the possibility that intervention motifs encode psychological thresholds rather than literal events. When symbolic cognition intensifies, humans begin to experience themselves differently. They develop narratives of origin, guilt, destiny, and cosmic order. They look for explanations that match the magnitude of their own transformation. The leap into symbolic life can feel like a gift or a burden. It can feel like being awakened. In that psychological framing, the "interveners" are not external beings but the mind's way of personifying an internal change. The teacher is the symbol of the new faculty itself—language appearing as a god, memory appearing as an ancestor, the moral sense appearing as a lawgiver. The myth says, "something happened to us," and it gives that "something" a face.

But the recurrence of intervention narratives does not only reflect metaphor and psychology. It also reflects a real pattern in how innovations spread. When a cultural system reaches a tipping point, change can be rapid. A new technique can cascade through networks. A new symbolic tradition can spread quickly if it solves coordination problems. A new story can unify groups and expand alliances. In such cascades, the time scale of transformation can be short—within a few generations. For people inside that transformation, it can feel as if a new world has arrived. Later, when memory looks back, it may compress that cascade into a single moment of origin. That compression produces a founder figure: a hero, a god, a teacher. Intervention becomes the narrative shape of a cultural tipping point.

So far, this remains within mainstream plausibility. But the book's larger thesis presses further, not by declaring certainty, but by asking whether the web model adds a deeper layer. If human evolution involved repeated interbreeding with archaic lineages and the absorption of lost populations, then human capacities could have been shaped not only by internal mutation and selection but by the exchange of genetic and cultural variation between groups with different histories. This is not intervention in the supernatural sense. It is intervention in the demographic sense: the encounter between lineages can change what a population can do. The Denisovan altitude case has already shown that a borrowed genetic variant can profoundly affect survival. The question is whether similar processes could have influenced cognitive traits—whether admixture, population structure, and refuge dynamics could have introduced variations that affected learning, social cognition, or language readiness.

Here we must be especially careful. Unlike hypoxia adaptation, cognition is difficult to map to specific genetic variants, and the evidence does not allow a clean claim that archaic introgression caused a cognitive leap. The responsible position is that it is possible in principle, but not established. Still, the hypothesis matters because it offers a naturalistic bridge between "intervention" myths and evolutionary complexity. The myths speak of non-human teachers or hybrid origins. The science confirms that humans are, in fact, a hybrid lineage in a broad sense—shaped by admixture. The connection is not proof of literal watchers, but it suggests that the ancient intuition of mixed origins is not entirely wrong, even if its narrative form is mythic.

There is another, more grounded way to approach the intervention question: intervention as specialization. As societies become more complex, knowledge becomes concentrated in certain individuals and roles—shamans, priests, astronomers, metalworkers, navigators. These specialists often appear, in myth, as semi-divine because their knowledge seems beyond ordinary comprehension. A person who can predict seasonal shifts, navigate by stars, or produce metal from rock can look like an intermediary between worlds. In prehistory, the earliest specialists may have emerged not in cities but in mobile communities where knowledge was scarce and precious. If certain lineages preserved specialized knowledge through refuge periods—highlands, isolated valleys, stable coastlines—then when they reconnected with other groups, they could appear as bearers of "given" knowledge. The intervention motif could therefore be the mythic memory of specialists re-entering broader networks after periods of isolation.

What, then, do we do with the most literal reading—the idea of non-human intelligences guiding human development? The disciplined response is to treat it as hypothesis, not evidence. The recurrence of the idea shows that humans repeatedly interpret discontinuity as external agency, but recurrence does not establish reality. If one wishes to argue for literal intervention, the burden would be to produce testable predictions and evidence that cannot be explained by known mechanisms. In the absence of such evidence, the safest position is that intervention myths are cultural instruments for explaining rapid change, encoding contact, specialization, catastrophe survival, and the human need to assign origin to something larger than ourselves.

And yet, the question still matters, because it tells us something about the human condition. Humans do not accept that they are accidental. They seek narrative proportionality: big changes require big causes. When the mind turns from animal life into symbolic life—when it begins to build worlds of meaning and moral order—it naturally imagines a source worthy of that change. In that sense, intervention narratives may be less about what happened in prehistory and more about how humans experience themselves once symbolic consciousness is fully awake. The story of teachers is the story of the mind recognizing its own power and asking where it came from.

This chapter therefore ends not with an answer that collapses mystery into certainty, but with a clarified frame. The cognitive leap can be approached through mainstream mechanisms: thresholds, demography, culture, and selection. The intervention motif can be read as mythic compression of those mechanisms—contact, teaching, specialization, refuge survival, and tipping-point cascades. The web of human evolution adds a further layer, suggesting that "given" traits can be inherited through admixture, making the lineage genuinely composite even when no external intelligence is invoked. And beyond all of that, the intervention question persists as a human narrative instinct—our habit of explaining our own becoming as a gift from beyond.

From here, the book moves into a more concrete domain where myth and evidence collide in a different way: the traditions of giants, mighty ones, and anomalous bones. There, the danger of overclaim is even greater, and the need for careful separation between pathology, misinterpretation, and genuine anomalies becomes essential. But the theme remains consistent: the human story contains chapters that are messy, hybrid, and contested, and the most honest approach is to keep the boundary between evidence and interpretation visible even as we explore what the boundaries might conceal.

CHAPTER 6 — GIANTS IN MYTH AND BONE

Few ideas are as widely distributed in human storytelling as the giant. From mountain valleys to coastal islands, from desert kingdoms to forest cultures, people tell of beings larger than ordinary humans—mighty ones who stride across landscapes, lift stones no one else can move, dominate battles, eat too much, laugh too loudly, and leave behind the feeling that the world used to be inhabited by something more immense than the present. These stories are so common that they almost feel inevitable, like fire myths or flood myths. That ubiquity is the first fact we must take seriously, because it means the giant is not a local curiosity. It is a global motif. And whenever a motif becomes global, the question shifts. The question is no longer, "Did a single culture invent a strange story?" The question becomes, "What human experiences repeatedly generate this kind of story, and what—if anything—might be remembered within it?"

In many traditions, giants are not just big. They are early. They belong to the beginning of things, to the age before the world became settled and measured. They are framed as ancestors, rivals, or fallen beings who existed when boundaries were looser—between humans and gods, between nature and culture, between heaven and earth. Sometimes they are portrayed as monstrous, violent, and chaotic, embodying an older wildness that must be defeated so order can emerge. Sometimes they are portrayed as the first builders or the first rulers, whose greatness is both literal and symbolic. Sometimes they are half-divine, born of unions that violate ordinary categories, and their size is a visible sign of that boundary-crossing. In these narratives, giantness is rarely a random trait. It signals something: power, threat, antiquity, or sacred difference.

The phrase "mighty ones" matters here as much as "giants." Many mythic texts and oral traditions do not strictly describe giants as towering beings in the way modern imagination does. They describe them as mighty, renowned, heavy, strong, ancient, or "great" in a moral and social sense. Over time, greatness in reputation often becomes greatness in size. A hero who is said to be incomparable can be imagined as physically incomparable as well. A warrior whose violence feels superhuman can be remembered as larger than human. A lineage of rulers who dominated a region can be imagined as a larger breed, as if political power and physical magnitude belong to the same category. In this way, myth often translates social inequality into bodily difference. It gives the powerful a different scale.

That translation is not a trick; it is a deep human habit. The human mind understands power through metaphor, and one of the oldest metaphors is vertical. Leaders are "above." Ancestors are "greater." The sacred is "higher." When those metaphors are carried into story, they can become literalized. A being who is "above" others in authority becomes a being who is literally above them in height. The giant is therefore a natural mythic form for hierarchy. It is the body made into a political symbol.

Yet giants also appear in contexts where hierarchy alone does not explain the pattern. In many stories, giants inhabit the margins: mountains, forests, deserts, caves, and islands. They occupy the places where ordinary people do not want to live, or the places where the world feels older. This suggests another function. Giants are boundary guardians. They mark the edges of safe human territory. They personify the danger of wilderness and the unknown. To say there are giants beyond the ridge is to say the ridge is a threshold, and crossing it is not merely travel but transgression. In a premodern world where survival depends on careful territorial knowledge, such personifications can be adaptive. They warn children away from cliffs, caves, and hostile valleys. They encode caution into narrative form.

There is also an ecological reading that does not require literal giant humans. Early human communities lived alongside megafauna—enormous animals that would have seemed almost mythic even in life. In some regions, these animals persisted into the human memory horizon long enough to be woven into stories. When you have seen a creature far larger than any human, when you have hunted it or fled from it, your mind becomes familiar with the category of enormous living beings. It becomes easier, in story, to populate the ancient world with giants. The giant becomes a human-shaped version of the world's lost bigness—the sense that the past was crowded with immense forms now gone. This is not a definitive explanation, but it is a plausible contributor to why the giant feels natural to the imagination.

Another contributor is the landscape itself. When you live among megaliths, ruins, and stoneworks whose construction methods are not obvious, the mind searches for proportional causes. If the stones are enormous, then the builders must have been enormous—or must have had help from beings who were. People do not need written engineering theory to reason this way. They use the simplest intuitive model: big things require big makers. Even when a society can build impressive stone structures itself, it may still treat older structures as the work of giants if the cultural lineage of knowledge has been interrupted. The giant becomes a narrative solution to the engineering mystery. It preserves awe in a form that makes sense to ordinary experience.

This is one reason giant myths often attach themselves to specific places. Certain boulders are said to have been thrown by giants. Certain hill formations are said to be giant bodies turned to stone. Certain ancient walls are said to be the work of giant hands. The landscape becomes a canvas for memory, and the giant becomes the brushstroke that explains anomalies. In this role, the giant is not primarily a biological claim. It is a local interpretation of scale: an attempt to account for the outsized.

Across cultures, giants also play a moral role. They are frequently framed as earlier beings who failed a test—becoming violent, arrogant, cannibalistic, or defiant of the sacred order. Their destruction then becomes a lesson: the world is not only physical but moral, and when beings become too great in pride or appetite, they are brought down. In these narratives, the giant is a warning about power without restraint. The size is not just spectacle; it is the embodiment of excess. A giant who consumes too much and dominates too much is simply the human vice of excess magnified until it becomes visible.

At the same time, not all giant traditions are purely moralized. Some are nostalgic. They remember an age of heroes, when people were stronger, braver, and closer to the gods. In these stories, the giant may not be evil but simply greater—an ancestor stock that has dwindled. Here, giantness becomes a symbol of decline, the feeling that the present is weaker than the past. This is another deep human tendency: to imagine origins as more potent than outcomes. Whether that tendency reflects reality or psychological pattern, it shapes myth everywhere. The giant is one form of the "golden age" impulse made bodily.

These different functions—hierarchy symbol, boundary guardian, ecological echo, engineering explanation, moral warning, golden age nostalgia—are not mutually exclusive. A single tradition can blend them. A giant can be both a primeval ruler and a wild monster. A "mighty one" can be both a lineage of elites and a race of semi-divine beings. A story can shift over generations as it travels, adopting new meanings in new landscapes. This is why the global distribution of giant myths does not point neatly to one cause. It points to a family of causes that produce similar narrative forms because the human mind, under similar pressures, tends to craft similar solutions.

The most careful way to handle this, especially within the thesis of this volume, is to treat giant myths as potential carriers of layered memory rather than as literal testimony. Myth can preserve real experiences—encounters with unusually tall individuals, the impression of foreign peoples who seemed large, the awe of monuments, the fear of wilderness, the echo of megafauna, the reality of social hierarchy—while converting those experiences into a single archetype. The archetype then travels farther than any single event. It becomes part of the human mythic toolkit. In this way, the recurrence of giants may tell us less about a secret species and more about what humans repeatedly confront: scale, power, and the feeling that the past was inhabited by forces larger than ordinary life.

Still, it would be dishonest to pretend that the giant motif remains purely symbolic in the modern imagination. Many people treat giant stories as historical claims. They point to reports of oversized skeletons, to alleged suppressed discoveries, to local legends of burial mounds, and to photographs that circulate outside scholarly contexts. Those claims are part of the contemporary landscape of this topic, and they cannot be ignored if the goal is to explore the "missing chapters" narrative responsibly. But the correct order matters. The myths come first as cultural data: they are real, widespread, and meaningful regardless of literal truth. The next step is to examine the physical claims—bones, burials, and alleged finds—with the stricter discipline that physical evidence demands.

Because once bones enter the conversation, the standard changes. A story can hold multiple meanings without collapsing. A skeleton either exists in the record in a verifiable way, or it does not. A measurement either holds up under scrutiny, or it falls apart. The transition from myth to bone is therefore the place where this chapter must become more careful, more forensic, and less forgiving of wishful interpretation.

Once giants move from story into bone, the entire conversation changes temperature. Myth can be elastic. It can carry moral meaning, cultural identity, and memory without needing to be falsifiable in the strict sense. Skeletal claims are different. They are physical. They are measurable. They demand provenance, context, and repeatability. And yet the modern giant narrative often arrives wrapped in a very specific frame: not simply that oversized skeletons existed, but that they were found and then hidden, confiscated, or suppressed by institutions. This is the "suppressed finds" claim—an idea that has become almost as central to modern giant lore as the giants themselves. If it were true, it would be a dramatic indictment of scientific culture. If it is false, it reveals something else equally important: how the human mind and media ecosystem generate the feeling of hidden history even when the evidence does not support it.

The first thing to understand is why the suppressed-finds idea is so persuasive. It draws power from two real features of archaeology and institutions. One is that archaeological interpretation does change. Museums reclassify items. Collections get reorganized. Bones once described in sensational terms may later be understood as pathological, mismeasured, or misidentified. Another is that institutions can be opaque. Many collections are not on display. Many artifacts sit in storage. Many excavation reports are not widely read. From the outside, this can look like concealment even when it is normal curation. Add to that the long history of human remains being handled unethically—collected, moved, displayed, and sometimes withheld—and you have a fertile environment for suspicion. Suppression feels plausible because institutions are not always transparent, and the public often encounters archaeology through fragments and headlines rather than through method.

But persuasion is not proof. When we shift from plausibility to evidence, the suppressed-finds narrative runs into a hard problem: extraordinary skeletal claims require extraordinary documentation. An oversized skeleton is not like a mysterious object that could plausibly vanish into a drawer. Human remains are among the most documented categories of archaeological material precisely because they are ethically sensitive and scientifically valuable. A genuinely anomalous skeleton would be studied, measured, photographed, and contextualized. Its discovery would usually involve multiple witnesses: field crews, local officials, sometimes journalists, sometimes community members. It would have an excavation context: stratigraphy, associated artifacts, a burial structure, an orientation, a location. It would have a chain of custody. The more dramatic the claim, the more documentation is required to treat it as real. This is not a bureaucratic preference; it is the only way to separate fact from rumor.

The suppressed-finds tradition, however, often relies on the opposite of chain-of-custody evidence. It favors anecdotes and secondhand reports. It often appears as a story that "someone's grandfather" saw bones, or "workers uncovered a skeleton," or a "museum took it away," followed by a lack of traceable records. Sometimes it appears as photographs, but photographs themselves are not self-validating. Images can be staged, miscaptioned, or manipulated. Without context—without a verifiable place, date, and excavation record—a photograph is a narrative device, not evidence. The modern ecosystem of giant stories thrives on this ambiguity because ambiguity allows belief to fill the gaps.

This brings us to the second driver: misinterpretation and scale illusion. Humans are prone to error when judging size, especially in photographs. A skull held closer to a camera can look enormous. A bone placed beside a small tool can appear giant. A dig site photographed at an angle can distort proportions. Even without deliberate deception, scale can mislead. Archaeology also deals with fragmented remains. A large femur fragment can be mistaken as belonging to an exceptionally tall individual when it is actually from a large animal, or from a human with a different body proportion, or simply misidentified. Most people are not trained in osteology, and even trained people require careful measurement to avoid mistakes. In this environment, an oversized-skeleton claim can emerge easily from error, excitement, or sensational framing.

A third driver is pathology. Human bodies can become unusually large through known medical conditions. Gigantism and acromegaly can produce extreme height, and other disorders can create unusual bone shapes that appear monstrous in isolation. Such individuals can become the seed crystal around which myths thicken. A community that encounters an unusually tall person might remember them as a giant. If that person's skeleton is later discovered, the story can escalate. In this scenario, no suppression is needed. The story grows because the body is genuinely unusual, and the narrative does what narratives do: it expands a striking fact into a world-explaining claim.

A fourth driver is the long history of sensational journalism and promotional archaeology. In the nineteenth and early twentieth centuries, newspapers frequently published exaggerated or poorly verified archaeological stories, often tied to local pride, tourism, or the thrill of the strange. “Giant skeleton found” was the kind of headline that sold papers, the way “mystery object” sells clicks today. Those early reports still circulate, sometimes detached from their original context and treated as if they were formal scientific announcements. The suppressed-finds narrative often grows out of this gap between old media sensationalism and modern expectations of scientific documentation. When later researchers cannot find supporting records for a sensational claim, believers sometimes interpret the absence as proof of suppression rather than as evidence that the claim was never robust in the first place.

A fifth driver is deliberate hoax. It would be naive to pretend hoax is rare in anomalous archaeology. Human beings fake things for profit, attention, ideology, or simple mischief. In the giant arena, hoaxes can range from crude—misidentified animal bones—to sophisticated—constructed images, staged digs, fabricated “museum letters.” The problem is not merely that hoaxes exist. The problem is that hoaxes are sticky. Once a striking image or story enters circulation, it becomes part of the mythology. Retractions rarely travel as far as the original claim. Over time, the ecosystem of giant narratives becomes layered: a few plausible kernels, many distortions, and a set of emotionally satisfying explanations—suppression, conspiracy, forbidden knowledge—that prevent the story from collapsing under scrutiny.

Now, none of this means that unusually tall humans never existed. They did and do. Nor does it mean that institutions have never mishandled evidence or controlled narratives. They have. But the specific claim of widespread suppression of giant skeletons faces a practical challenge. Suppression at the scale implied by popular narratives would require coordination across time, countries, researchers, and communities, and would leave traces: contradictory documentation, whistleblowers with hard evidence, collections with missing catalog entries, excavation records pointing to vanished remains. The more ambitious the suppression claim becomes, the more evidence it should generate, not less. In reality, suppressed-finds stories often behave like fog: the closer you approach, the more the details dissolve.

A responsible approach, therefore, is to separate three categories that are often blended together. First are genuine cases of exceptional height, which can be documented medically and osteologically and do not require any grand theory. Second are ambiguous claims—local legends of burial mounds, reports of "large bones," stories of museum involvement—that may reflect misinterpretation, exaggeration, or partial truth but lack sufficient documentation. Third are clear fabrications and recycled media artifacts that circulate because they are emotionally and visually compelling. The tragedy of the suppressed-finds narrative is that it tends to treat all three as one category, flattening complexity into certainty.

And yet, the suppressed-finds narrative persists because it satisfies a deeper hunger that runs through this entire volume: the hunger for missing chapters. If the human story feels too neat, then a hidden race of giants becomes an ideal disruptive force. It provides drama, conflict, and an explanation for monumental architecture, elite power, and mythic memory all at once. Suppression becomes the mechanism that keeps the neat story intact. In that sense, the suppressed-finds motif is not only a claim about archaeology. It is a claim about authority. It suggests that official narratives are curated, that truth is managed, and that the extraordinary must be hunted like contraband.

This is why the correct response is not to mock the impulse but to discipline it. If someone wants to argue for oversized skeletons as evidence of a distinct population, the argument must stand on verifiable remains with documented context, measurable anatomy, and careful comparison to known human variation and known pathologies. It must explain why the evidence is not already integrated into mainstream osteology. It must show what is unique, what is repeatable, and what cannot be explained by error, exaggeration, or disease. Without that, "suppressed finds" becomes a narrative that cannot be tested, and anything that cannot be tested becomes a mirror reflecting whatever the believer wants to see.

In the next subsection, we turn directly to the heart of that disciplined inquiry: how to interpret unusually large human remains when they do appear, and how to distinguish between pathology, natural variation, measurement error, and the possibility—however constrained—of a distinct population pattern. The question is not whether stories exist. The question is what the bones can actually support once we treat them as evidence rather than as mythic fuel.

If you strip away the noise of rumor, headlines, and conspiracy framing, the question that remains is actually straightforward. When unusually large human remains are reported—whether in old accounts, local traditions, or legitimate osteological collections—what are we looking at? Are we looking at ordinary human variation pushed to extremes? Are we looking at pathology, where known medical conditions produce exceptional height or unusual bone morphology? Or are we looking at something population-level: a pattern that would suggest a distinct human group or branch with systematically larger bodies? The discipline of this subsection is to show why the first two explanations are far more common and far more compatible with what we know, while also defining what evidence would actually be required to take the third seriously.

Human height varies widely even within a single population, and it varies between populations due to genetics, nutrition, and environment. In any large enough population over enough time, you will find outliers—people who are unusually tall or unusually large-framed. You do not need a separate species or branch to produce a person who stands far above average. This is basic statistics. Outliers are rare, but they exist. The trouble arises because myths and sensational accounts often treat outliers as proof of categories: one tall skeleton becomes “a race of giants.” But biological inference does not work that way. A category claim requires a sample, a distribution, and context. Without those, the outlier remains what it most likely is: an outlier.

Pathology is the next layer, and it is the most important to address because it can produce exactly the kind of “giant” bodies that become legends. Gigantism, typically associated with excess growth hormone during childhood, can lead to extraordinary height. Acromegaly, associated with growth hormone excess after growth plates close, can produce enlarged hands, feet, and facial bones, altering the skull and skeleton in ways that can look almost non-human to the untrained eye. Other conditions can influence growth patterns, joint morphology, and bone thickness. In skeletal remains, signs of such conditions can be present in characteristic ways—though interpretation requires expertise and careful comparison. The key point is simple: known medical phenomena can produce bodies that match the imaginative template of a giant, and they do so without requiring a lost species.

Pathology also interacts with memory. A person with extreme height in a premodern community would be unforgettable. They would be narrated. They might be feared, revered, or mythologized. Their life could become a story, and their story could become a lineage legend. If that person was also a warrior, a leader, or simply socially dominant, the myth could spread beyond the individual and become a symbol: "in those days, people were bigger." Over generations, the story can detach from the individual and become a claim about the past itself. This is one plausible mechanism by which real biological outliers seed cultural traditions of giants.

There is also the simple problem of misidentification. Archaeology frequently encounters fragmented remains. Without careful analysis, large animal bones can be mistaken for human bones. Even within human remains, bones can be mixed or disarticulated, and earlier excavations were often less meticulous about context than modern standards demand. A site might contain multiple individuals, and bones could be combined unintentionally. A femur from one person and a tibia from another could be interpreted as one unusually tall individual. Measurement errors can compound the mistake. Historical accounts are full of enthusiastic extrapolations from partial remains, and once a number is printed, it tends to persist. A "seven-foot skeleton" becomes "nine feet" in retelling, and the exaggeration is rarely corrected.

If we want to ask whether unusually large remains could represent "another branch," we must define what that would actually mean in scientific terms. Another branch would not be established by one or two enormous individuals. It would require repeated findings in a definable region and time range, showing a consistent pattern of body size significantly outside known human variation, and ideally accompanied by other anatomical differences that are not readily explained by pathology or nutrition. It would require population-level evidence: multiple skeletons, consistent measurements, and a coherent archaeological context—settlement patterns, burial practices, associated material culture. It would also require genetic evidence if preservation allows, because genetics is now the most direct tool for testing ancestry claims. A "branch" is not a vibe. It is a detectable pattern across multiple lines of evidence.

This is where the giant narrative often collides with reality. The known fossil record already contains multiple human forms—Neanderthals, Denisovans, Homo erectus, and others—showing variation in body proportions, robustness, and cranial morphology. If there had been a widespread branch of humans consistently taller by several feet than typical Homo sapiens, it would likely have left a stronger footprint. Height is not subtle. It changes joint loading, bone proportions, and the practical mechanics of movement. A population of true giants would also have significant energetic requirements. Large bodies require more calories. They would need ecological conditions that could support them reliably, and their material culture might reflect those needs. Again, none of this makes a giant branch impossible in principle, but it makes it demanding. It would require a coherent ecological and archaeological framework, not only sensational bones.

It is also important to recognize that "giantness" is not the only way myths encode difference. A group could seem "giant" because they were taller on average than neighboring groups due to nutrition, genetics, or both. In a world where many populations experienced periods of undernutrition, a well-fed, robust group could appear enormous by contrast. This contrast effect is powerful. Even a modest height difference can be mythologized when groups are in conflict or when one group dominates another. The defeated describe the victors as huge; the victors describe their ancestors as larger than life. In this way, "giants" can sometimes be the memory of real intergroup differences amplified through narrative and emotion.

We should also consider the role of selection and ecology. Certain environments favor larger bodies—colder climates, for example, can correlate with larger body mass in broad biological patterns, though human variation is complex and influenced by many factors. Highland or rugged environments can favor endurance and robustness rather than extreme height, but local patterns of body size can still emerge. Again, the key is that such patterns would be gradients and distributions, not a separate species. The web model of human evolution suggests that populations could mix and exchange traits, producing regional patterns that look unusual without representing a separate branch.

So where does that leave the question? It leaves it with a hierarchy of explanations. At the base is ordinary variation: some humans are simply very tall. Above that is pathology: some humans become extremely tall or unusually shaped through medical conditions, and those individuals can seed myths. Above that is misidentification and error: bones can be mixed, mismeasured, or misclassified, producing inflated claims. Only after those are addressed do we reach the possibility of the population-level pattern that would justify talk of another branch. And even then, the branch would need to be defined by repeatable evidence across multiple sites and ideally supported by genetics.

Within the thematic frame of this volume, this hierarchy is important because it preserves the boundary between mystery and method. The "missing chapters" impulse can be productive if it pushes us to look at overlooked data and ask careful questions. It becomes destructive when it treats absence of evidence as proof of conspiracy, or when it treats isolated claims as population history. The giants topic, more than almost any other in this volume, tests our ability to keep that boundary intact.

The next subsection turns to why giant narratives persist so strongly in relation to monumental construction. Across continents, people have looked at megaliths, cyclopean walls, and massive temple platforms and concluded that ordinary humans could not have built them. Giants become the intuitive builders of the old world. We will examine that logic carefully—what it gets right about human psychology, what it gets wrong about engineering and labor, and what it might still preserve as cultural memory even when the literal claim cannot be sustained.

There is a particular kind of awe that rises when you stand before stone that seems too heavy to belong to human hands. It is not just admiration. It is cognitive dissonance. The mind knows what a body feels like. It knows what a man can lift, what a team can carry, what ropes can drag across earth. And then it meets a block that seems to mock those intuitions—a wall of fitted stones whose joints are tighter than the blade of a knife, a monolith taller than a house, a platform whose foundations disappear into the ground as if it were anchored to bedrock by sheer will. In that moment, a simple explanatory reflex often takes over: if the stone is gigantic, the builders must have been gigantic. This reflex is ancient. It repeats across continents. Cyclopean walls, megalithic temples, giant statues, and mysterious ruins are attributed again and again to giants, to mighty ones, to beings of an earlier age. In many places, that attribution is not a fringe idea. It is the default local explanation.

The logic is easy to understand because it is proportional. Humans use proportional reasoning constantly. Big effects come from big causes. A small animal cannot leave a footprint like a boulder. A child cannot raise a roof beam alone. If a structure looks beyond the capacities of ordinary life, then ordinary life is not credited. Something larger is. In societies without written engineering tradition, the simplest "larger" is not advanced machinery. It is advanced bodies. Giants are the most intuitive machine the human imagination can invent because they require no theory of technology. They require only the idea that the past contained beings more powerful than the present.

This is why "giants built it" is so common where cultural discontinuity exists. When a later population inherits monumental structures built by a previous culture, the knowledge chain is broken. The inheriting population sees the result but does not possess the methods. They can see the stones, but they cannot reconstruct the organizational system that moved them: the labor coordination, the quarrying techniques, the rope systems, the ramp designs, the rollers, the levers, the seasonal workforce, the political authority that mobilized thousands, the religious incentives that made people work for a project that might outlive them. Without that context, the stones look impossible. Giants become a narrative bridge over a missing engineering culture.

In this sense, giant-builder myths are not necessarily claims about biology. They are claims about discontinuity. They signal that the present community perceives the structure as belonging to another world. That other world may have been a real previous society, a previous dynasty, a vanished population, or simply an earlier phase whose memory has been mythologized. The giant is the mask worn by a lost tradition. Where historians see cultural succession, myth sees monstrous ancestry. Where archaeologists see labor systems, myth sees superhuman force.

But there is another reason giant-builder stories are powerful: they preserve awe in a morally satisfying form. If the builders were giants, then their works can remain beyond ordinary scrutiny. The monument is not simply impressive; it is sacred or uncanny. It belongs to a time when the world was different. This elevates the structure into a moral landscape. Many giant-builder stories place the giants in a primeval epoch that ends violently—giants destroyed by gods, drowned by floods, defeated by heroes. The ruins then become relics of a fallen world. The monument is explained not only by giant strength but by giant fate. The stones stand as proof that greatness once existed and then was judged. In that framing, architecture becomes theology.

The problem is that this narrative, while psychologically and culturally coherent, does not align neatly with what we know about human capacities. Humans do not need to be giants to move large stones. They need organization, time, leverage, and social motivation. The brute strength of a single body is rarely the primary factor in monumental construction. Levers multiply force. Rollers reduce friction. Ramps change the direction of effort. Coordinated teams can apply steady power over long durations. A workforce can be fed and housed by an agricultural surplus. Authority can mobilize labor. Ritual can justify it. When these pieces align, humans can achieve feats that feel impossible to later observers who only imagine individual strength.

This is why many megalithic structures, once studied in detail, reveal practical logic rather than supernatural impossibility. Quarries are found nearby. Tool marks show extraction methods. Transport routes become plausible when terrain is analyzed. Construction stages are visible. The stones may be enormous, but the engineering principles are often simple, relying on friction management and incremental movement rather than sudden lifting. In other words, what looks impossible in a single mental snapshot becomes possible when expanded into process: a thousand small actions repeated with discipline.

Yet the persistence of giant-builder myths still matters, because it points to something true about the human experience of ruins. People are not just trying to solve an engineering puzzle. They are reacting to the presence of a past that feels superior or alien. That reaction can include an accurate intuition: the builders often did possess capacities that the later population lacked, not in terms of biology, but in terms of social organization, political cohesion, and technical tradition. A society that cannot coordinate large-scale labor will naturally struggle to imagine that another society could. The limit is not the body; it is the system. Giant myths encode a system-level difference in a body-level metaphor.

There is also an elitist dimension that intersects with this volume's broader themes. Monuments are often built by hierarchical societies. They reflect centralized authority, specialized labor, and social stratification. For those living under or near such structures, the builders can seem like a different kind of human—not because they were taller, but because they wielded the power to compel thousands. In later memory, the elite builders become "mighty ones." Their might is political, economic, and religious. Over time, that might can become literalized into physical size. The giant-builder myth can therefore encode the memory of elite classes: larger-than-life rulers, priest-kings, dynasties that claimed divine descent. The "giants" are not a separate species; they are the past's concentrated power remembered as a different scale of being.

And in some cases, giant-builder myths may also encode the memory of outsiders. Many monumental traditions spread through contact and migration. A new group arrives with new building forms, new religious models, new technical expertise. To those already present, the newcomers can seem like beings from elsewhere—strange, powerful, imposing. Their structures reshape the landscape. Over generations, the newcomers become mythic. Their foreignness becomes otherworldliness. Their ability becomes superhuman. In this way, giant-builder stories can be the fossilized memory of contact events: moments when one group's expertise overwhelmed another's expectations.

This is where the "missing chapters" thesis finds a disciplined foothold. Not in the literal claim that giants existed and moved stones, but in the recognition that myths cluster around real discontinuities. If people repeatedly attribute monuments to giants, it suggests that monumental traditions often outlast the memory of their methods. It suggests that cultural collapse, migration, and knowledge loss are common enough that later populations inherit stone without inheritance of explanation. Those are missing chapters in the mundane but profound sense: the chapter of labor, organization, and technique vanishes, leaving only stone and awe. Myth fills the gap.

The risk, of course, is that myth can also mislead by turning awe into biology. If we mistake giant-builder stories for literal data about human types, we can lose the more valuable insight: that human civilizations can rise to extraordinary organizational capacity and then fall, leaving behind monuments that later people cannot easily reproduce. The "giant" is often the memory of a civilization's peak, not the memory of a different species.

And yet, to remain honest to the thematic arc of this book, we cannot entirely dismiss the possibility that unusual human bodies contributed to elite narratives. Tall individuals, robust lineages, or populations with distinctive traits could have been elevated into leadership roles, and their physical presence could have fed mythic exaggeration. But even in such scenarios, the builders of megaliths would still have relied on labor systems, not on personal height. A very tall ruler does not move a hundred-ton stone alone. A system does.

The next subsection turns to the deeper symbolic logic that makes giants such a persistent mythic form in the first place: why size equals authority across cultures. If we can understand that logic, we can better understand why giant stories attach themselves so naturally to power,. and why they persist even when the physical evidence for a giant race remains elusive. Size is not just a measurement in myth; it is a language of domination, origin, and sacred distance.

Across cultures, authority is rarely imagined as a small thing. It is imagined as height, weight, radiance, enlargement—something that takes up more space in the world than ordinary life does. Kings sit higher than others. Gods dwell above. The sacred mountain rises. The hero stands out. Even in everyday language, greatness and bigness are intertwined: a "great" leader, a "high" office, a "big" name. Myth takes these metaphors and makes them literal, because myth is not concerned with subtlety. Myth is concerned with power made visible. That is why giants persist. A giant is authority with a body. It is dominance translated into scale.

This translation begins in the human organism itself. Humans are primates, and primates are intensely sensitive to hierarchy. In many primate species, size correlates with dominance—not always, but enough that the brain carries ancient associations between physical magnitude and social power. Height can intimidate. Broad shoulders can signal strength. Deep voices can command attention. Even in modern societies, where legal systems and institutions regulate power, people still respond instinctively to physical presence. In small-scale societies, where personal dominance can matter more directly, these cues can be even more significant. Myth builds on this ancient wiring. It amplifies what the nervous system already understands: bigger is harder to challenge.

But size in myth is not only about intimidation. It is also about distance. Authority is what stands apart. It is what cannot be approached casually. A giant is separated from ordinary humans by physical distance. You cannot look a giant in the eye without craning your neck. You must physically acknowledge the hierarchy. This is a perfect mythic symbol for sacred difference. Gods are not merely rulers; they are other. They are not just in charge; they are in a different category of being. Size accomplishes that categorical separation quickly. The giant is not simply a stronger man; it is a being whose scale marks it as belonging to another order.

This is why giants often appear in the earliest layers of mythic time. They belong to beginnings, to primordial ages. Myth often treats the origin of the world as an age of excess: bigger mountains, longer lifespans, stronger heroes, more violent forces. In such an origin age, the categories we live by—human and divine, nature and culture—are not fully separated. Giants fit perfectly into this liminal zone. They are human-shaped but beyond human. They are the bridge between gods and men, between the raw forces of nature and the domesticated order of society. Their authority is the authority of the first era, the time when power was not yet tamed.

Authority in myth is also often generational. Ancestors are imagined as greater than descendants. The first beings are imagined as more potent than later beings. This is the "decline from origins" pattern that appears in many cultures: a golden age gives way to weaker ages. In that framing, giants are the bodies of ancestors made literal. They represent the idea that the past had more vitality, more substance, more presence. Even when a culture's lived experience suggests that technology improves over time, myth often insists that the deepest power is behind us, not ahead. The giant is a fossil of that worldview: the past was bigger.

There is also a political function here that is easy to miss. Elites often legitimize their authority by claiming extraordinary ancestry. They claim descent from gods, from heroes, from founder figures. In many traditions, the rulers are “larger” in a metaphorical sense—closer to the sacred. Myth can literalize this by describing the founders as physically larger, or by describing dynasties as carrying giant blood. Once such a claim exists, it becomes a social technology: it marks the elite as a separate kind of human. It justifies hierarchy by embedding it in biology or divine origin. The claim does not have to be true to be effective. It only has to be believed.

This connects to a deeper mythic logic: size as legitimacy. In myth, rightful rule often belongs to those who can bear the weight of the world. A king is not simply elected; he is chosen by strength, by divine favor, by an ability to withstand forces that would crush ordinary people. Giants embody that capacity. They can carry stones, fight monsters, lift gates, carve rivers. These are not merely feats; they are proofs of fitness for rule. In a mythic world where the cosmos is a contested territory, authority is not a matter of paperwork. It is a matter of force, and force is expressed as size.

There is also the cosmological dimension. Many mythologies treat the world as layered vertically: underworld below, earth in the middle, heavens above. Authority is naturally associated with the upper layers. The mountain is sacred because it rises. The sky is sacred because it is above. When myth introduces a being that is physically tall, it is not only making the being stronger; it is aligning the being with the vertical structure of sacred space. A giant's head approaches the sky. The giant becomes a living axis between earth and heaven. This is why giants and mountains often appear together, and why giant stories so frequently situate giants in high places, on peaks, near clouds, in regions where the sky feels close. Size makes the being cosmologically relevant.

At the same time, giants are frequently portrayed as threats to cosmic order. Their size makes them difficult to restrain. They are often too hungry, too violent, too arrogant. This is not accidental. Myth uses giants to explore the danger of unbounded power. If authority grows too large, it becomes tyranny. If appetite grows too large, it becomes cannibalism. If pride grows too large, it becomes rebellion against the gods. The giant becomes the embodiment of power without limits. And because myths often function as social instruction, the giant's defeat becomes a moral lesson: order requires boundaries, and even the mighty must be constrained.

This moral layer also explains why giants are sometimes associated with catastrophe. In some traditions, giants are destroyed by floods, by fire, by divine punishment. Their fall marks a reset of the world. The implication is that the old age was too extreme to continue. The world had to be resized. In that sense, the giant is not only authority; it is the old scale of reality. Its destruction is the act of bringing the world down to human measure. The myth says: the cosmos once contained beings too large for a stable order, and the present world is the result of their removal.

Yet the emotional power of giants persists even when their moral role is negative because the giant also carries nostalgia. Even when giants are villains, they are impressive villains. They remind humans of what it would mean to be unrestrained—stronger, freer, closer to the raw forces of nature. This is why giant tales often oscillate between fear and admiration.

The giant is dangerous, but it is also magnificent. The psychological charge comes from that ambivalence. Authority is always ambivalent: it protects and it threatens. The giant is the mythic figure that contains both poles.

When we bring this back to the question of bone, the implications become clearer. The mythic equation of size with authority means that giant traditions will persist even if giant skeletons do not. Because the function of the story is not primarily biological description. It is social mapping: identifying who had power, who threatened the group, who built the old works, who belonged to the origin age, and why the present world is smaller. That mapping can be generated by many sources—real tall outliers, elite dominance, foreign conquerors, monument awe, ecological memory—and it will still produce the same archetype because the archetype is structurally useful.

And this is where the chapter must make its final pivot. If size in myth is authority and otherness made visible, then giant traditions might preserve, in distorted form, the memory of "different humanity"—not necessarily different species, but different social orders, different populations, different lineages of elites, different eras of organization and collapse. The last subsection turns toward that possibility: not the simplistic claim that giants were real as a separate race, but the subtler possibility that giant myths may be the cultural residue of human diversity and discontinuity—memory of a world in which the boundaries of "human" were experienced differently than they are today.

If we treat giants as literal biology, the story quickly becomes a battlefield of claims and counterclaims: photographs without context, sensational headlines, skeletal anecdotes, and the recurring assertion of suppression. But if we treat giants as cultural memory—memory shaped by metaphor, magnification, and the human need to explain difference—then the motif becomes more interesting and, in a sense, more plausible. It becomes a lens through which societies remember that the human past was not uniform. It was populated by varieties of people, by unfamiliar bodies, by foreign customs, by elites who seemed almost divine, by rivals who felt monstrous, by ancestors who loomed larger in imagination than their descendants ever could. In this reading, "giants" may not be a separate species standing ten feet tall across continents. They may be the mythic shorthand for a different humanity: the remembered experience of discontinuity in human types, power structures, and worlds that vanished.

The most basic form of "different humanity" is simply human diversity experienced as otherness. For most of our species' existence, humans lived in small groups. A person could live their entire life within a limited territory, encountering only a small range of body types and cultural practices. In such a world, the arrival of outsiders could be shocking. A neighboring group might be taller on average due to diet and genetics. Another might be more robust due to ecological pressures. Another might carry distinctive cranial shapes, scarification practices, or clothing that altered the perceived outline of the body. To modern eyes, such differences may not seem extreme. To small-scale societies, they can feel profound. The mind responds by categorizing: they are not like us. Myth then gives that category a stable form. The "other" becomes a giant, an ogre, a mighty one, a people of old who stand apart from ordinary humans.

This logic becomes even stronger when outsiders are associated with dominance. Conquest and hierarchy are powerful myth-generators. If a group is defeated repeatedly by a rival, the rival can become mythologized as larger and stronger than life. Physical intimidation matters in conflict, and tall or robust warriors do make an impression, but the mythic enlargement often reflects political power more than actual anatomy. The defeated remember the victors as huge because the victors felt huge—an overwhelming force. Over time, "they were stronger" becomes "they were bigger," and "they were bigger" becomes "they were not fully human." In this way, giant traditions can be the fossilized memory of historical power imbalance.

The next layer of "different humanity" is elite separation. Many early societies, even before full urbanization, developed hierarchies. Certain lineages claimed sacred descent. Certain rulers or priestly classes distinguished themselves through ritual, ornamentation, and controlled knowledge. Over time, these elites can come to be described in myth as semi-divine—children of gods, bearers of special blood, mighty ones who ruled when the world was young. Their "giantness" is often less about physical height than about distance: a class separated from ordinary life by privilege and access. Yet myth prefers bodies to abstractions. It makes class difference visible by exaggerating the elite's scale. The rulers become giants because their authority felt like a different order of being.

There is also the possibility of real biological outliers shaping collective memory. A single individual with extreme height or unusual morphology can become legendary in a community. If they are associated with leadership, warfare, or ritual roles, their presence can be woven into origin narratives. Over generations, the story can expand from "our ancestor was unusually tall" to "our ancestors were giants." This is not a trick of storytelling; it is how storytelling preserves salience. Humans remember what stands out. What stands out becomes representative. The exceptional becomes archetypal.

But the most consequential form of "different humanity" in the context of this volume is the one we already know is real, even if it does not map neatly onto giant folklore: the fact that Homo sapiens did not evolve alone. We lived alongside other human lineages. We met them. We interbred with some of them. We absorbed fragments of them into our genomes. This is not myth. This is established in the broad sense: the human lineage is composite. Even if we cannot connect that composite reality directly to giant legends, the existence of multiple human forms in deep time provides a powerful background against which myths of "mighty ones" and "other humans" become more than fantasy. They become culturally expected. If the ancient world contained different kinds of humans, then stories of different kinds of humans are not inherently absurd. They may be distorted, yes, but distortion is what cultural memory does when it stretches across millennia.

This is where we must hold the boundary carefully. The known archaic humans we can document do not fit the popular "giant" template. Neanderthals, for example, were robust but not towering in the sense of ten-foot myths. Denisovans are still largely known through genetics and fragments, and we cannot responsibly claim they were giants. The fossil record contains variation, but it does not give us a confirmed global race of enormous humans. So we cannot treat giant myths as straightforward memories of an archaic species. However, we can treat them as a narrative channel through which societies remember, in a mythic register, that humanity was not always one clean category. The "different humanity" might be the remembered existence of others—others who were physically distinct, culturally distinct, and sometimes terrifying.

The monument dimension returns here as well. When people inherit structures they cannot easily reproduce, they experience the builders as different humans—people of superior strength, superior knowledge, or superior organization. Myth transforms that difference into biology. The builders become giants. In this form, the giant is not a memory of a different species but a memory of civilizational discontinuity: the loss of techniques, the collapse of labor systems, the vanishing of specialized knowledge. The "giant age" is the age before the collapse, the time when humans could do what later humans cannot. In that sense, the giant story can preserve a true historical pattern—loss and decline—while being wrong in its literal explanation.

This is also why giant myths are often paired with themes of judgment and disappearance. Giants are destroyed, banished, drowned, turned to stone. They vanish from the world, and the world becomes smaller. That pattern mirrors something real in human history: populations vanish. Lineages are absorbed. Worlds end through climate shifts, conflict, and collapse. Even when the specific events in the myth are not literal, the structure of the story reflects a recurring reality: the past contains peoples who are gone, and their absence leaves a gap that demands narrative explanation. Giants fill that gap. They give shape to vanished others.

Seen this way, the giant motif becomes a diagnostic tool for the broader project of this volume. It reveals where human memory tends to deposit its strongest feelings about discontinuity: in bodies that are bigger than life. It reveals that humans repeatedly imagine the past as populated by beings who were not merely earlier, but categorically different. That instinct may be partly psychological—our tendency to magnify origins—but it may also be anchored in the genuine complexity of our ancestry and the repeated experience, across continents, of encountering and replacing others.

The responsible conclusion, then, is not that giant myths prove a hidden branch of human evolution. It is that giant myths preserve the human experience of difference and loss. They are the mythic echo of human diversity—biological outliers, foreign peoples, elite classes, vanished builders, and perhaps the deep-time presence of other humans whose existence is now confirmed in genetics but whose lived reality is largely lost. The myth says, “They were bigger.” The evidence says, “They were different.” The translation from difference to bigness is the mind’s oldest shorthand for power and otherness.

CHAPTER 7 — MEGALITH BUILDERS AND BIOLOGICAL OUTLIERS

Megaliths force a particular kind of question because they sit at the intersection of physics and imagination. A carved stone can be admired as art. A painted wall can be understood as symbolism. But a monolith weighing tens of tons imposes itself as a problem of bodies—bodies that quarried it, dragged it, lifted it, and placed it with intent. However sophisticated the planning, the work still resolves into muscle and breath. This is why megalithic construction so often generates speculation about unusual builders. If the stones seem beyond ordinary capacity, then ordinary bodies are doubted. The chapter's task is to approach that doubt without feeding it into fantasy. The real story of megaliths is already extraordinary, and it begins not with giants or hidden species but with what human bodies can do when endurance, coordination, and social organization are pushed to their limits.

The first reality to establish is that megalith building is not a test of maximum individual strength. It is a test of sustained collective effort. A single person's peak lifting ability matters far less than a group's capacity to apply steady force over hours, days, and seasons. The difference is crucial. Most megaliths were not "lifted" in the way modern people imagine—cleanly raised and placed in one motion. They were moved by incremental techniques that convert time into power: levering, rocking, hauling, rolling, sliding on prepared tracks, and raising with ramps. These techniques do not require superhuman strength. They require patience, engineering intuition, and the social discipline to keep many people working in coordination.

That discipline depends on endurance more than brute force. Endurance is the capacity to repeat effort without collapse. It is the ability to keep pulling when the body is tired, to keep coordinating when the mind is strained, to keep returning to the work site day after day. Endurance can be trained. It can be culturally embedded through ritual, obligation, and identity. A society that builds in stone often does so not because it has stronger individuals, but because it has stronger systems for sustained labor: food surplus to feed workers, leadership that can organize schedules, and cultural incentives that make the work meaningful. The "strength" of megalith builders is therefore often systemic strength—an ecology of cooperation.

Yet systemic strength still expresses itself through bodies, and those bodies leave traces in the kinds of tasks they performed. Moving stone long distances over uneven terrain requires not only muscles but joints that can withstand repetitive load, and a cardiovascular system capable of steady output. Quarrying and shaping stone requires upper-body endurance, fine motor skill, and the ability to work with tools for long periods. Transport requires coordination and rhythm, often using songs or chants to synchronize pulling and stepping. Placement requires careful attention and nerve, because a single misstep can injure or kill. When such work becomes habitual within a community, it selects—not genetically in a strict immediate sense, but socially—for certain kinds of people: those resilient enough to endure, those disciplined enough to coordinate, those experienced enough to avoid mistakes. Over generations, the identity of the community can become tied to these traits. They become "stone people" not because their DNA is different, but because their lifeways sculpt their bodies and behaviors.

This is where the concept of "biological outliers" begins to enter in a grounded way. Before we ever reach cranial anomalies or elite traits, we should acknowledge a simpler form of outlier: the occupational body. Bodies adapt to work. Skeletal remains can show this, not in dramatic monstrous forms, but in patterns: robust muscle attachments, joint wear, and signs of repetitive strain consistent with heavy labor. A population engaged in intense physical work can appear, osteologically, different from a population living a more sedentary life. This difference is not a separate species; it is the long-term imprint of labor on bone. If megalith-building societies produced a workforce conditioned by hauling, quarrying, and construction, their bodies might show a distinctive robustness compared to neighboring groups. That robustness, in turn, could feed myths of mighty builders and larger-than-life ancestors.

Endurance is also tied to environment. Communities living in rugged landscapes—highlands, islands, coastal cliffs—often develop strong locomotor adaptation through daily life: climbing, carrying, navigating uneven ground. They may be smaller-bodied in some contexts and more robust in others, depending on diet and ecology. What matters is not size but functional strength: grip strength, stability, and stamina. It is easy for modern observers to assume that bigger bodies are required to move big stones, but the mechanics of hauling and levering reward technique, rhythm, and coordination as much as mass. A well-organized team can outperform a disorganized group of larger individuals. In this way, megalithic achievement does not automatically point to unusual biology. It points to unusual social cohesion.

But social cohesion itself has biological preconditions. Humans can only sustain large coordinated projects if they can cooperate at scale. That cooperation depends on trust, shared identity, leadership legitimacy, and the ability to manage conflict. These are cognitive and social traits, not only physical. The megalith, therefore, is not merely a monument of strength. It is a monument of coordination. It implies that a community could keep people aligned—motivated enough to work, disciplined enough to follow plans, committed enough to invest effort in something that might not pay immediate survival dividends. This is why megaliths often feel like the edge of civilization even in pre-urban contexts. They signal that humans had crossed a threshold in collective action.

Once a community crosses that threshold, it also develops specialists: planners, toolmakers, quarry masters, transport coordinators, ritual leaders who sanctify the work, and skilled artisans who finish stone surfaces. Specialization changes how labor is distributed, and it can change who holds authority. In many megalithic contexts, those who "knew how" would be a source of prestige. Knowledge becomes a kind of power, and power tends to separate itself socially. That separation is important because it becomes a breeding ground for later myths about elite lineages being different—closer to gods, stronger in blood, distinct in appearance. The stone project, in other words, can create not only monuments but hierarchies.

At this point, we arrive at the core tension of the chapter. Megaliths provoke biological speculation because they are physical feats, and physical feats invite physical explanations. But the deeper explanation often lies in the interface between body and culture: endurance shaped by lifeway, strength multiplied by technique, and labor made possible by social systems. If we want to talk about "outliers," we must first talk about how ordinary humans become extraordinary through training, repetition, and organization. The megalith builders were outliers in the sense that they lived in societies capable of sustained collective action and in bodies conditioned for heavy work—not necessarily because their genetics placed them outside normal human range.

And yet, the chapter title promises more. It suggests that certain populations and certain elites may have displayed traits that seemed unusual—cranial forms, non-local features, markers of difference that later cultures interpreted as signs of special origin. The transition from "strength and endurance" to those more contentious claims must be made with care. The correct next step is to examine one of the most frequently cited physical anomalies in ancient contexts—elongated skulls and cranial modification—and to distinguish between cultural practice, pathology, measurement error, and the rare possibility of unusual cranial morphology not explained by intentional shaping.

Few images in alternative history have had as long a life as the elongated skull. Photographs of stretched crania in museums, drawings of narrow-headed nobles, and dramatic comparisons between ordinary skulls and highly modified ones are used again and again to suggest that ancient elites were not merely socially distinct but biologically different—perhaps even non-human. The intensity of this claim comes from how the skull functions in the human imagination. The skull is not just bone; it is the housing of the mind. Alter the skull, and people feel as if the mind itself has been altered. That is why elongated skull narratives carry such emotional charge. They imply a different kind of consciousness. They suggest that the builders, the priests, or the rulers were a separate type of being.

The disciplined approach begins by separating "elongated skulls" into categories that are often collapsed into one. The first category is intentional cranial modification—head shaping practiced in many societies across the world. The second is pathology and developmental variation—conditions that can alter skull shape. The third is normal human variation and measurement perspective—where certain head shapes appear extreme because of photographic angles, selective examples, or unfamiliarity with population diversity. Only after these categories are distinguished can we ask whether any remaining cases warrant the language of "anomaly" in the stronger sense.

Intentional cranial modification is the most widespread and the most important category because it explains the majority of dramatic elongated skull examples. This practice is simple in principle: an infant's skull bones are not yet fused, and the head can be shaped gradually by applying steady pressure over time, often using boards, cloth bindings, or cradle devices. Because the shaping is done early and consistently, the skull grows into a new form. The resulting cranial shape can be striking—high, flattened, conical, or elongated. In many societies, this modification is not hidden. It is celebrated. It is an identity marker.

The fact that such practices existed in many regions matters because it dissolves a common modern assumption: that elongated skulls must be rare, secret, or unnatural. They can be the opposite. They can be a public signal of group belonging, beauty, rank, or lineage. In some contexts, cranial modification is associated with elites, which naturally fuels the suspicion that "the elite were different." But "different" here often means culturally marked, not biologically non-human. The skull becomes a billboard for identity, a visible declaration that a person belongs to a specific lineage or social class. In hierarchical societies, that declaration can be powerful. It makes status literally embodied.

This also explains why cranial modification often appears in contexts tied to power. If rulers and noble lineages want to present themselves as separate from the ordinary population, they use symbols that cannot be faked easily. Clothing can be stolen. Jewelry can be copied. A body modified from infancy is a lifelong credential. It announces that one's family had the resources and social legitimacy to shape identity at the deepest level—literally shaping the head. In this way, cranial modification can be understood as a political technology. It manufactures visible distinction.

But the presence of intentional modification does not mean every elongated skull is modified, nor does it mean that every modified skull can be recognized easily. Some modification styles are subtle. Some create forms that look like normal variation unless you know what to look for. Others are so dramatic they become the icons of sensational claims. The key is that osteological analysis can often detect modification patterns through characteristic flattening and deformation, especially when the practice follows consistent cultural methods. That is why, in professional contexts, elongated skulls are often discussed as modified rather than as mysterious anomalies.

The second category—pathology and developmental variation—matters because it introduces a different mechanism for unusual cranial shape. The skull can change shape due to premature fusion of cranial sutures, developmental disorders, or other medical conditions. Such cases are usually rare at the population level, but they can be visually dramatic. In skeletal remains, pathology must be diagnosed carefully because the signs can overlap with modification effects. The difference is that modification tends to produce consistent patterns across multiple individuals in a culture, while pathology tends to produce more idiosyncratic patterns. If a site yields many similarly elongated skulls, modification is more likely. If a single skull is bizarrely shaped in a unique way, pathology becomes more plausible.

The third category—normal variation and perspective—matters because modern viewers are often shown extreme examples without context. A museum display may contain the most striking skull in a collection because it is visually compelling. Photographs may emphasize dramatic angles. Comparisons may place a highly modified skull beside an average skull to amplify contrast. Without population context, viewers assume the extreme skull represents a distinct group rather than a special case. This is a familiar pattern in anomalous claims: a rare example is treated as representative, and representativeness is mistaken for category.

Once these categories are in place, we can address the most common leap: the claim that elongated skulls indicate non-human ancestry. This claim usually rests on two impulses. One is aesthetic: the skull looks "wrong" compared to modern expectations. The other is symbolic: because the skull houses the brain, a different skull shape is imagined to imply a different brain, a different intelligence. The first impulse is understandable but unreliable. Humans are poor at assessing biological variation across unfamiliar contexts. The second impulse is emotionally powerful but scientifically slippery. Skull shape can change without creating a fundamentally different kind of brain. The brain is plastic, especially in development, and many modified individuals lived normal lives within their societies. There is no responsible basis for claiming that cranial modification equals a non-human mind.

However, the skull question does not become trivial just because modification explains many cases. It remains significant because it intersects with the volume's core theme: elites as marked lineages, and the possibility that some elites carried non-local traits due to migration, intermarriage, and demographic complexity. Cranial modification can amplify that perception of difference. If an elite lineage is already genetically distinct in subtle ways because it intermarried with outsiders or maintained endogamy, and then it also modifies skull shape intentionally, the result is a doubled signal: biological difference plus cultural marking. Later observers might interpret this as evidence of a separate species when it is actually the layering of social strategy on top of ordinary human variation.

In addition, cranial modification often emerges in contexts where identity and cosmology are intense. The head is not merely a body part; it is the seat of spirit in many traditions. Modifying it can be framed as aligning the body with a sacred ideal—imitating ancestors, embodying beauty, or signaling closeness to the divine. This is where the skull becomes mythic material. When later cultures or modern audiences encounter such remains, they are not just seeing bone. They are seeing a deliberate attempt to become different. It is precisely this deliberate difference that invites the interpretation of otherness.

Therefore the responsible approach is neither dismissal nor sensationalism. It is to treat elongated skulls as evidence of cultural practice, identity engineering, and hierarchical signaling, while remaining open to the fact that some skulls may reflect pathology or rare variation. The question the evidence can support is not “were they aliens?” but “what social system made it valuable to alter the body, and what does that tell us about elites, identity, and power?” When the skull is understood in that frame, it becomes a window into the deep politics of appearance.

This sets up the next subsection directly. Once we admit that elites often engineer difference—through body modification, controlled marriage networks, ornamentation, and ritual knowledge—the question becomes how much of what we perceive as “non-local” or “anomalous” in elites is actually the result of mobility and admixture. If elites traveled, intermarried, and formed networks across regions, they could carry traits that stand out in local populations. The next subsection explores that dynamic: non-local traits in ancient elites, and how migration, alliance, and selective reproduction can create the impression of a distinct caste without requiring a separate species.

Across much of human history, elites have been mobile in a way ordinary people rarely were. They travel to marry, to negotiate, to conquer, to trade, to study, to serve as hostages, to claim legitimacy, and to build alliances that stretch beyond the horizon of local villages. This mobility produces a simple but powerful effect: elites can look different. They can carry physical traits that stand out in a region—differences in stature, facial structure, pigmentation, hair texture, or other features that communities read as "non-local." In a world where most people were born, lived, and died within a limited range, such differences could become socially charged. They could be read as signs of divine ancestry, foreign origin, or a separate kind of humanity. The purpose of this subsection is to show how non-local traits can arise through ordinary human processes—migration, alliance-building, endogamy, and network power—without requiring the leap to non-human bloodlines.

The first principle is that elites often form marriage networks rather than marrying randomly within a local population. Marriage for elites is not primarily about romance; it is about consolidation. It is about linking lineages, securing resources, and stabilizing political relationships. When elites marry across regions—sometimes far across regions—they introduce new genetic variation into their lineage. If they maintain a degree of endogamy afterward, marrying within a narrow social class, the introduced traits can become concentrated. Over generations, a ruling caste can become visually distinctive not because it belongs to another species, but because it is a selective subset of the broader population shaped by controlled reproduction.

This process does not require modern genetic knowledge. It emerges naturally from social strategy. If elite families repeatedly choose spouses from certain allied lineages—perhaps those known for height, for beauty, for charisma, for "noble" appearance—those preferences can shape the lineage's phenotype. A caste can develop a recognizable look: taller bodies, different facial proportions, lighter or darker features depending on alliances. These patterns are not absolute, and they fluctuate with political shifts, but they can be strong enough to be noticed. When noticed, they become mythic material. The elite begins to look "not like us," and the mind translates that into a deeper claim: they are "not from here."

The second principle is that elites often have better nutrition. This is often overlooked in speculative narratives because nutrition is mundane, but it is one of the strongest drivers of visible difference, especially in stature and robustness. In agricultural and early urban societies, ordinary people frequently lived close to nutritional limits. Elites, by contrast, had more reliable access to protein, varied diet, and calories. Over generations, better nutrition can produce taller and more robust individuals, even without major genetic differences. In a population where malnutrition is common, the well-fed can appear like a different breed. Again, myth readily turns this into a biological category. The ruler becomes physically larger, and physical largeness becomes proof of special origin.

The third principle is that elites often engineer their appearance intentionally. They use clothing, headgear, hairstyles, adornment, and body modification to mark status. Cranial modification is one dramatic example, but many subtler practices exist: dental modification, scarification, ritual posture and gait training, and aesthetic norms that alter how a person carries themselves. These practices can make elites look and feel different in a crowd even if their biology is not unusual. When combined with non-local ancestry and better nutrition, the engineered signal becomes overwhelming. The elite is not just socially distinct; they are visually and physically distinct. For common people, this can feel like evidence that the elite belongs to another order of being.

A fourth principle is that elites often control access to certain roles that are already associated with otherness: priesthoods, ritual specialists, navigators, metalworkers, scribes, astronomer-priests. These roles carry the aura of hidden knowledge. A person who can predict the seasons, interpret eclipses, or produce metal from stone appears to possess powers beyond ordinary comprehension. If such roles are monopolized by a lineage that is also visually distinctive, the impression of "non-local traits" becomes fused with the impression of supernatural authority. The body and the knowledge reinforce each other. The result is the archetype of the priest-king who is not fully human, or whose blood contains something sacred.

This is where mythology and social reality begin to overlap in a particularly potent way. Many mythic traditions describe ruling lineages as descended from gods, sky beings, or ancestral heroes. It is tempting to read these claims as literal genetic memories. But they can also be understood as political language: a symbolic justification for hierarchy. The elite claims divine descent to legitimize rule. The people see the elite's non-local appearance and accept the claim more readily because it "fits" the visible difference. The myth becomes credible because the phenotype provides a visual anchor. Over time, the myth and the phenotype merge into a single narrative: they look different because they are different.

We should also consider the role of conquest and replacement. When a new ruling class arrives—through invasion, migration, or political takeover—it can impose itself over local populations. Such ruling groups may have different physical traits, different languages, and different customs. Even if the takeover is not violent, the social separation can be sharp: a foreign elite intermarrying among itself, controlling resources, and maintaining distinct identity. To the local population, the new rulers can appear alien. Their difference can become exaggerated in memory. Their bodies become larger, more beautiful, more terrifying, more divine. This is one of the oldest engines of myth-making: the foreign ruler as a being from beyond.

In the context of megalithic builders, this dynamic becomes especially interesting because monumental construction often coincides with complex political organization. The ability to mobilize labor suggests authority, and authority often involves elites who claim special status. If those elites were part of wider networks—trading, marrying, moving—then their bodies could carry non-local traits. Their distinctiveness could then be projected backward onto the builders themselves: the builders were different humans. In this way, the impression of "biological outliers" can attach to monumental cultures even when the labor was performed by ordinary local bodies under organized systems. The elite face becomes the face of the entire civilization, and the elite's distinctiveness becomes a myth about the whole population.

Now, how does this intersect with the volume's broader theme of hybridization? Carefully. We know that human populations mixed repeatedly in deep time. We also know that later historical societies mixed through migration and alliance. "Non-local traits" are therefore expected; they are not anomalies. But the perception of those traits as anomalous is culturally significant. It reveals that many communities experienced contact and mixing not as a smooth blend, but as a confrontation with difference. That confrontation becomes story: gods came, strangers taught, mighty ones ruled, builders were not like us. In many cases, the underlying reality could simply be population movement and elite network formation.

It is also worth noting that "non-local traits" are not always accurately perceived. Humans are prone to stereotyping and to overreading small differences. A slightly different head shape, a different hair texture, a different style of dress can be interpreted as evidence of an entirely different origin. When such perceptions are tied to power relations, they harden into legend. The elite becomes an inherited other. This is why the same culture can describe its founders as "taller" and "brighter" and "more beautiful" even when the biological differences were modest. The mythic story is describing social distance, not providing a forensic measurement.

So the disciplined conclusion of this subsection is that elites can become biologically and visually distinctive through three ordinary mechanisms: selective marriage networks across regions, nutritional advantage, and deliberate body/appearance engineering. These mechanisms can produce "non-local traits" that feel like evidence of a separate human type. When layered with specialist knowledge and political authority, the impression becomes mythically amplified. The elite becomes a different humanity in the story-world of the culture, regardless of the actual genetic distance.

The next subsection confronts the most direct question that follows from this: were the builders physically different? Not the rulers, not the priests, not the legendary founders, but the people who moved stone.

That question forces us to separate elite myth from labor reality, and to examine what differences—if any—would be detectable in osteology if a megalith-building population truly had unusual physical traits. It also allows us to explore how myths about builders can arise even when the builders were ordinary humans working under extraordinary systems.

The question sounds simple, but it hides several questions inside it. Were the builders physically different from neighboring populations in a way that would be visible in skeletal remains? Were they different in the sense of training and occupational conditioning—bodies shaped by labor rather than by lineage? Were they different because they belonged to a distinct migrating group that introduced new traits into a region? Or were they "different" only in myth, with physical difference serving as a story device that explains monumental achievement? To answer responsibly, we have to keep these meanings separate, because the word different is the doorway through which speculation enters.

Start with what is most plausible and most consistent with how humans actually build large things: the occupational difference. A population engaged in quarrying, hauling, and construction will often show robust physical adaptations—not genetic, but skeletal signs of repetitive strain and muscle use. Bones remodel under stress. Muscle attachments can become more pronounced. Joints can show wear patterns consistent with heavy lifting and repetitive motion. If megalith building was part of a community's sustained life rather than a one-off event, then it is reasonable to expect that some skeletal remains would show the imprint of that life. Such bodies could appear "stronger" than those of more sedentary groups. Over time, that difference could become part of cultural identity. People could be known as builders, carriers, stone workers. Later storytellers might then translate that occupational robustness into a narrative of a physically different people.

The next plausible layer is selection within a population. Even without genetic divergence, a society can channel certain individuals into labor roles based on physical aptitude. Taller or stronger individuals may be chosen for hauling. Those with better balance may be assigned to placement work. Those with endurance may become carriers. This does not create a separate branch of humanity, but it can create a visible labor class: the builders as a subset of the population. If such selection is repeated over generations within the same community, the "builder" identity may become associated with certain body types. Again, the difference is social and functional rather than species-level.

Now consider migration and admixture. Megalithic traditions often appear in clusters or spread along coastlines and trade routes. Where a building tradition spreads, people often assume ideas traveled without bodies. Sometimes that is true; techniques can diffuse. But in many cases, bodies travel with ideas. Builders migrate, specialists relocate, marriages join networks. If a migrating group contributed disproportionately to monumental projects—either as organizers or as ritual specialists—their traits could become overrepresented in the social memory of the builders. In the archaeological record, this could show up as non-local burial practices, different isotopic signatures indicating mobility, or shifts in skeletal population profiles. The key point is that even if mobility contributed to visible difference, it would still be human difference: population movement, not non-human ancestry.

This is where the "physically different" question often becomes inflated. People see one or two elite burials with unusual features and assume the builders were a separate type. But monuments are not built by elites alone. They are built by labor forces, often local and often numerous. If a monument reflects the identity of an elite class, the labor force might still be drawn from ordinary local populations. In such a scenario, the builders' physical difference would not necessarily be extreme. The elite might be distinct; the workforce might not. Yet myth tends to flatten distinctions. It assigns the elite's qualities to the whole civilization, and the civilization's achievements to the elite's body. "They built this" becomes "they were different."

So what would we actually need to see to justify the claim of a physically distinct builder population in a stronger sense? We would need repeated osteological evidence across multiple sites showing a consistent profile: significantly different average stature, distinctive body proportions, or other anatomical traits not readily explained by nutrition and occupational stress. We would need those traits to be present not only in elites but in the general burial population associated with the building community. We would need coherent dating and context, showing that the difference is not an artifact of sampling or mixing populations from different periods. And ideally, we would want genetic evidence consistent with a distinct ancestry component. Without these lines of evidence, the "different builders" claim remains a narrative inference, not a demonstrated biological fact.

It is also important to understand why the physical-difference claim is psychologically attractive. Monumental stonework creates a mismatch between modern imagination and ancient process. Most modern people do not routinely witness coordinated manual labor on a massive scale. They do not live in societies where thousands can be mobilized to build a temple without modern machines. As a result, they underestimate what organized human labor can accomplish. When they see the stones, they assume the builders must have had unusual bodies or unusual technology. This underestimation is a recurring error, and it is one reason giant-builder myths arise so naturally. The human body becomes the explanatory lever because the true lever—social organization—feels abstract and unfamiliar.

But social organization is not abstract to the people who lived it. For them, building in stone could be a seasonal rhythm, a collective rite, a political obligation, a sacred duty. The bodily experience of hauling and raising stone could be embedded in community life. In such contexts, endurance and strength become culturally cultivated. People train by living. They become strong because their daily life demands it. Their bodies then become evidence of their culture, not of a separate species. The megalith is therefore the fossil of a social order: a society capable of binding bodies into one coordinated intention.

If we keep that frame, we can interpret "physically different" in a way that remains both powerful and responsible. The builders may have been different in that their bodies were conditioned by labor, their endurance was trained by lifeways, and their communities were organized around projects that gave physical work sacred meaning. Their difference may have been real and visible, but it would not necessarily point to non-human lineage. It would point to a human capacity that modern life often hides: the capacity to become extraordinarily capable through collective discipline.

And yet, the persistence of myths about non-human builders and strange elites suggests that something else is going on as well: the collision between mythic description and osteological reality. People describe builders as giants, as mighty ones, as not quite human. Bones, when they can be studied, often suggest ordinary variation plus cultural modification. The next subsection explores that collision directly. It asks how mythic descriptions map—or fail to map—onto actual skeletal evidence, and why the mismatch itself may be the most important clue to how cultural memory works. The question is not whether myth is "wrong," but what myth is doing when it insists on bodies larger than life.

Myth speaks in bodies. Osteology speaks in measurements. When the two are placed side by side, the contrast can be jarring. Myth tells us the builders were mighty ones, giants, semi-divine ancestors with skulls like towers and strength like storms. Osteology, when it has enough material to work with, often tells a more restrained story: ordinary human variation, occasional pathology, evidence of strenuous labor, and cultural practices that intentionally shaped appearance. The tension between these two languages is not a mistake to be corrected. It is a phenomenon to be understood. Because the mismatch between mythic description and skeletal reality is itself a clue to how human societies remember power, origin, and discontinuity.

The first reason myth and osteology diverge is that myth is not trying to report anatomy. Myth is trying to report significance. When a tradition says the founders were larger, it is often saying they were more authoritative, closer to the sacred, or capable of organizing the world. Size is the symbol of that capacity. A myth that describes the builders as giants may not be describing tall bodies at all. It may be describing the perceived magnitude of their achievements—moving stone, building in alignment with sky cycles, establishing law, defeating chaos. In such narratives, bodily size is a moral and cosmic measurement. It is the way the story signals that these beings were not ordinary. Osteology, by contrast, does not measure significance. It measures bones.

This difference in purpose means that even if we found a cemetery associated with a megalithic culture and measured its individuals, the results would not necessarily "disprove" the myth. The myth could remain true in its own register: the builders were giants in impact, not in centimeters. This is a difficult idea for modern minds because modern discourse is trained to treat claims as either factual or false. Myth operates differently. It makes claims that are true in meaning even when not literal in form. When we forget that, we make the classic error: we treat myth as failed science rather than as symbolic history.

The second reason for divergence is selection. Myths rarely describe whole populations. They describe exceptional figures: founders, heroes, priest-kings, civilizers, adversaries. The bone record, when it is available, often represents a broader cross-section—though even burial populations are biased by status and practice. If myths exaggerate elite bodies, and if elites were sometimes taller, better nourished, or intentionally modified in appearance, then the mythic "body" is not the average body. It is the elite body made archetypal. Osteology might show that elites had different diets, different mobility patterns, and sometimes different body modification practices, while the general population did not. The myth then becomes a magnifying lens: it takes the elite pattern and projects it onto the entire age.

This is why the skull becomes such a focal point. Elongated skulls, where present, are often tied to elite status and identity marking. Mythic language about "strange heads" or "shining foreheads" can attach to these visible markers. But the osteological interpretation may still be cultural modification rather than biological anomaly. In this way, myth and osteology can both be "right" in their own domains: the elite really did look different, and that difference really did matter, but the cause was social engineering rather than non-human ancestry.

The third reason for divergence is that the osteological record is incomplete and biased. We do not have skeletons from most people who ever lived. Burial practices vary, preservation is uneven, excavation is partial, and in many places the dead were treated in ways that leave little trace. If we expect osteology to settle mythic questions definitively, we will often be disappointed. Absence of evidence is common. This creates an opening for speculation, because where the record is thin, imagination rushes in. Myths feel like they fill the gap. Sometimes they preserve genuine memory. Sometimes they simply supply meaning. Either way, the incompleteness of osteology means it can rarely provide a total answer.

This incompleteness also means that extreme cases—unusual skulls, tall individuals, robust skeletons—can receive disproportionate attention. A single dramatic specimen in a museum can become the centerpiece of a whole narrative. Yet osteology demands population context. A single specimen cannot establish a distinct human type. It can establish an outlier. The mythic mind does not care about population context. It cares about archetype. This difference in reasoning is a major source of conflict. Myth is comfortable generalizing from the exceptional. Science is not.

The fourth reason for divergence is translation across time. Myths travel. They move from one language to another, from one culture to another, from oral performance to written text to later retelling. In this travel, descriptive words shift. “Mighty” becomes “giant.” “Great” becomes “tall.” “Heavy” becomes “huge.” “Shining” becomes “otherworldly.” These shifts are not necessarily deliberate distortions; they are the natural drift of metaphor across generations. A myth that began as a statement about authority can become, centuries later, a statement about anatomy. When modern readers encounter the later form, they assume the earliest form was equally literal. Osteology then appears to contradict myth. In reality, we may be comparing bone to a metaphor that hardened into literal language over time.

The fifth reason is that human bodies can be made to look extraordinary through cultural practice without becoming biologically extraordinary. Beyond cranial modification, societies use posture training, ritual movement, clothing silhouettes, and ornamentation to reshape how bodies appear. A tall headdress and an elongated hairstyle can make a person seem taller and more imposing. A rigid posture can communicate dominance. Ritual regalia can alter the perceived proportions of the head and shoulders. Such practices can feed mythic description, especially when later observers encounter statues and depictions rather than living bodies. If a culture depicts its elites with elongated forms—stylized heads, tall crowns, exaggerated features—later interpreters can mistake artistic convention for anatomical evidence. Osteology will not match art, because art is not always portraiture; it is often ideology.

Yet we must not swing too far and treat all mythic bodily difference as pure symbol. Human history really did contain people who were physically unusual, whether through pathology or variation. Human history also contained populations with different average traits due to diet, ecology, and ancestry. There is no need to deny that. The disciplined point is that such differences, even when real, are rarely as extreme as mythic description. Myth enlarges. It turns difference into category, category into cosmic struggle, and cosmic struggle into bodily scale. That is what myth does.

So how do we use osteology and myth together without collapsing one into the other? The answer is to treat each as a different kind of evidence.

Osteology tells us about bodies. Myth tells us about perceived difference, power, and social memory. When myth insists on different bodies, it may be pointing to one of several realities: elite separation, cultural body modification, contact with outsiders, occupational robustness, or the psychological need to explain monuments through superhuman makers. Osteology can test which of these are plausible by examining skeletal patterns where available. It can confirm occupational stress, nutritional differences, and modification practices. It can identify pathology. It can sometimes detect population movement indirectly. But it cannot confirm the mythic register of significance, because significance is not stored in bone.

In the context of this chapter's larger arc, the key insight is that mythic descriptions and osteology often do not align because they are describing different things with the same language. The myth says the builders were different. Osteology may show ordinary human bodies conditioned by extraordinary labor and marked by deliberate cultural practices. Both can be true at once: the builders were different in experience, identity, and social system, even if they were not a separate species. The "outlier" may be the culture, not the skeleton.

And yet, the final subsection of this chapter will push into the most charged version of the claim: the idea of a hybrid aristocracy—an elite caste that was not only socially distinct but genetically distinct through controlled marriage and possibly through intergroup admixture remembered as divine or non-human origin. This is where the web model of human evolution meets the political reality of elite lineage formation. The question will not be framed as proof of non-human blood, but as a disciplined exploration of how composite ancestry and social strategy can create the enduring impression—mythically and sometimes physically—of elites as a separate kind of humanity.

If there is one place where myth, biology, and politics lock together with almost no separation, it is in the origin story of elites. Across civilizations, ruling lineages are rarely content to be ordinary. They claim descent from gods, from sky teachers, from heroic founders, from beings who arrived from elsewhere. They frame their authority not as an accident of force but as a property of blood. And when a lineage is described as different in blood, the modern imagination inevitably asks whether that difference might have been real in some biological sense—whether certain elites were, in effect, a hybrid aristocracy: a caste shaped by selective marriage, mobility, and admixture that made them visibly and socially distinct over generations.

The danger is obvious. This idea can be turned quickly into crude narratives: aliens, giants, non-human kings, hidden races. The disciplined version is narrower and more defensible. It does not require non-human ancestry. It requires only three premises, each of which is historically plausible. First, elites tend to form endogamous networks, marrying within a narrow social sphere and often across regions to secure alliances. Second, elites often have nutritional advantages that influence stature and robustness. Third, elites frequently cultivate visible distinction through body modification, ornamentation, and ritual regalia. If these premises hold, then a caste can become both biologically and culturally distinctive relative to surrounding populations. It can look like a different kind of humanity without being outside the human range.

To ask “hybrid aristocracy” responsibly, we should clarify what hybridization means in this context. It can mean admixture between regional populations that would otherwise be more isolated. It can mean repeated intermarriage between different ethnic or linguistic groups during periods of conquest and alliance. It can mean the fusion of lineages from distant zones into a ruling network. It does not require hybridization between species. It requires hybridization in the demographic sense: the weaving of diverse ancestries into a narrow elite line. In a world where most people remained local, elites were often the mechanism by which distant genes moved across landscapes.

This is not merely theoretical. Elites historically exchange spouses as diplomatic instruments. They take foreign partners to legitimize claims. They send children to other courts. They incorporate local noble families to stabilize rule. They bring in specialists from afar—scribes, architects, astronomers, priests—and those specialists can become integrated into elite circles. Over generations, such practices can create lineages that carry traits uncommon in local populations. If those traits become associated with prestige, they may be reinforced deliberately through preference. The lineage becomes a curated phenotype as well as a curated myth.

The myth then does what myth always does: it explains curated difference as sacred origin. A ruling caste that looks different and claims divine descent becomes self-validating. The people see difference and accept the claim more readily. The rulers see acceptance and invest further in difference—through ritual, regalia, and controlled reproduction. Over time, the line between political strategy and sacred narrative disappears. Blood becomes destiny. This is the essence of aristocratic ideology: the claim that authority is inherited not merely socially, but biologically and cosmically.

Cranial modification fits neatly into this dynamic. If a lineage alters the head shape of its children from infancy, it is not only creating a status marker. It is manufacturing a visible boundary between ruler and ruled. The boundary is embodied. It cannot be removed like clothing. It becomes a lifelong declaration of belonging to an elite order. If that elite also maintains marriage restrictions, the signal becomes even stronger. People begin to speak of a "different stock," a "different blood," a "different people" within the same society. The mythic language of hybrids—semi-divine children, heroes born of gods and women, kings whose veins contain sacred essence—can be understood as the narrative expression of this social reality.

Now, to connect this back to the megalithic world, we should be cautious about projection. Many megalithic cultures are pre-urban or early-complex societies where hierarchy existed but may not have been as formalized as later states. Still, the presence of monumental projects implies leadership and shared ideology. Someone planned. Someone coordinated. Someone justified the labor. Where leadership and ideology exist, elite formation becomes likely. And where elite formation exists, claims of special origin often follow. The hybrid aristocracy hypothesis, in this setting, is not a claim about alien kings. It is a claim that certain lineages—organizers, ritual specialists, founders—may have become socially distinct and possibly biologically distinctive through their network connections.

This is where the "web" model of human history becomes a useful interpretive bridge. If humans have always been a composite lineage at the species level, then composite lineage formation at the social level is an echo of the deeper pattern. Elites, in many periods, may have functioned as the junction points of human networks, bringing together genetic and cultural material from multiple sources. They were not only rulers; they were conduits. That does not make them non-human. It makes them networked. It makes them biologically and culturally mixed in a way that ordinary villagers might not be. In a small-scale world, that networked character could appear uncanny.

But we must also keep the limits clear. A hybrid aristocracy, even if real in this demographic sense, would still fall within human variation. It would not generate a separate species. It would not automatically explain megalithic engineering as a biological mystery. Monumental building remains primarily a story of organization and technique. The aristocracy's distinctiveness would be a social effect: how power is embodied, justified, and remembered. It might contribute to myths of non-human founders and mighty builders, but it would not be the mechanical explanation for moving stone.

So why does the "hybrid aristocracy" idea feel so compelling in the broader alternative narrative ecosystem? Because it offers a synthesis. It allows myths of gods mating with humans to be read as memories of elite intermarriage with outsiders. It allows stories of sky teachers to be read as memories of specialist classes entering societies through contact. It allows claims of different blood to be read as reflections of controlled lineage and network ancestry. And it does so without requiring us to assert literal non-human intervention. It is a way to treat myth as encoding social and demographic reality rather than as either pure fiction or literal fact.

At the same time, this interpretive synthesis must remain honest about what it can and cannot prove. We can reasonably argue that elites often became visually and culturally distinctive. We can reasonably argue that marriage networks and mobility created non-local traits. We can reasonably argue that myths of divine descent are political technologies that piggyback on visible difference. What we cannot responsibly claim—without specific genetic datasets and direct evidence—is that any given elite group was biologically extraordinary beyond the human range. The hybrid aristocracy hypothesis is strongest as a model of how difference is produced and then mythologized, not as a claim that elites were a different species hiding in plain sight.

This chapter's purpose, then, has been to anchor the "biological outliers" theme in mechanisms that are both human and historically plausible: endurance shaped by labor, bodies marked by culture, elites distinguished by networks, and myths that translate social distance into bodily otherness. With that foundation, we can move into the next chapter, where hybridization is no longer a metaphor about elites but an explicit mythic theme: gods mating with humans, heroes born of mixed origin, and the recurring narrative that exceptional beings emerge from boundary-crossing unions. There, the task will be to keep the same discipline: separate established evidence from interpretation, treat myth as meaningful without forcing it into literalism, and use the web model of human ancestry as context rather than as proof.

CHAPTER 8 — HYBRIDIZATION MYTHS

Across the world's mythic traditions, the boundary between divine and human is rarely as firm as later theology insists it should be. Again and again, stories return to the same charged image: a god, a luminous visitor, a being from the sky or the deep, crossing into the human realm and taking a human partner. The result is never ordinary. Sometimes it is a hero. Sometimes it is a dynasty. Sometimes it is a catastrophe. But the structure repeats with such persistence that it begins to look less like an isolated fantasy and more like a human obsession—a narrative mechanism through which cultures explain the origin of exceptional individuals and exceptional orders.

The first thing to acknowledge is how universal the motif is. You find it in stories of storm gods and mortal women, of river spirits and human lovers, of mountain deities and shepherd girls, of sky fathers and queens, of shining strangers and chosen brides. The specific names change, the moral framing shifts, the details of the encounter are colored by each culture's cosmology, but the central claim remains recognizable: the divine has reproductive access to the human. That claim is extraordinary, and its recurrence demands explanation—not necessarily a literal one, but an explanatory frame that is strong enough to account for why so many societies independently return to it.

One explanation is social rather than biological: these stories are a narrative technology for legitimizing hierarchy. If a ruler can claim divine parentage, then rule becomes naturalized. Authority is no longer merely coercion or inheritance; it is cosmic order. The king is not simply the strongest man with the largest army. He is the child of heaven. In such a system, the myth of divine-human mating is not decorative; it is political infrastructure. It makes power feel inevitable. It discourages rebellion by sacralizing the ruler's body. It turns the lineage into a living altar. When we look at the breadth of the motif, it is difficult to avoid the conclusion that many cultures used it to explain why some people were allowed to command others.

But this political function does not exhaust the motif's meaning, because divine-human unions also appear in stories that are not primarily about kings. They can be about the origin of crafts, the arrival of law, the founding of cities, the birth of monster-slaying heroes, or the establishment of sacred knowledge. In these narratives, the union is often less about domination and more about transmission. The human partner becomes a conduit for something that cannot be generated from purely human stock. The child embodies that transmission. They are not merely a person; they are a bridge. In mythic logic, hybridization is the way the world upgrades itself.

That word—upgrade—must be handled carefully, because it can tempt us into modernizing the myth too literally. Yet the underlying idea is clearly present: the hybrid child is always described as exceptional, possessing strength, wisdom, charisma, or destiny beyond ordinary humans. This is not random. It is the reason the motif exists. The union between god and human is not erotic entertainment in its deepest form; it is an origin mechanism. It explains why the world contains beings who seem to exceed the normal limits of human capacity.

To understand the motif's emotional power, consider what it implies about human self-perception. If humans are merely one animal among many, then the appearance of genius, leadership, and creative force is difficult to explain in an intuitive way. Most people do not experience their own talents as the outcome of slow evolutionary processes; they experience them as gifts, as sparks, as mysterious differences between individuals. Myth takes that lived intuition and projects it into the cosmos. The exceptional person is exceptional because they carry a fragment of the beyond. The divine-human union becomes the most direct narrative for turning mystery into lineage. It is a way of saying: what we admire most in ourselves comes from somewhere higher than ourselves.

The motif also encodes a deeper tension: the divine is dangerous. Many divine-human union stories contain violence, coercion, deception, or imbalance. The god is not always a gentle lover; often the god is a force. The human partner can be overwhelmed, chosen without consent, or drawn into a relationship that alters their life irrevocably. This is not incidental. It reflects a fundamental mythic insight: power crossing into human life is not safe. It produces greatness, but it also produces disruption. Hybridization, in these narratives, is not a domestic marriage; it is a collision of categories. It can create heroes, but it can also create monsters. It can found dynasties, but it can also trigger flood, war, or divine punishment. The boundary is crossed, and the world pays a price.

This is where the motif begins to resonate with the broader themes of this volume. Hybridization is a fact of human evolution in a biological sense: our ancestry is not a single line. But mythic hybridization is not describing genetics in a laboratory way. It is describing the experience of boundary-crossing: the arrival of outsiders, the merging of lineages, the mixing of social orders, the shock of contact, the sudden appearance of new capacities and new hierarchies. In that context, a divine-human union can be read as a mythic encoding of contact events—moments when a community experienced strangers as beings from another world.

In small-scale societies, this perception is understandable. A group that arrives from afar with unfamiliar language, technology, or ritual expertise can appear superhuman. Their knowledge seems given by the sky. Their tools seem enchanted. Their confidence seems godlike. If such newcomers intermarry with local women—especially high-status women—the resulting children may indeed hold special status and special opportunities. Those children might become founders of new lineages, elites, or hero traditions. Over generations, the story of outsider intermarriage can become sacralized. The outsiders become gods. The women become chosen. The children become demi-gods. The myth preserves the memory that something outside the ordinary entered the community and changed its destiny.

This interpretation does not require us to claim that gods literally walked among humans. It requires only that humans repeatedly encounter the unfamiliar and interpret it through sacred categories. The "god" in myth is often what humans call a power that cannot be locally explained. Sometimes that power is nature. Sometimes it is another human group. Sometimes it is a specialist class whose knowledge seems uncanny. Sometimes it is the psychological eruption of new social complexity. Myth compresses these forces into characters and relationships because narrative is the mind's primary tool for making sense of change.

There is also a symbolic level that runs deeper than politics and contact. Divine-human unions often mirror the fusion of opposites that myth uses to explain creation itself: sky and earth, spirit and matter, order and chaos. The god represents the transcendent principle; the human represents the embodied world. Their union produces a being who can operate in both realms—a mediator, a hero, a king, a shaman. In this sense, the hybrid is a cosmological necessity. The world needs bridges between layers. Myth supplies them through lineage. The child of god and human is the living axis of the story's universe.

This is why these motifs persist even in societies that do not emphasize dynastic kingship. The hybrid is not only a political figure; it is a metaphysical figure. It embodies the idea that humans can participate in the divine, that we are not sealed off from larger forces, that the cosmos can enter the bloodline. Whether or not this corresponds to literal events, it corresponds to a deep human desire: the desire for meaning in our origin, and for continuity between our ordinary lives and the vastness above us.

Yet the same motif also carries anxiety. If gods can mate with humans, then humans are vulnerable to cosmic intrusion. The boundary of the body is not safe. The lineage can be altered. This anxiety appears in stories where hybrid children are both revered and feared—too strong, too strange, too uncontrollable. It appears in stories where the divine fathers are punished, banished, or restrained, suggesting that such unions are violations that must be contained. In other words, myth often treats hybridization as both the source of greatness and the source of instability. That ambivalence is one of the motif's strongest signatures, and it may reflect a real social memory: when lineages mix and new elites arise, societies can be transformed—but transformation can be traumatic.

So what do we take from the universality of gods mating with humans? We take, first, that cultures everywhere have used hybrid lineage as an origin explanation for exceptional individuals and exceptional authority. We take, second, that the motif likely encodes multiple realities at once: political legitimation, contact and migration memory, specialist power, and cosmological symbolism. And we take, third, that the motif is the mythic form of a truth that biology now makes undeniable in its own way: humans are not a pure line. We are shaped by mixture—whether mixture of genes, mixture of cultures, or mixture of worlds as experienced by the imagination.

The next subsection moves from the act of union to its product: the hero. If gods mating with humans is the engine, the hero is the output. The heroic figure is the most persistent form of hybrid in mythology, and the place where the motif becomes most narratively detailed. We will examine how heroes function as semi-divine hybrids, what traits they consistently carry, and why their mixed origin is not an ornamental detail but the key to their role in the mythic world.

Across mythic traditions, the hero is rarely presented as merely talented. The hero is presented as a different kind of being—born with a weight of destiny, marked from the beginning, and able to do what ordinary humans cannot. This is why heroic origin stories so often reach for the same solution: the hero is a hybrid. Their bloodline is mixed. One parent belongs to the human world of hunger, fear, time, and death. The other belongs to a realm beyond the ordinary—divine, ancestral, spirit, or sky. The mixed origin is not a flourish. It is the justification for the hero's role. It explains why they can step into the cracks between worlds and survive where others fail.

In many traditions, the hero's hybrid status is expressed directly: the hero is called semi-divine, god-born, a child of heaven, a descendant of luminous ancestors. In others, the language is more symbolic: the hero is conceived under strange circumstances, receives a divine sign at birth, is protected by spirits, or is raised by beings outside ordinary society. But the structural message is the same. The hero is not fully of this world. They are a bridge. And because they are a bridge, they can do the story's central work: they can mediate between realms that ordinary humans cannot reconcile—chaos and order, wilderness and settlement, the monstrous and the civilized, the old age and the new.

The hybrid hero almost always appears at a threshold moment. This is another signature. Heroes arise when a world is unstable: when monsters roam, when giants threaten, when floods or famines loom, when kingdoms need founding, when laws must be introduced, when the sacred must be re-established after disruption. Ordinary humans, the myths insist, are insufficient at these thresholds. They are too fragile, too fearful, too entangled in everyday life. The hero must be more. Hybridization supplies that "more" in narrative form. It gives the hero extra strength, extra knowledge, extra legitimacy, or extra protection. The hybrid is the mythic answer to crisis.

Yet the hero's mixed origin also creates tension, because a bridge is never fully at home on either side. The semi-divine hero is often portrayed as restless, dangerous, or doomed. Their presence disrupts normal social order. They may be too violent, too proud, too passionate, too haunted. They may accomplish what must be done and then die young, ascend, vanish, or withdraw from ordinary life. In this way, hybrid origin explains not only the hero's capacity but also the hero's alienation. The hero belongs to the world as a function, not as a citizen. They appear, they act, they depart. The hybrid is the mythic form of a necessary outsider.

This outsider quality is especially clear in the hero's relationship to law. Many semi-divine heroes are founders of order: they slay monsters, clear wild spaces, establish boundaries, and create the conditions for society. But they themselves are often boundary-breakers. They commit transgressions. They challenge kings. They violate taboos. They are too large for the rules they create. This paradox is not accidental. It reflects a deep mythic logic: the force that creates order must be strong enough to break the old disorder, and that strength cannot be fully domesticated. The hero's hybrid blood is the narrative device that allows the story to hold both truths at once: the hero is necessary, and the hero is dangerous.

In some traditions, the hero's hybrid status is expressed through explicit bodily markers—unusual size, unusual strength, unusual hair or eyes, a radiance, a mark that sets them apart. In others, it is expressed through feats that imply bodily difference: wrestling giants, lifting stones, traveling between worlds, enduring wounds that would kill ordinary men. These markers and feats are important for this volume because they show how myth turns social and psychological difference into physical language. To say a hero is god-born is to say that their body and will cannot be measured by ordinary standards.

But we should also notice how often the hero's hybrid origin is tied to knowledge as much as to strength. Many heroes do not merely fight; they know. They receive instructions from divine parents, learn secret crafts, discover sacred laws, or carry ritual authority. In some cases, the hero is as much teacher as warrior. They introduce agriculture, metallurgy, navigation, writing, calendars, or moral codes. When this happens, hybridization is not simply a power boost; it is an explanation for cultural innovation. The hero becomes the narrative vehicle for the sudden appearance of complex knowledge. The myth is saying: this did not arise gradually from ordinary life; it was imported into the human world through a boundary-crossing lineage.

This is precisely where the hero motif begins to echo the volume's larger thesis about missing chapters. From a mainstream perspective, culture can change rapidly through networks, demographic thresholds, and cumulative innovation. From a mythic perspective, sudden change is often too abrupt to be explained by ordinary human discovery. It must be given. The hero, as a hybrid, becomes the courier of that gift. The myth is not necessarily wrong in perceiving that knowledge often arrives through contact and transmission rather than isolated invention. But it translates contact into divinity, and transmission into bloodline. The hero is the personification of cultural diffusion.

Seen through that lens, semi-divine heroes can be read as remembered junction points in human networks. A charismatic leader from outside a community marries into it and becomes a founder. A stranger introduces a new craft and is mythologized as a god's child. A lineage forms around an individual who carried non-local knowledge and non-local traits. Over generations, the story crystallizes: he was not fully human. His father was a god. His mother was chosen. The details vary, but the narrative function remains: the hero explains why the world changed.

This interpretation gains strength from one consistent feature of hero myths: the hero's story often includes travel. The hero moves between regions, between islands, between mountains and lowlands, between courts and wilderness. Travel is the signature of contact. It is the mechanism by which knowledge spreads, alliances form, and lineages mix. When a hero is always on the move, the myth is preserving the idea that transformative power comes from beyond the local. The hero is not merely strong; the hero is connected. Hybrid blood, in myth, is the symbol of that connection.

Yet the myths insist on something more intimate than travel. They insist on mating. This is important. Mating is the ultimate form of integration. It is not simply exchange of goods or ideas; it is exchange of blood. In myth, the transformation of the world requires not just learning but lineage. The new order must be embodied in descendants. This is why heroes are often founders of dynasties or ancestors of tribes. The hybrid child becomes the seed of a new people. The hero is not only a person; the hero is a genetic and cultural pivot.

In the disciplined frame of this book, we treat that pivot as metaphorically resonant with real processes. Populations mix. Lineages form. Elites curate ancestry. Cultural systems can shift rapidly when network nodes appear—leaders, specialists, migrants, innovators. Myth gives these processes a human face and a sacred origin. It turns demographic complexity into family drama. The hero's semi-divine parentage is the mythic way of saying: this person did not come from nowhere; they came from beyond the ordinary, and their descendants carry that beyond into the future.

The next subsection focuses on the most persistent claim attached to hybrid heroes: that hybrid children are always exceptional. We will examine why myths insist on this pattern, what functions it serves in legitimizing authority and explaining genius, and how it interacts with the human tendency to turn rarity into destiny. The question is not whether exceptional children exist—they clearly do—but why myth repeatedly insists that exceptional ability must come from crossing the boundary between human and non-human.

Myth almost never produces a hybrid child who grows up to live an ordinary life. The child of a god and a human is not a quiet farmer, not a modest craftsperson, not a person who simply inherits land and dies in peace. The hybrid child is always marked—stronger, wiser, more beautiful, more dangerous, more destined. Even in stories where the hybrid becomes tragic or monstrous, the exceptional quality remains. The hybrid is never average. This consistency is one of the most revealing features of hybridization myths, because it tells us what the motif is doing. It is not describing romance. It is generating explanations for extremes.

Human societies are always confronted by extremes. A person appears with extraordinary strength, extraordinary charisma, extraordinary intelligence, extraordinary violence, extraordinary luck. A dynasty rises with improbable speed. A city becomes a sacred center. A new craft spreads like fire. These events can be explained, in principle, by social conditions and historical processes. But myths are not written as academic analyses. They are written as meaning machines. They turn extremes into narrative. And one of the most efficient narrative engines for producing exceptional beings is hybrid origin. The child is exceptional because the child is mixed. The child carries a power that did not exist in the ordinary human line.

This "exception through mixture" pattern serves several functions at once. The first is legitimization. If a society wants to justify why a particular person should lead, why a particular family should rule, or why a particular lineage should be treated as sacred, it must explain why they are different. The easiest explanation is that they are not purely human. Their origin is outside the ordinary pool. Hybrid birth turns privilege into destiny. It tells the community: this person is exceptional not because we decided to elevate them, but because the cosmos did. By shifting the source of authority from social choice to divine biology, the myth hardens hierarchy into something that feels natural.

The second function is narrative compression. Hybrid origin is a shortcut that allows a story to skip the slow work of training and social context. A hero can be born with capacities already installed. The story does not need to show years of learning or the gradual accumulation of skill. It can simply state: he is the child of a god. In a mythic world, this is a sufficient explanation. It is the equivalent of a modern story giving a character superpowers. But unlike modern fiction, myth uses superpowers as social explanation. The hybrid child's exceptional ability is the story's way of making the hero's role credible.

The third function is psychological. Humans have a strong tendency to search for origins that match outcomes. When we see extraordinary ability, we intuitively believe it must have an extraordinary source. Ordinary sources feel unsatisfying. "He trained hard" does not fully capture what it feels like to witness genius or destiny. "She is the product of a particular educational environment" does not capture what it feels like when a person seems to outpace their peers with ease. Myth answers that emotional mismatch by providing an origin that feels as large as the outcome. If the child is extraordinary, their parentage must be extraordinary. The hybrid is the mythic balance sheet of awe.

The fourth function is social cohesion. Hybrid children often serve as shared ancestors. They become the origin point of tribes, clans, or nations. In this role, the hybrid's exceptional nature is not just personal; it is collective. The people become exceptional because they descend from an exceptional being. This kind of story forges identity. It tells a community: we are not random; we are chosen. Our origin is sacred. Our place in the world is meaningful. Hybrid ancestry therefore becomes a way of binding people together through a shared mythic pedigree.

But if hybrid children are always exceptional, why are they sometimes also dangerous? Because myth uses hybridization to express ambivalence about power. Hybrid children embody the crossing of a boundary, and boundary-crossing always carries risk. A being that is not fully human may not be fully governable by human rules. This is why many hybrid heroes are impulsive, violent, or cursed. Their greatness is inseparable from their instability. In some traditions, the hybrid must be tamed, taught, or controlled. In others, the hybrid's fate is to be destroyed or removed because their power threatens the social order. This ambivalence is crucial. It shows that hybridization myths are not simply flattering stories for elites. They are also warnings about the cost of importing too much power into the human realm.

This brings us to the question of why hybrid children are depicted as exceptional in specific ways. Across traditions, the exceptional traits cluster into a few categories. One is physical dominance: the ability to overpower opponents, lift stones, slay monsters, endure wounds. Another is cognitive or magical knowledge: the ability to understand hidden laws, read omens, craft tools, navigate seas, speak with spirits. Another is charisma and beauty: the ability to command loyalty, inspire armies, seduce, or radiate authority. These are precisely the traits that matter most in early societies when survival depends on strength, when leadership depends on persuasion, and when innovation depends on specialist knowledge. In other words, hybrid myths make exceptional children exceptional in ways that matter for social transformation. The hybrid is the mythic lever for changing history.

In the context of this volume, the most intriguing aspect of the pattern is that it aligns with what real hybridization can do in a biological sense—though not in the simplistic way myths suggest. In biology, mixture between populations can produce novel trait combinations. It can increase genetic diversity. It can introduce adaptations from one group into another. We already know, from the broad story of human ancestry, that admixture contributed to modern human biology in real ways. But mythic hybridization is not reporting population genetics. It is capturing the intuition that mixture can produce difference. It encodes that intuition through the language of gods and humans.

If we apply this intuition socially rather than biologically, it becomes even more persuasive. When different groups mix—through migration, intermarriage, conquest, alliance—the resulting children can indeed occupy exceptional social positions. They may inherit networks from both sides. They may become bilingual. They may have access to multiple traditions. They may become mediators and leaders precisely because they can move between worlds. Their "exceptional" role is not purely genetic; it is structural. Hybrid children, in the social sense, often become bridges. And bridges are valuable.

This is why so many myths cast hybrid children as mediators: they negotiate between gods and humans, between tribes, between wilderness and city, between old order and new. In real societies, children of mixed ancestry or mixed affiliation often do become mediators, whether formally or informally, because they embody connection. Myth elevates that practical role into cosmic destiny. The child becomes the one who can speak to heaven and earth because, metaphorically, they belong to both.

Now consider how this pattern would function in a society that is building monuments and forming hierarchical states. Elites may curate marriage alliances across regions. Foreign specialists may be incorporated into ruling circles. Children who embody these alliances may become symbolically elevated as the proof of legitimacy. Over time, the elite's mixed origins—mixed in geography, mixed in culture, mixed in network—can be mythologized as mixed in species. The "god" becomes the outsider or the sacred specialist; the "human" becomes the local lineage; the child becomes the founder. The story preserves the reality that power arrived through connection and mixture, but it frames that reality as divine reproduction.

This leads naturally into the next subsection's theme: kingship as genetic memory. If hybrid children are always exceptional, and if those children often become founders of lineages, then kingship itself can be framed as the inheritance of hybrid blood. The king is not merely a political office; the king is the embodiment of a mythic ancestry claim. The next subsection explores how societies turn lineage into legitimacy, how "blood" becomes a political language, and why the idea of hybrid origin becomes so tightly linked to the institution of rule.

In many ancient societies, kingship is not presented as a job. It is presented as an inheritance of essence. The king does not simply govern; he embodies a lineage claim about what kind of being is permitted to govern. That claim is often expressed in the language of blood—sacred blood, royal blood, divine seed, the right line, the chosen house. This language persists across cultures because it is one of the most powerful tools ever invented for stabilizing hierarchy. If authority can be framed as genetic memory—an inherited property carried in the body—then political order becomes naturalized. Rebellion becomes not merely illegal but cosmically wrong. The state becomes a family story, and the family story becomes history.

The phrase genetic memory must be handled with care. It does not mean that societies understood DNA. It means that societies understood heredity as a principle: children resemble parents, traits run in lineages, certain families consistently produce leaders, and ancestry can be invoked as proof of legitimacy. Ancient kingship systems often treat this heredity not as statistical tendency but as metaphysical law. Blood is imagined as carrying not only physical traits but destiny, charisma, and divine favor. In this way, "memory" is stored in the lineage: the accomplishments of the founders, the blessings of the gods, the rights to land and ritual authority. To be of the line is to inherit the world.

Hybridization myths fit seamlessly into this framework. If the founding ancestor is the child of a god and a human, then every descendant carries a trace of that boundary-crossing origin. The dynasty becomes a living artifact of contact between worlds. Kingship becomes the institutionalization of that contact. The king is not merely descended from a founder; he is descended from the moment the divine entered the human bloodline. This makes kingship far more than administration. It becomes sacred continuity.

The most revealing aspect of this is how often kingship mythology insists that rule must remain within the right line even when the right line produces incompetent rulers. A society will endure weak kings rather than abandon the lineage principle, because lineage is the core of legitimacy. The line is the anchor that keeps social order from dissolving into perpetual competition. If anyone can rule, then every strongman becomes a candidate, and civil war becomes a constant risk. Lineage is the solution: a narrow corridor through which power is permitted to pass. Hybrid origin myths strengthen that corridor by making it sacred.

This corridor is reinforced by ritual. Kingship is often accompanied by ceremonies that dramatize the king's unique nature: coronations, anointings, sacred marriages, rituals of renewal, rites that link the king to the fertility of the land and the stability of the cosmos. These rites do not merely celebrate the king; they manufacture the king. They make the lineage claim visible and emotionally real. The body of the ruler becomes a public symbol of cosmic order. In such contexts, the idea that kingship is carried in blood is not abstract; it is performed repeatedly until it becomes unquestionable.

The "sacred marriage" motif is especially relevant because it shows how reproductive imagery is used to legitimize rule. In some traditions, the king's union with a queen or priestess is framed as a reenactment of divine unions that founded the world. The king is not just a man taking a wife; he is repeating the cosmic act that binds heaven and earth. The dynasty's children are therefore not only heirs; they are the continuation of the sacred pattern. Hybridization myths often hover behind such rituals: the first king was god-born, so every royal union echoes the original crossing of boundaries. The marriage bed becomes a political engine.

In practice, elite marriage networks also create the conditions for real demographic distinction. Kings and nobles often marry into allied houses across regions. They bring foreign spouses into courts. They keep the elite pool narrow. Over generations, this can produce lineages that look different from surrounding populations in ways that acknowledge mobility and network power. When people see the court's faces and bodies as distinct, the myth of distinct blood becomes more credible. Social strategy and myth reinforce each other. The lineage becomes, in both perception and reality, a separate stream within the broader population.

This is why kingship often carries an obsession with purity, even when hybrid origin myths exist. The founding ancestor may be a hybrid of divine and human, but once the dynasty exists, it frequently becomes anxious about dilution. The sacred blood must not be mixed with the wrong line. The royal seed must not be contaminated. Such language appears not only in explicit rules but in taboo, scandal, and fear around succession. The paradox is revealing: the dynasty's legitimacy comes from boundary-crossing origin, yet its stability depends on controlled boundaries afterward. The myth celebrates the original crossing but polices later crossings. This is the politics of blood in its purest form.

From the perspective of cultural memory, this produces an effect that can last for centuries: the king becomes a different kind of human in the public imagination. He is portrayed as larger, more radiant, more powerful, more connected to the sacred. Artistic depictions may exaggerate these qualities—taller bodies, stylized heads, enlarged eyes, imposing posture. Court rituals may cultivate them through costume and performance. Body modification may reinforce them. Nutritional privilege may make them physically more robust. The result is a convergence: the king looks different, acts different, is treated different, and is said to be different in blood. Over time, this convergence becomes mythic certainty. Kingship is not a role; it is a species-like category inside the human species.

This is the point where the "genetic memory" framing becomes most valuable for the book's thesis. Hybridization myths can be read as narrative memory of admixture, but kingship myths can be read as institutional memory of how societies manage admixture and difference. The king is the node where multiple streams meet: ancestry, alliance, migration, ritual, and power. The dynasty becomes the archive of those streams, stored not in writing alone but in bodies—faces, features, and the visible continuity of the court. People may not remember the detailed history of alliances and migrations, but they remember that the ruling house came from somewhere, that it has always been distinct, that it carries a sacred origin. In that sense, kingship acts as memory: a living record of contact and consolidation.

This does not mean kingship myths are literally genetic records. They are not precise enough for that. But they are demographic stories written in sacred language. They remember that lineages matter, that ancestry is a form of power, and that mixing and boundary-crossing can produce new orders. In societies that experienced repeated population movement and elite network formation, the myth of hybrid kingship can preserve, in symbolic form, the reality that political legitimacy often arrives from beyond the local—through outsiders, alliances, and imported knowledge.

The next subsection turns to the most ambitious version of this interpretive move: myth as a record of interbreeding. Not interbreeding with gods in a literal biological sense, but interbreeding between populations, castes, and lineages in ways that later cultures encoded as divine unions. We will examine how far that reading can go without collapsing into fantasy, what kinds of historical patterns it can plausibly preserve, and where the evidence ends and interpretation begins.

To say that myth records interbreeding is to walk a narrow ridge. On one side is the temptation to literalize: to treat gods and humans as biological categories and to read ancient stories as veiled scientific reports. On the other side is the temptation to dismiss: to treat all such stories as pure invention with no relationship to lived history. The disciplined position is between these extremes. Myth is not a lab notebook, but it is not random noise either. It is a cultural memory system. It preserves what communities felt was important about origins, power, and difference. If a society repeatedly tells stories about beings from beyond mating with humans and producing exceptional offspring, it may be preserving—through metaphor and exaggeration—the memory of real mixing events: outsiders merging with locals, elite lineages formed through foreign alliances, or culturally disruptive encounters that were later sacralized.

Interbreeding, in the broad demographic sense, is one of the most consequential processes in human history. Populations move. They meet. They form alliances, trade networks, and conflicts. They take spouses. They absorb one another. Most of this happens quietly, without leaving a dramatic narrative trace. But some mixing events are socially explosive. They alter hierarchies. They produce new elites. They introduce new skills. They reshape identity. Those are the events myth tends to preserve, because myth preserves what transforms the world.

When a myth says a god took a human lover, it can be read as a symbolic way of saying that power entered the community from outside. The "god" may represent a foreign group with superior organization or technology, a conquering elite, a ritual specialist class, or simply a lineage that arrived from a distant place and became dominant. The human partner represents the local lineage, the land, the people. Their union produces a child who becomes founder, hero, or king—exactly what would happen in a historical scenario where outsiders integrate with locals and establish a new ruling line. The myth compresses a complex demographic and political process into a family story because family is the most intuitive model humans have for continuity.

This reading becomes more persuasive when we notice how often hybridization myths concentrate around founding moments: the first kings, the first cities, the first laws, the first crafts. These are precisely the moments when intergroup contact can be most transformative. A migrating group enters a region with new ways of organizing labor or new ritual systems. Local groups adopt and adapt. Marriage alliances stabilize the new order. The children of these alliances become symbols of unity and legitimacy. Over time, the foreigners are remembered not as merely foreign but as divine, because their power seemed to originate beyond ordinary experience. The myth does not say "we merged with outsiders and built a state." It says "the gods came, and our founders were their children." The content is different, but the structural memory—contact, mixing, new order—remains.

Yet we must also acknowledge a deeper reason myth uses mating as its language of contact: mating is irreversible. It is the point at which difference enters the lineage. Trade can stop. Alliances can dissolve. Technologies can be copied. But children are permanent. Once lineages merge, a community cannot simply return to its previous state. In this sense, the mythic emphasis on sex and reproduction is not gratuitous. It is a recognition that the most profound form of integration is biological. Myth is telling us: the boundary was crossed in a way that changed the bloodline. Whether the literal actors were gods or humans from elsewhere, the story insists that the change was embodied.

This is also why hybridization myths so often frame the children as exceptional. If a new ruling line emerges from intergroup mixing, those children will be socially exceptional by definition. They may inherit claims from both sides. They may become mediators. They may be given special training. They may be positioned at the center of political life. Their exceptional status then becomes interpreted as exceptional nature. The myth says they are extraordinary because they carry divine blood. The historical reality could be that they carried network power and social advantage. The effect is similar: the child is a pivot, and the pivot is made sacred.

However, not every hybridization myth needs to encode a literal demographic mixing event. Myth also uses interbreeding symbolically to describe the fusion of categories: sky and earth, spirit and matter, civilization and wilderness. In many traditions, divine-human unions are not about foreigners but about the world's structure. The god represents a cosmic principle, the human represents embodied life, and their union produces the mediator who binds the cosmos. In such cases, reading the myth as demographic memory would be a category error. The challenge, then, is to distinguish when the interbreeding motif is likely to be grounded in social history and when it is likely to be primarily cosmological symbolism.

One clue lies in how concretely the myth anchors itself to places, lineages, and political claims. Myths that attach divine-human unions to specific dynasties, named founders, and territorial legitimacy often have strong political functions. They are more likely to be entangled with real social processes—conquest, migration, alliance. Myths that function more as cosmic drama, detached from specific political lineages, may be less about demographic memory and more about metaphysical meaning. But even this distinction is not absolute. Politics and cosmology often merge, because rulers borrow cosmic language to legitimize rule.

Another clue lies in the myth's moral framing. Some traditions describe divine-human unions as violations that trigger punishment—floods, exile, cosmic reset. This can be read as social memory of disruptive mixing events that were perceived as destabilizing: outsiders entering, lineages changing, old orders collapsing. But it can also be read as a moral allegory about boundaries and transgression. Again, the same narrative can carry multiple layers: a warning, a legitimization, a memory, a cosmic model. Myth is not one-dimensional.

In the context of this volume, the key value of reading myth as interbreeding record is not that it proves any single event. It is that it provides a conceptual bridge between the known biological reality of human admixture and the cultural obsession with hybrid origin. Humans interbred with other human populations repeatedly. We also constructed myths that frame the most transformative mixing events as unions between worlds. This parallel is not trivial. It suggests that the human mind has long understood, at an intuitive level, that mixture changes destinies. It also suggests that societies often experienced mixing not as gradual blending but as boundary-crossing shocks—moments when the familiar world was invaded by difference.

This is where the "missing chapters" theme returns. If modern narratives of human origins once insisted on a clean line—one species replacing others without much mixing—then myths of hybrid origin looked like fantasy. But as our understanding has shifted toward a web of ancestry and admixture, myths of mixed origin begin to look like cultural expressions of a reality we now know exists in some form: humans are composites. Myth does not describe the mechanism correctly, but it may preserve the intuition that the human story is not pure.

The next subsection is the pivot to science. We will examine what modern genetics actually confirms about interbreeding and admixture, and how those confirmations can be placed beside hybridization myths without collapsing one into the other. The aim will be to show what science supports, what science does not, and why the overlap—limited but real—matters for understanding how humans remember their origins.

The most important shift in the modern understanding of human origins is not a new fossil in a museum drawer. It is the realization that ancestry is not a single line. For much of the twentieth century, popular presentations of evolution trained the public to imagine a ladder: one species turning into another, a clean replacement, a straightforward climb toward "modern humans." But genetic evidence has increasingly forced a different picture, one that looks less like a ladder and more like braided rivers. Different human populations met. They mixed. They exchanged genes. They left traces of one another inside the bodies of their descendants. In that sense—carefully defined and not sensationalized—science has confirmed the basic intuition that makes hybridization myths so persistent: the human story includes mixture.

This confirmation, however, is narrower than myth, and it must be kept narrow if we want to remain honest. Science does not confirm that gods mated with humans as literal events. It does not confirm that hybrid children were semi-divine beings. It does not confirm that ancient elites carried non-human blood. What science does confirm is that interbreeding among human lineages occurred and that it mattered. It also confirms that the genetic legacy of such interbreeding is not evenly distributed. Some populations carry more of certain ancestry components than others. Some inherited specific genetic variants that conferred advantages in particular environments. This is where the overlap with myth becomes meaningful: not as proof of divine unions, but as proof that mixture can produce real biological consequences that would be experienced as exceptional.

The simplest scientific point is that modern humans carry genetic material from archaic human groups that did not survive as separate populations. This implies repeated contact and integration rather than total replacement. In the older, overly simple story, a "modern human" emerged and swept across the world, eliminating all others. In the newer story, modern humans expanded, but they also absorbed. They did not merely replace; they incorporated. That incorporated legacy is precisely what the volume means by the human species as composite. It is not poetic; it is biological. It is also an uncomfortable idea for any narrative that wants human origins to be clean and singular. Complexity is harder to teach, harder to compress into a timeline, and harder to turn into a reassuring identity story. Yet complexity is what the evidence increasingly demands.

A second scientific point is that admixture is not only an ancient event; it is a recurring process. Human populations have mixed repeatedly throughout prehistory and history. Each migration, each trade corridor, each conquest, each alliance can alter the genetic landscape. Most of these events are subtle at the scale of individuals, but across centuries they can change populations. This matters because it means that hybridization is not an anomaly. It is normal. If myths obsess over hybrid origin, they may be reflecting an underlying human condition: humans are always crossing boundaries, and boundaries are always porous.

A third scientific point—one that resonates strongly with myth—is that some inherited variants from other human groups appear to have been beneficial in certain contexts. In high-altitude environments, for example, specific adaptations related to oxygen processing became crucial for survival, and some of these adaptations are linked to ancient admixture. In immune function, too, variants can be shaped by contact with new pathogens, and admixture can introduce diversity that helps populations respond to unfamiliar disease landscapes. The relevance to myth is not that a god gave a gift, but that mixture can look like a gift when it confers survival advantage. If a community thrives in an environment where others struggle, it will be interpreted as chosen, blessed, protected by powerful ancestors. The biology can create the lived experience that feeds the myth.

But science also complicates the mythic picture in a way that is easy to miss. Myth often treats hybridization as a rare, dramatic event: a god descends, a union occurs, a hero is born, history turns. Genetics suggests something different: admixture can be diffuse, repeated, and distributed across many individuals. It can happen through multiple contacts over long periods. It can be minor in percentage but major in impact if certain variants matter. In other words, hybridization does not necessarily produce a single obvious hybrid child. It produces a population-level mosaic. Myth compresses mosaic into a person because stories work best with persons. Science expands persons into patterns because evidence works best with patterns.

This is why the relationship between myth and science is best described as parallel rather than identical. Myth preserves the meaning of mixture: the sense that boundary-crossing creates exceptional outcomes and destabilizes old orders. Science preserves the mechanism: gene flow across populations and lineages. Where myth says "the gods gave us heroes," science says "admixture contributed to variation and sometimes adaptive advantage." Both frameworks acknowledge that mixture changes destiny. They differ in their explanatory language and evidentiary standards.

A fourth scientific point is that the concept of "species" becomes harder to apply cleanly in deep time. The more we learn about ancient humans, the more we see that lineages we once thought were separate may have had overlapping ranges and occasional interbreeding. This does not mean everything interbred with everything. It means the boundaries were not absolute. This aligns surprisingly well with the mythic intuition that the world once contained beings who were not quite human and not quite divine, entities that crossed categories. Myth externalizes the ambiguity of boundaries into characters—giants, demigods, watchers. Science internalizes the ambiguity into population models and gene flow. Again, the overlap is not literal. It is structural.

A fifth point is that genetics has revealed the existence of "ghost" contributions—signals of ancestry from populations we have not yet clearly identified from fossils. This is not a license for fantasy; it is a reminder of incompleteness. The fossil record is partial. DNA preservation is uneven. There were likely populations whose bones have not been found or not been recognized, yet whose genetic traces persist. For this volume's thesis, this matters because it shows that the human past contains missing chapters in a concrete sense: populations that existed, interacted, and then vanished as distinct groups, leaving only fragments. Myths of vanished beings—mighty ones, ancestors, teachers, others—can be read as cultural ways of grappling with such disappearance, even if they do not map neatly onto specific populations.

At the same time, science sets boundaries that myth does not. It does not support the claim that ancient elites were biologically non-human. It does not support the claim that megalith builders were a separate species. It does not support the claim that exceptional individuals require non-human ancestry. Human exceptionalism can arise within Homo sapiens through cultural complexity, social structure, and individual variation. Genetics can explain some aspects of adaptation and ancestry, but it does not replace culture as the engine of civilization. This distinction is crucial, because hybridization myths often become a way of denying human agency: "we could not have done this without gods." The evidence does not require that conclusion. Human beings, organized and motivated, are more capable than myth gives them credit for in one direction, and less in another.

So what does science actually allow us to say that feels like "myth remembered"? It allows us to say that the deep intuition behind hybridization myths—that the human story includes mixture, that boundaries were crossed, that lineages absorbed one another, that vanished populations left descendants—has a real analog in the genetic record. It allows us to say that admixture can introduce traits that feel like gifts, especially in harsh environments. It allows us to say that the human past is more complex than the clean, linear stories once told in classrooms. And it allows us to say that the "missing chapters" are not merely gaps of imagination; they are gaps in our evidence, gaps we know exist because the genome itself contains traces we cannot fully assign.

CHAPTER 9 — THE WATCHERS AND THE SKY TEACHERS

If hybridization myths are about blood, the Watcher myths are about knowledge. They tell a different kind of origin story—one that does not begin in the womb but in the mind. Across continents and eras, cultures repeatedly claim that the most transformative knowledge was not patiently discovered through ordinary trial and error. It was given. It arrived from elsewhere. It was taught by beings who were not quite human: sky visitors, luminous instructors, divine messengers, ancestral civilizers, serpent sages, mountain spirits, star-born teachers. The names change, the costumes shift, the moral tone varies from reverent to fearful, but the structure holds. Humanity becomes human—fully human, in the mythic sense—only after instruction from something beyond itself.

The universality of this motif is not trivial. It appears in traditions that never contacted one another directly, in environments that could not be more different, in societies ranging from small-scale communities to imperial civilizations. This suggests that we are not dealing with a single historical rumor that spread globally. We are dealing with a recurring way humans frame the mystery of cultural leaps. When a society looks back at the sudden appearance of agriculture, navigation, metallurgy, calendars, sacred law, monumental architecture, or complex ritual, it often experiences the leap as too steep to be purely human. The mind asks: who taught us? And myth answers: they did—those who came from above, from the deep, from the old world, from the realm of gods.

To understand why this answer recurs, begin with the basic nature of human cultural memory. Most people do not experience civilization as a gradual curve. They experience it as a set of finished facts: we have fire, we have tools, we have crops, we have boats, we have metal, we have writing. The origins of these things are often invisible, stretched across generations, scattered across regions, and lost to time. When later societies inherit the results, they cannot easily imagine the long chain of incremental improvements that produced them. They therefore compress invention into a moment and personify it. The "teacher" becomes the narrative container for a process that is too slow and distributed to remember.

But this still does not explain why the teacher is so often non-human. Why not simply wise ancestors? Sometimes the myth does use ancestors. Yet again and again it insists on otherness. The instructor is described as shining, descending, winged, serpent-like, giant, pale, dark, masked, or otherwise marked as not fully human. This is the crucial pattern. It reveals that the myth is not merely commemorating learning. It is commemorating a boundary-crossing event. The teacher belongs to another realm. The knowledge therefore belongs to another realm. The origin of civilization is framed not as internal development but as contact.

One reason for this is political. Knowledge is power, and power needs legitimacy. If a priesthood claims that its astronomical calendar, ritual law, or sacred metallurgy came from gods or sky beings, that knowledge becomes unquestionable. It cannot be challenged by ordinary people because its source is beyond ordinary dispute. The myth of non-human instruction therefore functions as an authority shield. It protects specialist knowledge from skepticism. It also protects the social order built around that knowledge. If law came from sky teachers, then law is sacred. If writing came from divine messengers, then scribes become custodians of cosmic order. In this way, non-human instructor myths often support the institutional stability of civilizations.

But the motif cannot be reduced to propaganda, because it also appears in contexts where there is no state machinery to enforce it. It appears in oral traditions that treat the instructors with ambivalence or even hostility. In some stories, the teachers are not benevolent gods but suspicious strangers. They bring tools that are dangerous. They teach arts that disrupt the old ways. They demand sacrifices or impose rules. The knowledge is double-edged. This reveals something important: the myth may be encoding the lived experience of contact with outsiders whose skills were transformative but destabilizing. The outsider becomes non-human because the outsider's knowledge rewires the community's world.

Contact, in this sense, does not need to be interplanetary to feel cosmic. In a world without global maps, a stranger arriving from beyond the horizon can feel like a being from another realm. Their language is incomprehensible. Their tools appear magical. Their confidence looks divine. Their ability to predict seasons or navigate waters seems supernatural. When such outsiders teach, they become remembered as sky teachers. Their human reality dissolves into a mythic identity because the community's interpretive categories are sacred categories. The god is what you call a power you cannot localize.

There is also an internal psychological reason the motif recurs. Humans struggle with the problem of origin when the origin is complex. How did mathematics begin? How did music arise? How did agriculture become reliable? How did humans learn to smelt metal? These are not single inventions. They are networks of practice, often developed across different places and times, often refined through failure. Myth refuses that complexity. It wants a beginning. It wants a giver. The non-human instructor is the giver that satisfies the mind's demand for a clean origin story. It is the narrative that turns distributed emergence into a moment of revelation.

This demand becomes even stronger when the knowledge feels sacred. Astronomy is not merely a tool for planting; it is a map of the gods. Metalworking is not merely craft; it is transformation of earth into weapon and ornament. Writing is not merely record; it is the power to bind speech into permanence. When knowledge is sacred, its origin must also be sacred. A human inventor feels too ordinary as a source for cosmic order. A non-human teacher solves the problem. It places the origin where sacred things are believed to come from: beyond the human world.

Yet the motif also contains a subtle confession: that humans do not feel fully responsible for what they became. When a culture says "we were taught," it is admitting that civilization feels like something that happened to us as much as something we did. This sense of being changed by an external force may reflect real historical processes—waves of migration, technological diffusion, cultural replacement—where communities experienced sudden transformation without controlling it. The Watcher myth becomes the story form of that experience: the world arrived, and we were remade.

The Watcher theme also has a distinctive moral tension. Non-human instructors are often portrayed as ambiguous beings. Sometimes they are benefactors. Sometimes they are transgressors. Sometimes they teach forbidden knowledge. Sometimes the gods punish them for giving humans what humans were not meant to have. This tension suggests that the myth is also wrestling with the costs of knowledge. Fire warms but burns. Metal builds but kills. Writing preserves but controls. Civilization stabilizes but enslaves. By framing knowledge as a gift from outsiders, myth can assign blame for civilization's darker consequences. The teachers brought it. They are responsible. Humanity becomes both beneficiary and victim.

In the framework of this volume, the universal myth of non-human instructors is therefore not a claim we accept literally. It is a pattern we take seriously as evidence of how humans remember rapid change, contact, and the strange feeling that our most powerful tools arrived as if from elsewhere. It is also a pattern that complements the genetic story: just as our biology is composite, our culture is composite. Much of what humans are was assembled through exchange and absorption. Myths of sky teachers and Watchers may be the symbolic form of that cultural absorption.

The next subsection examines the most consistent feature of these traditions: the claim that knowledge was given, not discovered. That claim is not only mythic; it is a way of framing human agency. If we understand why cultures prefer the "gift" narrative, we can better understand how memory treats innovation—and why the idea of instructors persists even when the evidence points to gradual development.

The phrase appears in countless forms across mythic traditions: "they taught us," "they showed us," "they brought the arts," "they gave the laws," "they revealed the calendar," "they instructed the first priests," "they opened the mysteries." The underlying claim is remarkably consistent. Humanity did not climb toward civilization solely by stumbling forward through trial and error. Civilization arrived as a package of teachings delivered by beings who already knew. Knowledge was given, not discovered. This claim is not a minor detail in the Watcher theme. It is the core mechanism by which myth explains why culture can change faster than memory can account for.

To understand why this claim persists, start with the gap between lived experience and deep history. Most people experience knowledge as something transmitted, not invented. A child does not rediscover language; a child is taught language. A young artisan does not reinvent metallurgy; they learn it from a master. A novice does not derive the calendar from first principles; they inherit it as a social fact. For most individuals, "knowledge" is received. Even for the rare innovator, innovation builds on inherited scaffolding. In daily life, human knowledge feels like a gift from elders. Myth expands that everyday truth to cosmic scale. If a community experiences knowledge as transmitted from those who came before, then at the origin of civilization it makes sense—mythically—to place a teacher.

But myth goes further. It often insists that the teacher is not merely an elder but an outsider, a being from beyond the ordinary. This is because some forms of knowledge appear too complex, too precise, or too transformative to be explained as the product of gradual local tinkering. The calendar aligns with the sky. The metallurgy turns rock into blade. The rituals seem to bind the seasons. The monuments align with cardinal points. When later cultures encounter these achievements without knowing the long chain of incremental steps that produced them, they experience them as revelations. Revelation implies a revealer. Hence the teacher.

The "given" narrative is also a way of preserving humility before complexity. A society that believes sacred knowledge was received from higher beings is less likely to treat that knowledge as casually alterable. The claim that laws were given, not invented, places law beyond debate. It transforms social rules into cosmic rules. In that sense, the narrative is not only explanatory but regulatory. It stabilizes social order by making its foundations sacred. If the law is a gift from the sky, then the law is not a human negotiation. It is destiny.

This stabilizing function is especially strong in priestly systems, where specialized knowledge can translate directly into authority. The ability to predict seasonal change, track celestial cycles, and administer ritual calendars confers prestige because it appears to control the invisible structure of time. When such knowledge is framed as discovered by human trial, it remains potentially accessible; someone else could, in theory, replicate it. When it is framed as given by non-human instructors, it becomes exclusive. The priesthood becomes the custodian of an inherited revelation rather than the practitioner of a learnable skill. The myth therefore supports the monopolization of knowledge, which in turn supports hierarchy.

Yet the claim that knowledge was given is not always framed as benign. In many traditions, the gift is dangerous. It is forbidden. It is premature. It breaks a divine boundary. The instructors are punished or cast down for sharing what humans were not meant to have. This pattern reveals another reason the "given" narrative persists: it offers a moral explanation for the double-edged nature of knowledge. Humans can look at the costs of civilization—war, exploitation, environmental depletion, social stratification—and tell a story in which the origin of these costs is external. The knowledge came from outside. It was a gift that should perhaps have been refused. By framing knowledge as given, the myth can both honor civilization's achievements and critique its consequences.

There is also a cognitive reason the "given" narrative is so attractive: discovery is hard to imagine retrospectively. Incremental invention is messy. It involves failure, partial solutions, forgotten experiments, regional variation, and long plateaus. When humans remember, they tend to remember peaks, not plateaus. They remember the moment something worked, not the countless moments it didn't. The story of discovery therefore collapses into the story of a single breakthrough. A single breakthrough naturally invites personification. Someone did it. Someone showed it. Even in modern history, we often compress complex developments into named inventors. Myth performs the same compression, but it shifts the inventor outside the human world to satisfy the scale of the achievement.

Contact between groups adds another powerful layer to the "given" narrative. When a community receives a new technology or a new ritual system through exchange, it literally was given, in a practical sense. The knowledge came from elsewhere. It was not locally discovered. In such cases, myth may be preserving a real directional flow of culture. The "sky teachers" may be the mythic form of travelers, traders, refugees, or specialists who carried skills across regions. The receiving community experiences the arrival of knowledge as a gift because it is. But because the donors are foreign and the knowledge is transformative, the donors are mythologized as non-human. The more dramatic the cultural gap between donor and recipient, the more likely the donor will be framed as belonging to another realm.

This is why many civilizing figures in myth are described as arriving from the sea, descending from mountains, emerging from caves, or appearing from the sky. These are not only cosmological directions; they are also geographic metaphors for "outside." Sea routes connect distant worlds. Mountains are zones of refuge and mystery. Caves are liminal spaces. Sky is the ultimate elsewhere. Myth uses these directional symbols to encode the simple social reality: the knowledge came from beyond us.

In the context of this volume's thesis, "knowledge given, not discovered" aligns with the broader critique of linear progress narratives.

Just as human evolution is not a single line, human culture is not a single isolated invention stream. It is a web of transmission. Knowledge spreads through networks. It accumulates and recombines. Some communities become hubs of innovation; others become recipients who adapt what arrives. The mythic "gift" narrative can therefore be read as a symbolic expression of cultural transmission in a world where the mechanisms of transmission are forgotten.

But the motif also does something more intimate. It creates a relationship between humanity and the beyond. If knowledge is given, then humans are not alone. We are in conversation with something larger—gods, ancestors, spirits, or the memory of vanished peoples. This relationship can be comforting: we were guided. It can also be terrifying: we were manipulated. Either way, the relationship gives meaning to the suddenness of civilization. It tells us that we did not emerge from darkness by accident. We were taught. We were chosen. We were altered.

The next subsection moves from this general claim into the specific domains that Watcher myths almost always emphasize: astronomy, metallurgy, and the clustered appearance of complex civilization. These domains are not arbitrary. They are exactly the kinds of knowledge that feel revelatory because they transform the relationship between humans and nature. Astronomy turns the sky into a clock. Metallurgy turns stone into power. Together, they turn survival into order and order into hierarchy. If any knowledge would be remembered as "given," it would be these.

When Watcher and sky-teacher traditions list the gifts that "came from above," they tend to cluster around a particular set of arts. The details vary, but the pattern repeats with uncanny consistency: the knowledge of the sky, the knowledge of metal, and the knowledge of how to organize human life around those powers. Astronomy provides time. Metallurgy provides tools and weapons. Together they provide the scaffolding for complex society—calendars for agriculture and ritual, blades for authority and war, ornaments for status, and a new class of specialists who mediate between hidden forces and ordinary life. If hybridization myths explain exceptional blood, these gifts explain exceptional civilization. They explain why humanity seems to pass through a threshold and enter a world where the sky becomes legible and stone becomes power.

Astronomy is the first pillar because it transforms the relationship between humans and uncertainty. Before systematic sky watching, seasons arrive as moods of nature. You feel the rains. You fear the drought. You follow animal migrations. You plant and hope. But once a society learns to read the sky with precision—tracking the sun's path, measuring the year, predicting solstices, observing the lunar cycle, noting the rising of particular stars—time becomes structured. Planting becomes planned. Ritual becomes synchronized. Authority becomes anchored in prediction. The person or class that can say "the rains will come," or "the festival must occur now," holds an invisible kind of power: they seem to command time itself.

In myth, this power is rarely described as the outcome of generations of observation. It is described as revelation. The sky teachers point to the heavens and explain what the lights mean. They establish the calendar. They define sacred days. They teach the names of stars. They connect constellations to gods and ancestors. The implication is that the sky contains a code, and only certain beings can read it. This framing is mythic, but the social reality behind it is clear: astronomical knowledge is specialist knowledge, and specialist knowledge creates hierarchy. Once time is formalized, those who control the formalization can control labor, taxation, ritual obligation, and legitimacy.

Astronomy also provides orientation. When a society learns cardinal directions, uses stellar navigation, or aligns structures to celestial events, space itself becomes organized. Sacred architecture—temples, ceremonial avenues, horizon markers—turns geography into cosmology. The built environment becomes a map of the sky on earth. This is one reason megaliths and early monumental sites are so often associated with celestial alignments in popular imagination. Even when claims are overstated, the deeper point holds: early complex societies often embed sky knowledge into their monuments because the sky is their most stable reference system. The heavens provide permanence in a world of change.

Metallurgy is the second pillar because it transforms the relationship between humans and matter. Stone tools and bone tools can be refined to a high degree, but metal introduces a different order of possibility. With metal, edges become sharper, tools become more durable, and weapons become more lethal. Metal also introduces an aura of transformation. To take dull rock, heat it, and extract a shining substance that can be hammered into form is to perform what looks like magic. The metallurgist becomes a ritual figure as much as a craftsperson. They know secrets of fire, air, and earth. They can make objects that change the balance of power in a community. It is no accident that many traditions place metalworking among the gifts of sky teachers or forbidden instructors. Metallurgy is not merely technology; it is social disruption compressed into craft.

In mythic accounts, metallurgy is often framed as a stolen or forbidden gift. The teachers reveal the art, and the gods punish them. This moral framing reflects the ambivalence of metal. Metal can build plows, but it can also build swords. It can craft ornaments that stabilize hierarchy through display, but it can also enable conquest that shatters old orders.

The society that first gains metalworking gains an advantage over those without it. That advantage can look like superhuman power. In a world where most people's experience of material transformation is limited to cooking and basic craft, metallurgy feels like contact with a higher realm.

When astronomy and metallurgy appear together, civilization accelerates. This is the third element in the motif: the sense of suddenness. "Sudden civilization" in myth is rarely a claim about a specific date. It is a narrative perception: that something changed quickly, that complexity arrived in a burst, that cities and temples and laws crystallized as if a switch had been flipped. From a mainstream perspective, early civilization does develop over long periods, with gradual intensification of agriculture, settlement, and social stratification. Yet it is also true that once certain thresholds are crossed—population density, surplus management, administrative systems—change can become rapid. A region can move from villages to urban centers in a relatively compressed span compared to the deep timescales of human existence. That compression is what myth remembers as "sudden."

The sky-teacher motif provides an explanation for this suddenness that is emotionally satisfying. It says: we were instructed. We received a package. We did not slowly invent; we were taught. This does not need to be literally true to reflect a real dynamic. Knowledge often spreads in clusters. Innovations often recombine. A society might adopt agriculture, calendar systems, and new craft techniques in overlapping waves through exchange and contact. From the perspective of later memory, these waves blur into one event. The teachers arrive, and civilization begins.

This is where the motif intersects with migration and network theory in a grounded way. A "teacher" in myth can represent a class of itinerant specialists—astronomer-priests, metalworkers, architects, ritual experts—who moved between communities and established new cultic or administrative systems. Such specialists could be absorbed into elite networks, marrying into local lineages and becoming founders of new orders. Their knowledge would appear as a gift because it was imported. Their social position would appear extraordinary because they carried tools and concepts that locals did not. Over time, their human identity could become sacred. They become Watchers, because they watch the sky. They become sky teachers, because they teach the sky's code.

The clustering of astronomy and metallurgy also reflects a deeper truth: both fields require abstraction, discipline, and social investment. They are not casual crafts. They demand time, specialization, and institutional support. You need sustained observation to build an accurate calendar, and you need controlled high-temperature processes to work metal reliably. These requirements naturally create specialist classes and centralized authority structures. Myth remembers that centralization as a gift from beyond because specialization itself feels beyond ordinary life. The ordinary person does not become a metallurgist or astronomer by accident. They are initiated. They are trained. They enter a knowledge lineage. That lineage can feel like a separate species of person inside the human species.

The sky-teacher motif therefore captures a real pattern: transformative knowledge often arrives through specialists, and specialists often form semi-closed lineages or institutions. Those institutions can claim non-human origin to protect their status and to sacralize their authority. The "given" narrative becomes a political and psychological tool that stabilizes the new order.

Yet the motif also insists on something else: that these teachers are never just human experts. They are beings with one foot outside the human category. That insistence is the subject of the next subsection. Why do myths so often refuse to let the instructors be merely human? Why are they described as luminous, winged, serpent-like, or descended from the sky? What does that otherness do in the story, and what does it tell us about how humans perceive the origins of their most powerful knowledge?

In the logic of Watcher myths, it is not enough that the instructors are wise. They must be other. They must be marked as belonging to a category beyond ordinary humanity—shining figures, winged messengers, serpent sages, mountain beings, star-born ancestors, giants, masked strangers, those who "came down" or "emerged" with a presence that ordinary humans cannot sustain. This insistence on otherness is one of the strongest signatures of the motif. It is also the most revealing, because it shows what the myth is really trying to protect: the sacred distance between ordinary life and the source of transformative knowledge.

One level of explanation is simple and structural. If the teacher were fully human, the knowledge might be understood as fully human. It might then be available, reproducible, and contestable. A myth that frames civilization's core knowledge as a human invention risks destabilizing the authority built on that knowledge. If the calendar was invented by men, other men might improve it, dispute it, or claim ownership of it. If sacred law was drafted by a committee, it might be renegotiated. But if the calendar and the law were given by beings beyond humanity, they become fixed. They are insulated from debate. The otherness of the teacher is therefore a stabilizer. It places knowledge outside ordinary politics.

This stabilizing function becomes especially visible in priestly traditions, where the boundary between knowledge and power is thin. The priesthood's authority depends on being perceived as a conduit to the beyond. If the origin of their knowledge is mundane, their status becomes mundane. The myth refuses this mundanity. It insists that the origin is uncanny. In doing so, it protects the institutional continuity of specialist classes. The teacher must be non-human because the knowledge must remain sacred.

But there is another level that is not merely political. Otherness is also a way of expressing the phenomenology of contact with the unfamiliar. When a community encounters outsiders who carry unfamiliar technologies, unfamiliar rituals, unfamiliar cosmological frameworks, the experience can be psychologically overwhelming. The outsiders may seem to possess powers: predicting seasons, navigating oceans, producing metal, curing illness, organizing large projects. In a world where knowledge is localized and experiential, such powers look like miracles. The mind struggles to place them within familiar human categories. It reaches for sacred categories instead. The outsider becomes not just foreign but superhuman—because the outsider's competence exceeds what the local worldview expects a human to be.

This is why Watcher figures are often described as liminal. They are not merely "gods" in a distant heaven. They are beings who move between realms. They descend. They teach. They depart. Their mobility marks them as not bound by the constraints of ordinary life. In narrative terms, mobility itself becomes a sign of otherness. To be able to arrive from beyond the horizon and leave again is to possess a freedom that ordinary people do not. The teacher's otherness therefore encodes an experience of asymmetry: the asymmetry between a local community rooted in place and a traveling group or specialist class that moves through networks.

Otherness also expresses a deeper fear about knowledge. Knowledge can feel inhuman because it demands inhuman discipline. Astronomy requires patience and long attention. Metallurgy requires mastery of fire and risk. Writing requires abstraction and memory externalized into marks. Law requires the suppression of impulse in favor of rule. These are not natural states for the human animal in its raw form. They require training. They require constraint. They require institutions. In a sense, they require a partial transformation of human nature. Myth externalizes that transformation by attributing it to non-human teachers. It says: we did not become this by ourselves. We were changed. The teacher is other because the new human is other.

This logic becomes especially powerful when the teachings are framed as forbidden. Many Watcher myths suggest that humans received knowledge too early, or in a way that violated cosmic boundaries. The teachers are punished not because they were ignorant, but because they breached a rule that even gods must obey. In such stories, the teachers cannot be fully human because humans cannot be held responsible for cosmic violations at that scale. Only a non-human being can commit a transgression against heaven. The otherness of the teacher therefore makes the moral drama possible. It creates a cosmic courtroom where the gift of knowledge can be judged.

There is also a symbolic dimension. Myth often encodes psychological and cosmic forces as beings. The sky teacher can represent the sky itself—its order, its regularity, its overwhelming vastness. The serpent teacher can represent the earth's hidden forces—fertility, danger, transformation. The winged messenger can represent inspiration—thought arriving as if from nowhere. In these cases, the teacher is never fully human because it is not primarily a historical person.

It is an archetype: the personification of a force that shapes human life. The myth says "a being taught us" because it is easier to tell a story about a being than to tell a story about a force.

Yet even when the teacher is archetypal, the insistence on otherness still points to something socially real: the division between specialists and ordinary people. Specialists often appear as a different kind of human within a community. They speak differently, dress differently, eat differently, obey different rules, inhabit different spaces, and claim access to hidden knowledge. Initiation rituals intensify this separation by marking the specialist as reborn. The new identity is not merely learned; it is acquired through transformation. From the outside, this transformation can look like becoming another species. Mythic otherness is a symbolic reflection of that social separation. The Watcher is the initiated specialist imagined as literally not human, because their role places them beyond ordinary life.

This also helps explain why Watcher figures are so often described with mixed features—part human, part animal, part luminous, part monstrous. These mixed features externalize the idea of crossing categories. The teacher is a hybrid because the knowledge itself is hybrid: it belongs to nature and culture, to earth and sky, to life and death. The being who carries it must embody that crossing. In a mythic vocabulary, the easiest way to show category-crossing is to make the character physically category-crossing. Wings, scales, horns, radiance—these are not random decorations. They are symbols of otherness made visible.

In the disciplined frame of this book, we hold two truths together. First, the otherness of the teachers is not scientific evidence of non-human instructors in a literal biological sense. Second, the otherness is evidence of something persistent in human memory: that transformative knowledge is experienced as external, disruptive, and not fully assimilable to ordinary human identity. The myths insist on non-human instructors because they are trying to preserve the felt magnitude of what happened when humans became something more than foragers—when they began to map the sky, reshape the earth, and bind society into rule.

The next subsection turns to the darker arc that frequently accompanies this motif: fall, punishment, and disappearance. The teachers often do not remain. They are cast out, destroyed, banished, drowned, imprisoned, or removed by divine decree. This pattern matters because it suggests that cultures not only remembered the arrival of transformative knowledge but also remembered rupture—an end to the age of direct instruction. Whether that rupture reflects moral allegory, social conflict, catastrophe, or the fading of specialist lineages, the myth insists on one thing: the teachers leave, and the world is never the same.

In many traditions, the sky teachers do not simply arrive, educate humanity, and fade gently into legend. They fall. They are punished. They are removed. Their age ends in rupture. Sometimes they are bound beneath the earth or sealed within mountains. Sometimes they are cast from the heavens. Sometimes they are drowned in flood, consumed by fire, or annihilated in divine judgment. Sometimes they disappear, leaving only fragments of knowledge behind—partial arts, incomplete calendars, broken rituals, corrupted lineages. This arc is so persistent that it functions like a second law of the Watcher myth: the gift of knowledge triggers consequences, and the bearers of that gift do not remain among humans.

At one level, the pattern is moral drama. It tells a story about transgression and boundary. Knowledge is portrayed as powerful enough to threaten cosmic order. The teachers are punished because they gave humans what humans were not meant to have, or because they violated a sacred separation between realms. In this reading, the fall is an allegory about limits. It warns that not all knowledge is safe, and that the pursuit of power carries a price. The punishment restores balance. The teachers become examples of what happens when boundaries are crossed without permission. This moral structure is satisfying because it gives the universe coherence. It says: the cosmos has rules, and even luminous beings cannot break them without consequence.

But the persistence of the fall motif suggests that something more than morality is being preserved. The myth is also explaining absence. If the teachers were once present—if the tradition insists they taught and walked among humans—then why are they not here now? Why do we no longer see them? Why must humans struggle, forget, and rediscover? The fall narrative answers this. It provides a reason for the end of direct instruction. It tells the community that the world is now in a different phase, a diminished phase, a post-contact phase. Humans are left with remnants rather than guidance.

This "why are they gone?" function is crucial because it mirrors how cultures often experience the loss of knowledge in real history. Specialist traditions can collapse. Priesthoods can be destroyed by war. Trade networks can fail. Centers of learning can burn. A civilization can fragment into smaller societies that inherit only scraps of the old system. From the perspective of later generations, the loss can feel like a cosmic withdrawal. The golden age is over. The teachers have left. The world is colder. Myth translates institutional collapse into celestial exile.

Consider how knowledge actually behaves across time. It is not a steady upward climb. It can regress. It can be lost. It can become incomprehensible when the social structures that maintained it disappear.

A calendar system requires trained observers and a tradition of record-keeping. Metallurgy requires supply chains, fuel access, and specialized expertise. Monumental building requires organization and surplus. When these conditions collapse, the knowledge can fade within a few generations. Later people may still see the monuments, still hear fragments of the old myths, but they no longer possess the full method. This produces precisely the emotional landscape the fall motif describes: we inherited wonders, but we do not fully understand them. Something that once existed is gone.

The fall narrative also serves as a mechanism for explaining why human knowledge is partial and conflicted. If the teachers were punished, then their teachings may be incomplete. If they were banished suddenly, then the transmission may have been interrupted. If their knowledge was forbidden, then it may have been hidden and corrupted. These explanations allow a culture to account for contradictions and gaps in its own tradition without blaming itself. The fault lies not in human incompetence but in cosmic disruption. The teachers were removed. The world changed. We are living in the aftermath.

Another dimension of the fall motif is political and social conflict. In some interpretations, the "teachers" represent an elite class—ritual specialists, astronomer-priests, metalworkers—whose authority depended on monopolized knowledge. Such classes can become targets during upheaval. Populations can rebel against them. New rulers can dismantle their institutions. Rival priesthoods can erase one another. When a specialist caste loses power or is expelled, the collective memory of their departure can be preserved as mythic banishment. The teachers were cast out. The gods punished them. The knowledge remained, but the beings themselves vanished. Myth therefore becomes a way of narrating elite turnover and cultural discontinuity.

This interpretation also helps explain why some traditions portray the teachers with suspicion. The teachers may be framed as deceptive, prideful, or corrupting. Their gifts may be portrayed as dangerous—introducing weapons, sorcery, unnatural arts. In such stories, the fall is not tragic but necessary. The removal of the teachers is a purification. This moral framing could reflect real tensions around new technologies and social stratification. Metallurgy and centralized law can benefit society, but they can also enable domination. Those who wield new tools can become oppressors. A later tradition that experienced the costs may tell a story in which the teachers were punished for bringing such arts. The myth thus becomes a critique of the very forces that built civilization.

Yet the most haunting element of the fall motif is disappearance. Often the story does not end with a clear execution or exile. It ends with a vanishing. The teachers withdraw into the sea. They return to the sky. They become invisible. They are sealed away, sleeping or bound until a later time. This is not merely a conclusion; it is a suspended ending. It implies that the teachers are not gone forever. They are absent but not erased. The world is left in a state of waiting. This feature matters because it gives the myth a temporal depth: it divides history into ages—an age of direct contact, an age of withdrawal, and sometimes an age of return.

This age structure mirrors how people experience long-term cultural change. When a civilization collapses, later generations often speak of the past as a different world—an age of giants, an age of gods, an age when knowledge was easy and greatness was common. The present becomes a diminished era. The future becomes the hope of restoration. Myths of sleeping teachers and promised returns encode that longing. They allow people to believe that what was lost is not lost forever. Somewhere, the source still exists.

In the context of this volume's thesis, the fall motif also intersects with catastrophe narratives. Many traditions connect the removal of teachers with floods, fires, cosmic upheavals, or moral resets. Whether or not these correspond to specific events, they represent a memory of discontinuity—something that breaks continuity between ages. Genetic bottlenecks, climate collapses, and cultural fragmentation are all real phenomena in human history. Myths translate such discontinuities into divine judgment. The teachers are punished; the world resets; survivors inherit fragments.

What matters is not to treat these myths as literal eyewitness reports, but to treat them as structured memories of loss. They encode the idea that the past contained capacities the present does not fully possess, and that the transition was not smooth. Something ended. Knowledge became fragmented. Memory became myth. The teachers became absent. This pattern sets the stage for the final subsection of the chapter: myth as evolutionary memory. If the Watchers are not literal beings, what kind of memory are they? Are they memories of contact between populations, memories of specialist classes, memories of catastrophe, or memories of the psychological shock of becoming civilized? The next subsection will argue that the Watcher motif may be a compressed symbolic archive of all these forces, preserving the shape of missing chapters even when it cannot preserve their details.

The Watcher motif endures because it solves a problem that sits at the edge of human self-understanding. How did we become what we are? Not merely biologically, but cognitively and culturally. How did a primate become a creature that reads the sky, smelts metal, builds monuments, writes law, and imagines itself in relationship with the cosmos? Science can explain many pieces of this transformation through gradual change, selection pressures, population dynamics, and cultural accumulation. But myth is responding to a different need. Myth is trying to preserve the felt discontinuities—the threshold moments where humanity seems to step across a line. In that sense, the Watchers can be read as evolutionary memory: a symbolic record of how humans experienced their own becoming, stored in narrative rather than in DNA.

This does not mean myth remembers evolution as a factual timeline. It means myth remembers evolution as an existential event. The Watchers embody the sense that our intelligence did not feel like a smooth internal development; it felt like contact with something beyond. The "teacher" is the personification of that beyond. Sometimes the beyond is another human population. Sometimes it is a specialist caste. Sometimes it is a vanished culture. Sometimes it is the sky itself—its order, its vastness, its uncanny regularity pressing on the human mind until the mind reorganizes. The Watcher is the way myth gives this pressure a face.

From an evolutionary standpoint, humans are unique not only because of brain size but because of cultural transmission. The mind is shaped by what it inherits—language, ritual, tools, norms, symbols. Cultural inheritance can change rapidly compared to biological inheritance. A single generation can transform how the next generation thinks, simply by changing what it teaches. This creates a phenomenon that feels like intervention. A child's mind becomes a different mind because it is filled with a different world. If a society suddenly acquires new knowledge—new crafts, new rituals, new cosmological systems—the society's cognition can reorganize within a few generations. Myth frames this as teaching from non-human instructors because it captures the experience of rapid cognitive transformation: the world arrived, and we became new.

In that sense, the Watcher myth is a memory of transmission itself. It says knowledge comes from teachers. It says civilization is inherited. It says the human mind is not self-created but formed through contact. This is not metaphysics; it is a deep social truth. Humans are made by other humans. But because the earliest phases of cultural formation are lost to time, myth shifts the teacher beyond the human realm. It says: the first teacher was not one of us. The first teacher was something other. That is myth's way of acknowledging how strange it is that a mind can be reshaped so profoundly by what it learns.

The Watcher motif also preserves a memory of the division between ordinary and specialist cognition. In complex societies, knowledge becomes stratified. Most people do not know how the calendar works, how metal is made, how the rituals align with seasons, how law is formulated, how long-distance navigation is performed. Specialists hold these skills. Specialists often form lineages and institutions with initiation barriers. From the outside, this looks like a separate species inside the species: people who know what others do not, people who seem to touch hidden forces, people whose speech is charged with authority. Myth externalizes that social division by making the specialists literally non-human. The Watchers are the archetype of specialist cognition imagined as otherworldly.

But the motif goes deeper than social stratification. It also preserves a memory of contact across populations. Human history is full of encounters that create sudden cultural change: migrations, trade expansions, conquests, refuges after catastrophe. In such encounters, one group may absorb the practices of another quickly, especially if those practices confer survival advantage or political strength. The absorbing group experiences the change as a gift from outside. Later, the donors are mythologized. They become civilizers and sky teachers because their knowledge arrived from beyond the local world. This is a plausible way to read the Watcher motif as demographic memory: not as a record of aliens, but as a record of how cultures experience the arrival of new systems.

There is a reason Watcher myths so often include the idea of "forbidden arts" and punishment. Cultural evolution is not purely beneficial. Knowledge creates power asymmetries. Metallurgy creates weapons. Calendars create taxation and labor obligations. Writing creates bureaucracy and surveillance. Law can protect, but it can also control. When a society later suffers under the burdens that civilization introduces, it may retrospectively frame the origin as a transgression. The teachers were punished for giving too much. Humanity was not ready. This moral framing functions as a memory of cost: the cost of complexity, the cost of hierarchy, the cost of living in a world where the sacred is administered by institutions.

The fall and disappearance motif strengthens the "evolutionary memory" reading because it encodes discontinuity. Human cultural history includes collapses and bottlenecks, moments when complexity contracts and knowledge fragments. After such contractions, later populations often inherit artifacts they cannot replicate and traditions they cannot fully explain. The past becomes mythic by default. The teachers become absent. The age of direct instruction becomes a golden age that is no longer accessible. This is a psychologically accurate way of remembering collapse: it feels like exile from a world where things were clearer and stronger. Myth holds that feeling as a narrative of banishment.

If we integrate this with the volume's genetic themes, a broader pattern emerges. Biology and culture both preserve traces of missing chapters. In genetics, traces appear as inherited variants and ghost signals of vanished populations. In myth, traces appear as recurring motifs—hybrid blood, sky teachers, forbidden knowledge, catastrophic resets, lost golden ages. Neither system preserves the past perfectly. Both compress, distort, and recombine. But both can carry something real: the shape of complexity, the memory that humanity is not a pure line, and the intuition that the human story contains ruptures.

This is where the Watcher motif becomes especially powerful as a lens. It is not asking us to believe in literal non-human instructors. It is asking us to take seriously the human experience of becoming civilized and the human experience of discontinuity. It is saying that people in the past felt the origin of their knowledge as external. They felt it as received. They felt it as dangerous. They felt it as tied to beings who were not fully human. That felt reality is itself evidence—evidence of how dramatic cultural thresholds can be, and how deeply humans reach for "the beyond" when trying to explain why they are no longer what they once were.

The value of this reading is that it preserves the myth's psychological truth while honoring scientific constraint. We can acknowledge that humans interbred and absorbed one another biologically. We can acknowledge that cultures transmit knowledge through contact and specialist lineages. We can acknowledge that civilizations rise and fall, fragmenting memory. We can then understand the Watchers as the story form of these processes: the personification of contact, transmission, and rupture. Myth becomes evolutionary memory in the sense that it remembers the forces that shaped humanity, even when it cannot remember their literal names.

From here, the book moves toward a more explicit confrontation with specific myth systems that describe “engineering-like” creation narratives and the boundary between myth and memory. But before that, the next chapter will sharpen the problem: when creation myths begin to sound like design, why do they appear so suddenly, and what does it mean to take them seriously without turning them into fabricated history? The Watchers have prepared the ground by showing that human memory encodes transformation as contact. The next step is to examine the most provocative versions of that encoding.

CHAPTER 10 — THE ANUNNAKI PROBLEM

There is a reason certain ancient creation narratives provoke a very specific modern reaction. People read them and feel an uncomfortable friction: the story is wrapped in gods and cosmic poetry, yet parts of it sound oddly procedural. They describe making, assembling, dividing, mixing, shaping, assigning functions, designing roles, solving labor problems, and establishing systems that behave like infrastructure. To a modern ear trained on technology and biology, these passages can resemble engineering language—crude, metaphorical, and mythic, but nonetheless close enough to trigger interpretation. This is the core of what might be called the Anunnaki problem: not the claim that any one myth is a literal technical report, but the persistent feeling that some myths describe creation as if it were a project.

The immediate temptation is to treat this resemblance as proof of something extraordinary. If a myth "sounds like engineering," then perhaps it records real engineering. Perhaps the gods were advanced beings. Perhaps they manipulated life. Perhaps the story is memory disguised as theology. But this leap is precisely what must be resisted if the book is to remain disciplined. The right approach is to ask a more careful question: why do these myths sometimes adopt a procedural, design-like tone at all? What human needs and social realities would cause a culture to describe origins in the language of making rather than in the language of spontaneous emergence?

One answer lies in the difference between hunter-gatherer cosmologies and the cosmologies of early states. In small-scale societies, origin stories often center on animal ancestors, tricksters, landscape transformations, or the gradual shaping of the world through episodic sacred acts. In state societies, origin stories increasingly reflect administration, labor, hierarchy, and division of function. The universe begins to resemble a kingdom. The gods behave like rulers and managers. Creation becomes an act of governance. In such a worldview, it is natural to describe the making of humanity and society as a kind of planning process, because the society itself is a planned process. The myth is mirroring the structure of the civilization that tells it.

When a culture lives inside institutions—temples, bureaucracies, workforces, taxation systems—the most intuitive model for "how things come to be" becomes institutional. Something exists because someone with authority made it exist. Work happens because it is assigned. Roles exist because they are designed. Order persists because it is maintained. A creation story told in that context will naturally sound like engineering, because engineering is simply the application of authority to matter. The gods become the ultimate administrators, and the world becomes their project.

This "administrative cosmology" does not require literal advanced technology. It requires only that people project their own experience of organization onto the cosmos. If you live in a world where canals must be dug, fields must be irrigated, bricks must be made, temples must be raised, and labor must be coordinated, then "creation" itself begins to look like coordinated labor. The divine act becomes a planning act. The gods set up systems the way kings set up cities. To modern readers, this can resemble engineering because we recognize system-building language. But for the ancient mind, it is simply how power works.

Another answer is that many of these myths are not naïve folk tales. They are theological literature. They were shaped by priestly classes whose job was to impose coherence on the world. Priests, scribes, and court intellectuals tend to favor orderly explanations. They think in categories, functions, correspondences, and hierarchies. They map the cosmos like a temple: every being has a place, every force has a role, every ritual has a timing, every god has a domain. When such minds write about origins, they often describe creation as the establishment of a system—because system is how they understand reality. The procedural tone is therefore an intellectual style, not evidence of literal engineering.

A third answer, more intimate, is that humans themselves are makers. We understand the world through craft. Long before we can explain geology or genetics, we can explain how a pot is shaped, how a tool is knapped, how a house is built, how bread is made. Craft is the oldest human epistemology: the way we know what causes what is by doing. Myth leverages this epistemology. It explains the biggest questions using the most familiar model: making. The gods make humans the way humans make things. They mix ingredients. They shape forms. They breathe life. The story feels like engineering because craft is the precursor to engineering. It is the natural metaphor for causation in a world without modern science.

This craft model becomes even more pronounced when a myth is trying to explain not only that humans exist but why humans have specific social roles. Why do some people work the fields? Why do some rule? Why are some bound to temple labor? Why are some chosen as priests? In societies with strong labor stratification, creation myths often contain explanations for social structure. Humans were made for certain purposes. The gods assigned tasks. The cosmos itself is built on division of labor. In such a worldview, creation is inseparable from economy. The story of origins becomes the story of why work exists and why hierarchy is legitimate. That is why these myths can feel like engineering: they are not just explaining life, they are explaining the social machine.

The Anunnaki problem, then, begins to take a clearer shape. It is not that the word "engineering" appears in ancient texts. It is that certain myths describe creation as purposeful design tied to labor and governance. Modern readers, living in an age of biotechnology and systems engineering, naturally hear echoes of their own concepts in this language. They see "mixing" and think genetics. They see "design" and think engineering. They see "assigning functions" and think programming. The text becomes a mirror. The danger is mistaking the mirror for a window.

Yet it would be equally careless to deny the legitimacy of the reaction. The reaction exists because the myths really do contain a procedural sensibility. They really do describe the world as built. And they often do so with striking specificity: not merely "the gods created," but "the gods organized," "the gods apportioned," "the gods established," "the gods crafted," "the gods appointed." This specificity matters because it suggests that the writers were thinking about origin through the lens of institution and craft. Whether or not the account is literal, it is revealing of how early complex societies conceptualized power. They imagined the cosmos as a managed project because their own world was becoming a managed project.

In the broader architecture of this volume, this chapter sits at a critical intersection. Earlier chapters established that human ancestry is composite and that myths often encode boundary-crossing and transmission. The Watcher motif framed knowledge as given. Now the Anunnaki problem pushes further: it frames humanity itself as made with purpose. That framing is provocative because it touches the deepest human anxiety about agency. Are we self-made, or are we designed? Myth answers in a way that is both flattering and humiliating: humans are important enough to be made deliberately, but also subordinate enough to be made for service.

The next subsection goes straight into that uncomfortable implication. Many of these narratives do not portray humans as the crown of creation in a modern humanist sense. They portray humans as workers—made to relieve divine labor, to serve, to maintain the system, or, in some variants, to rule on behalf of the gods. This is where the “engineering” tone becomes explicitly political. Creation is not just an act of artistry; it is the establishment of an economy. And when the origin story is an economy, the line between myth and ideology becomes difficult to separate.

One of the most unsettling features of certain ancient creation narratives is not that they describe gods making humans, but why they say humans were made. The purpose is often practical. Humans are created because work must be done. The world-system requires maintenance—fields irrigated, temples supplied, offerings prepared, order upheld. In these stories, the gods are not distant abstract creators; they are rulers presiding over an economy. They need labor. They face conflict over labor. They establish solutions to labor. Humanity becomes that solution. To modern sensibilities, this can sound cold, even degrading: humans as a workforce designed to relieve divine burden. But the logic is revealing. It reflects a society in which labor is central, hierarchy is assumed, and purpose is defined in terms of function.

This is where the “engineering-like” tone becomes sharper. A system has a problem—too much work, too much disorder, too many tasks that must be performed to keep reality stable. A decision is made. A new component is introduced into the system: human beings. Their raison d’être is embedded in the system’s requirements. In mythic language, this is theology. In social language, it is the naturalization of labor stratification. The story tells people what they are for. It makes work sacred by making it primordial. If humans were created to serve the gods, then service is not merely economic compulsion; it is cosmic alignment.

Yet the narratives often leave space for another purpose alongside service: rule. Humans are sometimes portrayed as servants of the gods and simultaneously as the rulers of the earth. This apparent contradiction is central to the chapter's argument. It shows how creation myths can encode a hierarchical chain: the gods rule the cosmos, the king rules humanity, humanity rules animals and land. In such a chain, to "serve" and to "rule" are not opposites; they are adjacent positions in the same structure. The king can rule humans while serving the gods. The farmer can serve the king while ruling the field. The structure is fractal: authority at every level, obedience at every level. Creation myths that sound like labor contracts are therefore also political blueprints.

To see how this blueprint functions, imagine the lived reality of early agrarian states. These societies depended on coordinated labor—irrigation works maintained, harvests stored, canals cleared, monuments built, temple economies supplied. The state and temple were not ornamental institutions; they were engines that organized survival. In such a world, the question "why do we work?" is not philosophical; it is existential. A myth that answers "because that is what humans were made for" is not merely a story. It is a stabilizing ideology. It turns the burden of labor into a metaphysical necessity. It makes resistance feel like rebellion against the cosmos.

But the same myths often include hints of tension. The gods themselves may resent labor. They may complain. They may revolt against the burden. They may fight over who must perform it. This is significant because it reveals that labor is understood as a burden even at the divine level. Work is not romanticized as pure virtue; it is treated as obligation. Humans inherit that obligation. The myth is therefore honest in a way that later moralizing systems sometimes are not: work is heavy. The world does not run on its own. Maintenance is eternal. Someone must carry it.

This is why the creation of humans can be framed as an economic invention. In myth, humanity appears as an instrument to solve a labor dispute among gods. In historical reality, humans were increasingly organized into labor systems that supported elites and institutions. The myth naturalizes that organization by projecting it upward: if even the gods required a labor solution, then human labor systems are not invented by kings; they are echoes of cosmic administration. The temple becomes the earthly version of the divine household. The king becomes the manager of the gods' estate. Humanity becomes the workforce that makes the estate function.

Yet there is another layer. The claim that humans were made to work is also a way of making human existence meaningful in a hard world. Work, suffering, taxation, obligation—these can feel arbitrary without an origin story. A myth that embeds them in creation provides existential coherence. It tells people: your toil is not pointless; it is part of the world's design. That coherence can be comforting even when it is oppressive. It explains why creation myths that emphasize service can be embraced by the very people they constrain. Humans often prefer a meaningful burden to a meaningless one. Myth supplies meaning.

The "serve or rule" tension becomes even clearer when we look at how such societies treated kingship. Kings are often positioned as intermediaries—humans, yet not ordinary humans. They represent the gods to the people and the people to the gods. They enforce labor obligations, administer law, and oversee construction, but they also perform rituals that sustain cosmic order. In such a system, the king's authority depends on the claim that humans are part of a cosmic labor structure. The king is the highest human laborer, the one whose work is governance and ritual. He is servant in the divine hierarchy and ruler in the human hierarchy. This dual role is not accidental; it is built into the mythic model of creation.

This is why many myths pair the creation of humans with the establishment of kingship and temple order. Humans are not created into a neutral wilderness. They are created into an administrative cosmos. The world is already a city in divine form, with hierarchies, roles, tasks, and obligations. To be human is to be assigned. This can sound bleak, but it is also a realistic reflection of how early states experienced themselves. People did not live as autonomous individuals. They lived as members of households, kin groups, temple estates, and labor rosters. The myth reflects that embeddedness by making it primordial.

However, it is important not to treat these narratives as uniform or to flatten them into one cynical reading. "Humans made to serve" can also be framed as humans made to participate in order. Service can be presented not as slavery but as alignment with the sacred. Work becomes a form of ritual. Agriculture becomes cooperation with cosmic cycles. Building becomes devotion. The same system that constrains can also inspire. Myth works precisely because it can hold both: burden and meaning, obligation and participation, oppression and purpose.

For the purposes of this volume, the most important consequence of this theme is how it reframes the question of human origins in the language of design. When a myth says humans were made for a function, it implicitly invites modern readers to think in terms of engineering. Function implies design. Design implies intention. Intention implies a designer. And suddenly, the myth feels like a coded account of creation as technological project. But we can see now how this feeling arises naturally from social conditions. A society built on organized labor and hierarchical administration will imagine origins as organized labor and hierarchical administration. The gods will sound like managers because managers have become the dominant human image of power.

This prepares the ground for the next subsection, where the language becomes even more provocative: blood, essence, and what can sound like genetic vocabulary. Many creation narratives explain human making through the mixing of substances—clay, divine essence, blood, breath, name, spirit. In modern minds, "blood" and "essence" can easily be heard as biology. But the ancient meanings are layered: ritual, lineage, legitimacy, and metaphysical substance. The task will be to explore why such language appears, what it meant in its own context, and how it can be read as symbolic without dismissing the intuition that these myths are unusually concerned with the mechanics of making a human.

If the idea that humans were made for work gives certain creation myths their political edge, the language of blood and essence gives them their visceral force. These narratives do not describe humanity as appearing from nothing. They describe humanity as composed. Ingredients are gathered. Substances are mixed. A form is shaped. Life is activated. Within this process, blood often appears not merely as a symbol of vitality but as a medium of transfer—something that can carry identity, power, and function from one being into another. To a modern reader living in the shadow of genetics, this can feel startlingly close to biological thinking. Blood sounds like heredity. Essence sounds like inherited code. Mixing sounds like recombination. But to handle this responsibly, we must step into the older worldview where blood is not simply a fluid but a metaphysical substance that bridges body, ancestry, and sacred authority.

In ancient thought across many cultures, blood is the most obvious sign of life. It flows when the body is wounded. It marks death when it leaves. It is connected to birth, to fertility, to lineage. It is therefore a natural carrier of meanings that modern minds distribute across separate categories: biology, kinship, and spirituality. When ancient myths speak of blood as a creative ingredient, they are saying that life is transferred through life. They are also saying that identity can be transmitted materially. A being made with divine blood carries divine quality. A being made with royal blood carries royal legitimacy. Blood is the tangible proof that something of the parent remains in the child.

This is why blood becomes an ideal mythic mechanism for explaining why humans are simultaneously earthbound and god-touched. Many myths frame humanity as a composite: shaped from clay or earth—matter, mortality, the soil—but infused with something that originates beyond the soil—breath, spirit, name, or divine substance. In such stories, the human is a hybrid of realms. The clay anchors us to the ground, to hunger, to decay. The divine substance elevates us into consciousness, speech, and purpose. Whether the substance is called blood, breath, or essence, the structure is the same: the human is built as a mixture.

This mixture is one reason modern readers hear "genetic language." Modern biology explains the human as a coded mixture: inherited material carried in cells, patterned and activated, producing traits and capacities. When an ancient narrative describes making humans by mixing substances, assigning functions, and animating the result, it sounds—at a high level—like a pre-scientific intuition that the human is assembled rather than spontaneously given. But the key is to keep the level clear. Ancient texts are not describing DNA. They are describing a metaphysics of composition. The resemblance is structural, not technical.

The concept of essence is central here. Essence in mythic language is what makes a being what it is. It is not simply personality; it is ontological content. If a god's essence is mixed into humanity, then humanity carries a fraction of the god's being. That fraction can explain everything from intelligence to moral agency to the capacity for ritual. It can also explain why humans can serve as intermediaries: if you share essence with the divine, you can communicate with the divine. In this sense, "essence" is a metaphysical substitute for what modern readers might call "information." It is the invisible content that shapes the visible outcome.

Blood is often the most dramatic expression of essence because it is both material and sacred. It is literally inside you, and it is ritually potent. Blood can be offered. Blood can purify. Blood can stain. Blood can bind oaths. Blood can mark kinship. So when a creation story says humanity contains divine blood, it is doing several things at once. It is explaining why humans are alive. It is explaining why humans belong to the gods' world. It is explaining why lineage matters. And it is giving an origin to social hierarchy: if some humans are closer to divine blood—through kingship, priesthood, or chosen descent—then their authority feels cosmically justified.

This is where "blood" shifts from vitality to politics. In many societies, the idea of "royal blood" is not metaphorical. It is treated as an actual quality that differentiates rulers from ordinary people. The claim that the king has divine blood is the claim that the king carries more of the sacred substance that made humanity. In that framework, kingship becomes not just office but concentration: a greater dose of essence. When a myth says the gods mixed their blood into humans, it is laying the foundation for later claims that certain lineages preserve that mixture more purely or more powerfully. The creation story becomes the first argument for aristocracy.

At the same time, blood language can be used to express moral ambivalence. If humans are made from divine blood, why are humans violent, deceitful, rebellious? Some myths resolve this by introducing conflict into the ingredients: humanity contains both a sacred element and a flawed element; or humanity is made from the blood of a defeated being; or humanity inherits the temperament of the one whose blood became the seed. In such stories, blood is not only power but fate. It explains why humans possess greatness and corruption simultaneously. Again, modern readers may hear echoes of heredity—traits inherited from ancestral stock. The myth is not wrong to intuit that inheritance shapes nature. It simply expresses the intuition through sacred substance rather than through genes.

The language of breath and spirit also intersects with this "proto-genetic" feel, though in a different way. Breath is often portrayed as the animating force that turns shaped matter into living being. Breath is invisible, yet it moves. It enters and leaves. It is shared between beings in speech. In many traditions, breath is closely linked to word and mind. To give breath is to give consciousness. When a creator gives breath, it resembles the activation of a system: the body is assembled, then powered on. The procedural sequence—shape, mix, animate—again triggers the modern mind's engineering lens. But for ancient thinkers, breath is simply the most immediate sign of life and spirit.

All of this can lead to a disciplined reading of "genetic language" that does not collapse into literalism. The myths use blood and essence because these are the strongest available metaphors for continuity, identity, and transfer. They are the pre-scientific vocabulary for what modern science would describe in different terms: inheritance, information, and embodied function. When we say these myths "sound like genetics," we should mean that they treat life as something composed of transferable substance and that they place extraordinary emphasis on the mechanism of making. That emphasis is what distinguishes them from more purely poetic origin tales.

Yet the other side of discipline is to acknowledge what these myths do not provide. They do not specify how traits are encoded, how reproduction works at a cellular level, or how variation arises. They are not predictive models. They do not behave like science. Their "mechanics" are symbolic mechanics—true within the story world and socially powerful within the culture, but not empirically testable as biological claims. To treat blood and essence language as literal genetics would be to mistake metaphor for mechanism.

Still, within the arc of this chapter, the point stands: these narratives present a strikingly material, procedural view of human origin. Humans are mixed. Humans are infused. Humans are assigned roles. Humans are made to serve a system. This combination is what keeps provoking modern interpretations. It suggests that the creators—whether priestly authors or the traditions behind them—were intensely concerned with explaining why humans exist within hierarchy and why hierarchy feels written into our nature.

The next subsection asks a different kind of question: why do these myths appear suddenly in the historical record? Not as isolated stories, but as fully formed traditions with a confident voice—complex cosmologies, named divine councils, elaborate creation mechanisms, detailed accounts of why humans were made. What social and historical conditions produce this kind of myth-making, and why does it cluster around the emergence of early states? This question moves us from the content of the myths to the moment of their crystallization.

From the viewpoint of a modern reader encountering these narratives on the page, they can feel as if they appear out of nowhere. A tradition surfaces with a confident cosmology: councils of gods, administrative hierarchies in heaven, procedural creation sequences, humans designed for roles, blood and essence mixed into clay, a world organized like a state. Even when we know intellectually that no myth is born in a vacuum, the literature itself can give the impression of abrupt emergence. One era seems to have scattered local spirits and episodic sacred tales; the next era has a full cosmic government. The question is not whether these myths truly appeared suddenly, but why they crystallize in forms that feel sudden—why they arrive in history as finished architecture rather than as visible gradual construction.

The first and most grounded answer is that written evidence is not the same as cultural origin. Oral traditions can exist for long periods without leaving direct traces. When writing enters a society and becomes institutionalized—especially under temple and palace systems—myth suddenly becomes visible. The appearance of a myth in writing can therefore be an artifact of preservation, not an artifact of invention. The story may have been told for generations; we simply did not have a durable record until a particular point. This is a sober explanation, but it is crucial, because it reminds us that "suddenness" in the record is often a property of archaeology and archives rather than a property of actual cultural change.

Yet this explanation alone does not fully account for the distinctive character of these myths. Many early written creation narratives do not feel like casual folk tales that happened to be written down. They feel like formal theology. They feel curated, systematized, and aligned with institutional needs. This suggests a second answer: early state formation changes the kind of myth a society produces. When societies develop complex administration—temple economies, kingship ideology, legal codes, labor organization—they require a cosmology that can legitimize and stabilize the new order. A simple origin tale about a trickster animal may not be sufficient to justify why a king can command thousands or why a temple can own land and labor. Institutions demand narratives that match their scale. The cosmos must become a state because the state must become cosmic.

This is why the myths can feel suddenly "engineered." They are responding to an environment where engineering—system-building, labor coordination, resource management—is now central to survival. In that environment, the most persuasive sacred stories are those that explain the world as a managed system. Creation narratives shift from episodic wonder to administrative logic. They explain why hierarchy exists by placing hierarchy at the beginning. They explain why work exists by making work primordial. They explain why law exists by making law divine. When the state becomes the dominant shape of life, myth becomes the dominant justification of the state.

A third answer is that these myths often emerge through the interaction of multiple traditions. Early states are not closed worlds. They are hubs of trade and conquest, where languages, gods, rituals, and stories mix. When traditions collide, they do not simply coexist; they recombine. Gods are merged, ranked, and given new functions. Local spirits are incorporated into larger pantheons. Older myths are reinterpreted through new political realities. Over time, a complex composite cosmology forms. When that cosmology is written down, it can appear suddenly—because what we see on the page is the result of long syncretic processes that occurred before the moment of inscription.

This recombination effect can produce exactly the kind of myth that modern readers find most provocative: a narrative that feels simultaneously ancient and strangely rational, poetic and procedural. It is ancient because it carries deep archaic motifs—creation from earth, life through breath, divine councils, sacred blood. It is procedural because it has been reorganized by literate institutions that think in systems. The scribal mind, trained to list, categorize, assign, and administer, reshapes myth into architecture. The result is a creation story that reads like a plan.

A fourth answer is that periods of stress and transition often generate intensified myth-making. When societies undergo rapid change—urbanization, population growth, climate instability, political upheaval, war—people seek explanations that restore meaning and coherence. The question "why are we here?" becomes urgent when the world feels unstable. In such contexts, myths can crystallize into more formal, authoritative forms. A priestly class may gather and standardize traditions to preserve identity. A ruling dynasty may sponsor the writing of sacred narratives to legitimate itself. A community may define its gods more sharply in response to rivals. This pressure can cause theological systems to appear abruptly in the historical record because the moment demands consolidation.

This consolidation is especially likely when writing is tied to institutions. Writing is expensive. It requires training, materials, and stable patronage. When a temple or palace invests in writing, it often invests in authoritative texts that support its legitimacy. The myth that is written down is therefore not random; it is selected. It is curated. It is likely to be the version that best serves the new social order. This selection effect contributes to the impression of suddenness. What survives is not the messy diversity of earlier storytelling but the polished canon of institutional theology.

A fifth answer is that these myths are, in part, retrospective. They are not only describing origins; they are explaining the present by projecting it backward. If a society has kings, it will imagine the gods as kings. If a society has bureaucracy, it will imagine the cosmos as bureaucracy. If a society has labor obligations, it will imagine creation as labor allocation. This is not deception; it is the natural way humans think. We understand the unknown by analogy with the known. The more complex the society becomes, the more complex the analogies become. The result is a creation myth that feels suddenly modern in its system-thinking, because it is written by a society that has already developed system-thinking.

In the disciplined framework of this book, this point is critical. The "engineering" feel of certain creation narratives does not automatically imply that they preserve literal engineering events. It can be explained by the emergence of early states, writing, and institutional theology. The myths appear suddenly because the record suddenly preserves them, and because the institutions suddenly needed them in their most formal, systematized form.

Yet the sensation that these myths arrive as finished architecture still matters. It matters because it highlights how quickly human societies can reorganize their own meaning structures when conditions change. It also matters because it creates the interpretive tension that drives this chapter: if these myths are so procedural, are they purely symbolic? Are they ideological? Or do they preserve some memory of extraordinary events—contact, catastrophe, or discontinuities that were experienced as interventions?

That tension leads directly into the next subsection's question. If we refuse both extremes—pure literalism and pure dismissal—where do we place these narratives? Are they symbolic, literal, or something in between? The answer will require careful boundaries: acknowledging what mainstream evidence supports, acknowledging what it does not, and exploring why myths that sound like design can still be treated as meaningful archives of how ancient peoples understood power, labor, and origin.

The interpretive pressure surrounding "engineering-sounding" creation myths always pushes toward extremes. One extreme reads the narratives as literal reports: the gods were real beings, the procedures describe actual technical acts, humanity was designed in a biological sense, and the texts preserve a hidden history. The other extreme reads the narratives as pure symbolism or propaganda: poetic metaphors with no relationship to historical events beyond the social needs of the culture. Both extremes offer a kind of relief. Literalism relieves the discomfort of ambiguity by turning myth into secret history. Dismissal relieves the discomfort by turning myth into harmless story. But this chapter's central claim is that the most defensible reading sits between. These myths are symbolic in form, but they may preserve real structural memories: memories of social organization, of contact and transmission, of the felt shock of civilization, and of the ways early states understood human purpose.

To frame this responsibly, begin with what "literal" would actually mean in this context. Literal would mean that the narrative's agents—named gods and councils—correspond directly to physical beings, and that the described acts correspond directly to empirical procedures. That is a high bar. These myths were not written as technical manuals. They are embedded in religious language, ritual worldview, and symbolic cosmology. They often contain obvious mythic features—cosmic battles, personified forces, divine speech shaping reality—that cannot be mapped onto empirical history without distortion. If we treat them as literal, we risk importing modern fantasies into ancient texts.

At the same time, "symbolic" cannot mean "there is nothing here." Symbols are not empty. Symbolic narratives can encode real relationships and real experiences. A story can be metaphoric and still preserve memory. It can preserve memory of how a society was organized, what it feared, what it valued, what it perceived as disruptive. It can preserve memory of migration and contact in transformed form. It can preserve memory of catastrophe in moralized form. It can preserve memory of the emergence of hierarchy as a cosmic inevitability rather than a human choice. When these myths describe creation as design and labor allocation, they are almost certainly preserving a real memory: that early civilization experienced itself as organized labor under divine authority.

So the "in between" reading begins by treating the myths as truth-bearing, but not in the narrow scientific sense. Their truths are not laboratory truths; they are structural truths. The gods can be read as personifications of institutions—temple, palace, priesthood, kingship, labor bureaucracy—projected into the heavens. The divine council can be read as a mythic mirror of elite councils and political negotiation. The creation of humans for work can be read as a sacralization of labor exploitation and agricultural obligation. The mixing of blood and clay can be read as a mythic way of expressing human duality: animal mortality combined with sacred legitimacy and consciousness. None of these readings requires literal gods. They require only that myth reflects society.

Yet the “in between” reading can also accommodate another possibility, without overcommitting to it: that some myths may preserve echoes of real discontinuities—moments when new social orders, new technologies, or new knowledge systems appeared abruptly to those who experienced them. A population that suddenly encounters outsiders with metallurgy may remember that encounter as divine teaching. A society that transitions rapidly into urbanism may remember that transition as a designed project. A culture that suffers collapse may remember the loss as punishment and banishment. These are not extraordinary claims; they are plausible dynamics of memory. They suggest that myth can be a distorted record of real events, not because it intends to record them, but because humans naturally mythologize turning points.

This view makes sense of the strongest feature of the Anunnaki-type narratives: their obsession with function. Humans are made for a purpose. The cosmos is organized by task assignment. The language is managerial. That is not how a small egalitarian band naturally talks about life. It is how a state talks about life. The myth is therefore telling us something literal about the culture that authored it: that it lived inside a world of organized labor and hierarchical administration, and it used divine language to justify that world. In that sense, the myth is symbolic but also literal as social document. It literally reveals how power thought about itself.

The next interpretive boundary involves the “engineering” sensation itself. Modern readers often project biotechnology onto ancient “mixing” imagery. But mixing substances is one of the oldest human metaphors for making, because craft is how humans understand causation. Pottery, bread, beer, dyeing, metallurgy—all involve combining ingredients and transforming them through heat and process. A creation story that uses mixing language is therefore not necessarily foreshadowing genetics; it may simply be using the strongest available craft metaphor for how life could be made. The “engineering” feel arises because craft logic resembles engineering logic at a high level: inputs, processes, outputs. Ancient myth and modern technology share that basic causal grammar because human cognition prefers it.

But this does not mean the modern reaction is meaningless. It means the reaction is a clue to the myth’s structure. These narratives are unusually concerned with mechanism. They do not simply proclaim that humans exist; they narrate how humans were assembled and why. That mechanism-mindedness is historically significant. It suggests that the culture was grappling with origin in a more analytic, system-focused way than purely animistic tales often do. The question becomes: why did the culture become mechanism-minded? The answer, again, points to institutional complexity—scribal reasoning, administrative order, and the needs of a society that had turned life into managed systems.

So the “something in between” reading can be stated plainly. These myths are not literal scientific accounts. They are symbolic narratives shaped by state societies, crafted by literate elites, and designed to stabilize hierarchy and meaning. Yet they may also preserve deeper memories of transition—memories of contact, innovation clusters, and discontinuities that were experienced as external intervention. In other words, they are not false; they are transformed. They carry truth in the way myths carry truth: through pattern, role, and moral structure.

The difficulty, of course, is that once we accept “in between,” we must draw lines. We cannot let the interpretive freedom become an excuse for inventing hidden histories. The discipline is to distinguish between what the myth clearly reflects—state hierarchy, labor ideology, specialist power—and what the myth merely invites us to speculate about—non-human designers, literal genetic engineering. The myth can invite speculation because its imagery resonates with modern concepts, but invitation is not evidence. A responsible reading keeps speculation labeled as speculation and stays anchored to what can be supported: the social and cognitive functions of the narrative.

This prepares the final subsection, which is the real resolution of the Anunnaki problem. The boundary between myth and memory is not a wall; it is a gradient. Some parts of a myth may be pure symbolism. Some may preserve social history. Some may preserve echoes of real disruptions. The task is to read the gradient without collapsing it. The next subsection will define that boundary more precisely, not by offering a single verdict, but by offering a framework: how to treat these narratives as meaningful archives of the human story while refusing the twin temptations of literalism and dismissal.

The boundary between myth and memory is not a line you can draw once and never revisit. It is a living edge. It shifts depending on the culture, the genre of the text, the social role the narrative served, and the kind of "truth" the story was built to preserve. This is why the Anunnaki problem persists. Certain creation narratives feel as though they are trying to say something mechanistic about human origin—something that resembles design, function, and composition. That resemblance provokes modern literalism, while academic caution provokes modern dismissal. The only way through is to define a framework for reading the boundary: a way to recognize where myth is primarily symbolic, where it likely reflects social memory, and where it may carry echoes of real discontinuities without claiming more than the evidence allows.

The first principle of that framework is genre. Not all mythic texts are the same kind of speech. Some are ritual hymns. Some are theological treatises. Some are royal propaganda. Some are etiological tales—stories written to explain why a ritual exists, why a place is sacred, why a social rule holds. A hymn will amplify grandeur and metaphor; it is designed to evoke awe, not to preserve detail. A courtly narrative will shape origins to defend kingship. A ritual text will encode timing, function, and authority because those are needed for practice. The boundary between myth and memory therefore begins with a simple question: what was this text for? If it was for legitimizing labor and kingship, then its "mechanics" are likely social mechanics more than biological mechanics. If it was for ritual, then its insistence on procedure may reflect the procedural nature of ritual itself.

The second principle is institutional mirroring. Early state societies project their own structures into their cosmologies. Divine councils resemble human councils. Heavenly hierarchies resemble bureaucracies. Creation resembles organized labor because the society lives inside organized labor. This mirroring is not merely a metaphor; it is a cognitive habit. Humans understand the unknown by analogies drawn from the most powerful realities they live under. In early agrarian civilizations, the most powerful reality is the temple-state system—its taxation, its work assignments, its legal obligations, its sacred calendar. When such societies tell origin stories, they do not invent hierarchy; they recognize hierarchy as the natural form of order and then declare that it existed from the beginning. In this way, myth becomes a memory of how power works, even when it speaks as if it were describing the cosmos.

The third principle is compression. Cultural memory compresses long processes into symbolic moments. The invention of a craft across centuries becomes "a teacher arrived." The gradual establishment of social hierarchy becomes "humans were made to serve." The slow formation of a calendar becomes "the gods set the seasons." This compression is not a flaw; it is how narratives survive. Stories that preserve every intermediate step do not travel. Stories that preserve the shape of the transformation do. When a myth sounds like engineering, part of what we are hearing is compression: the long, distributed emergence of civilization has been condensed into a single design act. The myth is preserving the truth that civilization is built, not the details of how it was built.

The fourth principle is the language of substance. Ancient myths use blood, breath, clay, name, and essence as the vocabulary of transfer and identity. Modern readers hear "blood" and think genetics. Ancient writers heard "blood" and thought life-force, lineage, legitimacy, and sacred potency. These languages can overlap structurally because both are trying to explain continuity through material inheritance, but they are not the same language. The boundary here is clear: ancient substance language is metaphysical and social, not molecular. It can preserve the intuition that humans are composites and that lineage matters, but it does not preserve a technical mechanism in the modern sense.

The fifth principle is social function. Myths endure because they do work. They justify kingship. They explain labor. They anchor ritual. They define identity against rivals. They provide moral interpretation of suffering. When a creation narrative insists humans were made to work, the primary "truth" being defended is not a factual statement about biological origin; it is a moral and political statement about obligation. The myth is telling people where they belong in the social order and why that order should feel inevitable. This does not make the myth false; it makes it functional. Its truth is the truth of how a society wanted itself to be perceived—by the gods, by the people, and by history.

And yet, within these principles, there remains room for actual memory. The boundary between myth and memory becomes most porous where myth interacts with real historical discontinuity—migration, collapse, conquest, climate stress, and the abrupt arrival of new knowledge systems. A community that experiences rapid transformation often narrates that transformation as intervention. “They came and taught.” “They arrived and organized.” “They gave laws.” These are mythic forms, but they can preserve the directionality of change: knowledge came from outside; a new order replaced an old one; authority shifted; certain lineages became sacred. Such memories do not survive as accurate chronicles. They survive as motifs and roles. The gods become the outsiders. The divine council becomes the elite coalition. The creation of humans for labor becomes the institutionalization of labor obligation. The myth is not history, but it can contain history in symbolic form.

This is where the Anunnaki problem resolves into a more mature question: what kind of memory are we dealing with? Not “did aliens engineer humans,” but “what discontinuity is this myth trying to explain?” The myth is clearly preoccupied with why humans exist within a labor hierarchy and why certain authorities can claim divine mandate. It is also preoccupied with the mechanics of making—mixing substances, assigning function, activating life. These preoccupations point strongly to early state conditions: temple economies, kingship ideology, craft mastery, and the scribal habit of thinking in systems. That is the most defensible memory layer: a memory of the moment when human life became organized under institutions that felt as absolute as the gods.

From this perspective, the boundary between myth and memory is navigable. We do not have to flatten myth into literal engineering, and we do not have to flatten it into empty metaphor. We can read it as a cultural archive: a record of how early complex societies understood power, labor, and human purpose, expressed in the most authoritative language available—divine creation. We can acknowledge that the myths may preserve echoes of real transitions and contact, while refusing to treat them as technical reports. We can hold ambiguity without letting ambiguity become permission for invention.

This framework also prepares the next chapter's shift toward lost human types and ghost populations. The Anunnaki narratives, when read responsibly, highlight something central to the volume: humanity has repeatedly told stories that frame our origin as composite, functional, and shaped by forces beyond the ordinary. Genetics, in its own language, has also shown that our ancestry is composite and that vanished populations left traces. The book's task is to keep these two archives—myth and genome—in dialogue without forcing them into the same category. Myth can remember structure. Genetics can remember mechanism. The boundary between them is not a wall. It is the space where interpretation must remain disciplined.

CHAPTER 11 — LOST HUMAN TYPES

There is a quiet fact at the center of modern human origins research that rarely makes it into neat diagrams. We know more hominins existed than we can confidently name. The human past is not only populated by the familiar cast—Homo sapiens, Neanderthals, Denisovans, and a handful of earlier species—but by forms that flicker at the edge of definition. They are implied by scattered bones that do not fit cleanly, by fossil fragments that resist classification, by traits that appear in unexpected combinations, and by genetic traces that hint at ancestors we have not yet pinned to a skull. The result is a paradox: the more evidence accumulates, the more the old certainty of a tidy family tree dissolves into a landscape of partial shapes. We can see the shadows of lost human types, but we cannot always draw their faces.

This problem begins with the nature of the fossil record itself. Fossils are not a representative sample of past life. They are a survival bias. Most bodies decay. Most landscapes destroy bone. Only a small fraction of individuals died in conditions that allow preservation, and only a smaller fraction of those remains will ever be found. Even when remains are found, they are often incomplete—a tooth, a jaw fragment, a partial cranium, a few postcranial bones. Yet classification requires patterns. To define a "type," scientists need enough material to distinguish variation within a population from differences between populations. Without that depth, any label risks being an illusion built on a handful of bones.

There is also the problem of variation within hominin populations. Modern humans vary dramatically in skull shape, height, robustness, and facial structure. If future researchers had only a few modern skulls from widely separated regions, they could easily mistake them for different species. Now multiply that complexity across hundreds of thousands of years, across shifting climates, isolated refuges, and repeated contact between groups. Ancient hominins likely varied even more, not less. Some were adapted to cold, some to heat, some to high altitudes, some to forests, some to open grasslands. Their bodies would have responded to diet, activity, and disease. If we find a robust individual and a gracile individual from different regions and times, are we seeing different species or different expressions of the same lineage? In many cases, the honest answer is: we cannot be sure.

This uncertainty is compounded by the fact that "species" is a human category imposed on a continuum. In deep time, boundaries between lineages may have been porous. Populations could diverge and still occasionally interbreed. They could maintain distinct traits and still exchange genes at the margins. They could be "different enough" for some purposes and "similar enough" for others. If that is true—and much of the evidence from later hominin interactions suggests it is—then our desire to draw crisp lines around every fossil may be misplaced. Some of the lost human types may not have been neat species at all. They may have been regional populations, semi-isolated groups, or transitional forms that existed for thousands of years without ever becoming a separate branch that lasted.

This is where the phrase "we know existed but can't define" becomes precise. We know, for example, that the world was once inhabited by multiple human lineages simultaneously. We know that some of those lineages left genetic traces in later humans. We also know that there are fossils that do not fit neatly into the well-known categories. But to define a hominin as a distinct type requires enough evidence to establish its consistent traits, its geographic and temporal range, and its relationship to other lineages. When the evidence is sparse, a "type" becomes a hypothesis, not a settled identity.

The past decade has made this problem more visible rather than less. On the one hand, new finds have expanded the cast of known hominins and challenged old assumptions about where and when certain traits emerged. On the other hand, these same finds have highlighted how much we do not have. A single jawbone can shift the timing of a lineage's presence in a region. A handful of teeth can suggest a population that lived alongside others. A partial skull can show a mosaic of traits that refuses our categories. Each discovery illuminates a larger darkness. Instead of a clean narrative, we get a map of islands with vast unknown seas between them.

In popular storytelling, this uncertainty is often treated as a weakness, but scientifically it is a sign of maturity. Early fields often begin with confident simplification. They build a coherent model from limited data. As data grows, the model becomes more complex and less certain. Human origins research is in that phase. We are moving from a story that could be taught as a line to a story that must be taught as a network. In that network, "lost human types" are not a speculative flourish. They are a natural expectation. If multiple lineages existed, if the fossil record is incomplete, and if gene flow occurred, then we should expect populations that we can infer but not yet define.

This expectation is strengthened by the way fossils tend to cluster. We have rich windows—certain caves, certain regions, certain time periods—where preservation and excavation have produced abundant material. But between these windows are long stretches with little or nothing. A lineage could exist for tens of thousands of years in a region that leaves no accessible fossils, and we might never see it directly.

We might only detect it indirectly through its impact on other lineages—through a trait that appears unexpectedly, or through a genetic signal that cannot be matched to known fossils. The unknown is not empty; it is simply unpreserved.

This has important implications for how we think about human identity. If there were many hominin populations that we cannot define, then our current categories are provisional. "Neanderthal" and "Denisovan" may themselves be umbrella labels for diversity we have not fully mapped. The term "Denisovan," in particular, reminds us how thin a label can be. An entire lineage can be recognized largely through genetic signals and a small number of fossils. If that is possible, then it is also possible that other lineages existed—lineages we have not yet named because the fossils are missing or the genetic traces have not been interpreted clearly.

Yet the most disciplined way to frame this is not to inflate it into mystery for its own sake. The existence of undefinable hominins is not evidence of secret civilizations or non-human engineers. It is evidence of the normal conditions of deep time: incomplete preservation, complex variation, and porous boundaries. The "missing chapters" are missing because the archive is physically broken, not because someone locked it away. But the fact that the archive is broken matters deeply. It means we should be cautious about narratives that claim to have the whole story. It means our picture of the human past is necessarily partial. And it means that future discoveries—fossils, improved genetic analysis, better dating—may add new characters to the cast or merge characters we currently treat as separate.

This is where the chapter's logic becomes a pivot for the volume. If lost human types existed and cannot yet be defined, then the human story is more crowded than our textbooks once suggested. It becomes easier to understand why myths speak of many kinds of beings—mighty ones, teachers, others, those who came before. Those myths are not literal hominin catalogs, but they are cultural expressions of a correct intuition: the past held more than one kind of "human," and the present has inherited only fragments.

The next subsection moves from fossils to genetics. If fossils are the broken bones of deep time, genomes are the hidden archive inside living bodies. Some populations are known almost entirely from genetic signals—"ghost" ancestries that appear in patterns of DNA but cannot yet be tied to a defined fossil type. This is where the modern search for lost human types becomes both more powerful and more humbling: the body itself remembers what the earth did not preserve.

If fossils are the visible ruins of deep time, genetics is the buried archive—the record that survived not in stone but in living bodies. This archive has changed the human origins story in a way that is both clarifying and unsettling. It has clarified that interbreeding occurred, that ancestry is braided, and that some adaptive traits were inherited from archaic lineages. But it has also unsettled the old picture by revealing something we did not expect to find so clearly: traces of populations for which we have no clear fossil identity. These are the so-called ghost populations—ancestral groups implied by genetic patterns, present as signals in genomes, yet absent as defined species in our museums. In other words, we can detect them without being able to name them.

The reason this is possible is that genomes preserve relationships in statistical form. Every person carries a mosaic of inherited segments. Over generations, these segments are shuffled, broken, and recombined. Yet they still carry signatures—shared patterns, unusual divergences, stretches of DNA that look more similar to one lineage than another.

By comparing genomes across modern populations and, where possible, across ancient DNA, researchers can infer that certain segments came from a source that is not fully accounted for by known lineages. The inference is not mystical. It is simply the recognition that the observed variation is too structured to be explained by the populations we have already identified. Something else must have contributed.

This is the crucial point: ghost populations are not a claim that "unknown beings" existed in an imaginative sense. They are a claim that our known categories are incomplete. A genetic signal can indicate admixture from a lineage that diverged earlier than the lineages we have fossil names for, or from a regional population that remained distinct long enough to leave a recognizable imprint. The "ghost" is not supernatural. It is a missing label attached to a real demographic event: contact between groups we have not yet fully mapped.

The existence of these ghost signals is not surprising when we remember how incomplete the fossil record is. It would be more surprising if genetics did not reveal missing contributors. Fossils are rare, but mating is common. A population could contribute modest ancestry to others and then vanish, leaving no fossils we can confidently assign to it. Or its fossils may exist but be misclassified. Or they may lie in regions where preservation is poor and excavation limited. Genetics does not require the body to fossilize to preserve a trace. It requires only that the population interbred and that some descendants survived. In that sense, genes can remember what geology forgets.

Yet genes also have their own limitations. Genetic signals fade over time. Recombination breaks ancestral segments into smaller pieces. Drift and selection can erase or distort patterns. If admixture events are ancient enough, the remaining signal can be difficult to interpret. Models must be constructed, assumptions tested, and alternative explanations ruled out. Ghost populations, therefore, are not direct sightings. They are inferences. They are the shapes that emerge when multiple lines of genetic evidence point toward an extra contributor. Their reality is supported by pattern coherence, but their identity remains unresolved.

This unresolved identity has an important consequence for the book's theme. The human story is not merely a story of known actors. It is a story of interaction among many populations, some of which have been erased as distinct groups. The genome is telling us: you have ancestors you cannot name. You are made of lineages that disappeared. Your ancestry includes chapters that left no monuments. This is a powerful scientific counterpart to the mythic intuition of lost peoples and hidden contributors. Again, myth and science are not the same archive, but here they echo each other in a disciplined way: both imply that the past was more populated and more complex than the present remembers.

Ghost populations also complicate any attempt to treat "Neanderthal" or "Denisovan" as singular entities. Even within these categories, genetic diversity suggests that what we call a lineage may have included multiple regional branches. Denisovan ancestry, for example, appears in varying degrees across different parts of Asia and Oceania, suggesting that the Denisovan world was not a single small group but a broader, diverse presence. If one named lineage can be internally diverse, then ghost lineages could also represent clusters rather than single types. The "ghost" may not be one population; it may be a family of populations we cannot yet separate.

The most transformative aspect of this is how it shifts the meaning of extinction. We tend to imagine extinction as disappearance without trace. Genetics reveals that extinction can occur as absorption. A population can disappear as a distinct group while surviving as ancestry within others. In that scenario, the population is both gone and present. Its bones may never be found, but its genetic contributions persist. Ghost populations are therefore the scientific face of a haunting idea: some extinct humans live on inside us without a name.

This raises a subtle point about how we define identity. If a population's physical remains are lost and its culture is unknown, and it survives only as genetic segments scattered across descendants, what does it mean to say it "existed"? It existed biologically, certainly. It existed demographically, as a group that was distinct enough to leave a signal. But beyond that, we cannot say much. We cannot reconstruct its language, its rituals, its self-understanding, or its daily life from genetic fragments alone. Genetics can reveal that a population contributed to modern ancestry; it cannot, by itself, recover the lived reality of that population. This is where the scientific humility becomes essential. The genome gives us the outline of contact and mixture, but it does not give us the story.

Still, even outline is extraordinary. Before genomic tools, the absence of fossils could easily be mistaken for absence of populations. Now, absence of fossils is recognized as a limitation of preservation, not a proof of nonexistence. This reframes the human origins field. Instead of asking only, "what fossils do we have?" we also ask, "what populations must have existed to produce the genetic patterns we see?" In doing so, we turn living humans into witnesses of deep time. Our bodies become the archive.

The presence of ghost populations also changes how we interpret migration and adaptation. If modern populations carry genetic contributions from unknown groups, those contributions may have had effects—especially if certain variants provided adaptive advantages. Some inherited segments may relate to immune function, environmental tolerance, or metabolic traits. This does not mean ghost ancestry equals superior ancestry. It means that admixture is not just historical; it is functional. It can shape survival. In that sense, ghost populations are not only missing chapters; they are active pages still being read by the body.

But the existence of ghost signals also highlights how easily narratives can outrun evidence. It is tempting to fill the unknown with dramatic explanations. Yet the most defensible position is simpler: ghost populations are expected under conditions of incomplete fossils and frequent contact. They are not proof of hidden civilizations. They are proof that deep time was crowded and that extinction is often absorption. The mystery is not that unknown populations existed; the mystery is how many, where, and how they interacted.

This sets up the next subsection's central question: why is the record fragmentary in the first place? Not only in fossils, but in genetics, archaeology, and cultural memory. Why do we have such uneven access to the past, and why do certain regions and time periods dominate our evidence while others remain silent? The answer will bring together geology, climate, preservation bias, and the simple fact that humans lived through enormous environmental changes that wiped away traces. The ghost populations are not anomalies. They are the natural product of a world that does not preserve its history evenly.

The human past did not vanish because it was insignificant. It vanished because the planet is an aggressive editor. Over tens of thousands and hundreds of thousands of years, landscapes move, rivers shift, coastlines drown, deserts expand, forests swallow sites, and caves collapse. Bone decays. Wood disappears. Fire destroys. Water dissolves.

Even stone can be buried beyond reach or ground into dust. When we say the record is fragmentary, we are not merely admitting a gap in knowledge. We are describing a fundamental condition of deep history: most of what happened left no durable trace that survives to the present in a form we can recover.

The first reason is preservation bias. Organic materials—skin, hair, textiles, plant fibers, wooden tools—almost never survive on deep timescales except under rare conditions. Even bone, which can last far longer than soft tissue, is vulnerable to acidity, moisture, microbial activity, and physical disturbance. Many regions of the world are hostile to preservation. Tropical environments, for example, can be brutal for ancient DNA and skeletal remains. Acidic soils dissolve bone. High humidity accelerates decay. In such regions, entire populations could have lived and died for millennia leaving little for archaeology to find. This means absence of evidence is often just absence of preservation.

The second reason is geological and climatic transformation. Human history unfolded during a period of dramatic environmental volatility. Glacial cycles repeatedly expanded and retreated, reshaping habitats, altering sea levels, and shifting the distribution of resources. A camp that sat near a river during one climate phase may now be buried under meters of sediment or destroyed by later erosion. A coastal settlement from a lower sea-level period may now lie underwater, inaccessible or eroded. A valley that once served as a migration corridor may now be a desert. A region that once preserved bone may now be too warm and wet for DNA to survive. The planet did not simply keep the record and hide it; in many cases, it actively erased it.

Sea level is one of the most consequential erasers of all. During glacial maxima, sea levels were much lower than today, exposing vast coastal plains and land bridges that are now submerged. Humans and other hominins often preferred coastlines and river mouths because they offer rich resources. If this is true—and evidence suggests it often is—then many ancient sites may now be underwater. Underwater archaeology is difficult, expensive, and still limited compared to land excavation. Even when underwater sites are found, they can be heavily disturbed by currents and sediment shifts. This means large portions of human coastal history may be literally drowned, especially in regions where early dispersals occurred.

The third reason is that human populations were often small, mobile, and lightly footprinted. For most of our existence, humans did not build cities. They did not leave thick layers of architecture and trash deposits. They moved seasonally, used perishable materials, and left traces that could be scattered and thin. A temporary camp can disappear within a few centuries under the right conditions, leaving only a few stone tools that can be moved by water or buried. When the archaeological signal is small to begin with, it does not take much to erase it. Our expectations, shaped by later civilizations with massive construction, can mislead us. Deep prehistory may look silent not because it was empty, but because it was subtle.

The fourth reason is that discovery is uneven. Archaeology is not a neutral global survey. It is shaped by politics, funding, accessibility, and historical focus. Some regions have been excavated intensively for generations; others have barely been touched. Some countries have robust heritage institutions; others have been disrupted by conflict and instability. Some landscapes are easy to survey; others are remote or dangerous. Even within well-studied regions, the distribution of discoveries is partly a matter of chance: a cave is found, a quarry reveals a layer, a road cut exposes sediment. Our picture of the past is therefore partly a picture of where we have looked and what we happened to stumble upon.

The fifth reason is definitional uncertainty. Even when we find fossils and artifacts, classification can be ambiguous. A tooth can suggest a lineage, but teeth vary. A skull fragment can show a trait, but traits overlap. A stone tool industry can reflect cultural transmission, but tool types can be shared across groups and can converge independently. When evidence is fragmentary, interpretation becomes probabilistic. Labels become provisional. Some fossils may represent distinct populations that we have incorrectly merged into known categories. Others may represent normal variation that we have mistakenly split into new types. In either case, the record is not only fragmentary in physical terms but in conceptual terms. We do not just lack data; we sometimes lack the ability to categorize what we have without imposing modern expectations.

Genetics, for all its power, does not escape this fragmentation. Ancient DNA survives only under certain conditions—often cold, dry, stable environments. That means the genetic record is also biased geographically. We have far more ancient DNA from certain regions than from others, not because those regions were more important to human history, but because they were kinder to molecules. Moreover, genetic signals degrade over time. Recombination breaks ancestral segments. Drift can erase traces. The deeper the time depth, the harder it becomes to reconstruct detailed relationships. Ghost populations can be inferred, but their identity can remain elusive because the signal has been blurred by thousands of generations.

The combined result is a past that appears discontinuous even when it was continuous. Humans did not blink in and out of existence. Populations did not always vanish without descendants. Yet our archives—fossil, archaeological, genetic—are full of holes, and the holes are not evenly distributed. This creates the illusion of sudden appearances and sudden disappearances. A lineage seems to arrive suddenly when it first appears in the record, even if it existed earlier elsewhere. A lineage seems to vanish suddenly when the record goes quiet, even if it persisted in a refuge we have not found. Our narratives are therefore shaped as much by the structure of the archive as by the structure of history.

This is where the book's "missing chapters" framing becomes more than a rhetorical device. It is a sober description of the condition of evidence. The chapters are missing because the pages were destroyed, scattered, and sometimes never written in durable materials. When people speak about suppressed chapters, the disciplined response is to recognize that the biggest suppressor is time itself. The earth suppresses by erosion. The sea suppresses by drowning. The climate suppresses by decay. The archive is not hidden in a vault; it is broken in the open.

But fragmentation does not mean hopelessness. It means caution. It means we should be wary of overly clean narratives. It means we should expect surprises. And it means we should interpret myths and cultural memories with both humility and seriousness: humility because myth is not a fossil, seriousness because myth may preserve the shape of what the archive cannot.

The next subsection turns from fragmentation to consequence. If the record is fragmentary, what forces created the disappearances that left such gaps? Climate collapse and extinction are among the most powerful. Populations can shrink, fragment, and vanish when environments shift faster than adaptation can keep up. In deep time, these shifts were not gentle. They were world-rearranging. To understand lost human types, we must understand the climates that unmade them.

If the record is fragmentary because the earth is an editor, it is also fragmentary because the earth has repeatedly changed the story's cast. Human evolution and human dispersal did not unfold on a stable stage. They unfolded during dramatic climate oscillations that repeatedly reshaped habitats, redrew migration corridors, and turned refuges into death traps. In such a world, extinction is not an exceptional event; it is a recurring outcome. Populations appear, expand, fragment, and sometimes vanish—not necessarily because they were "inferior," but because the environment that supported them collapsed or because they were trapped between shifting ecologies and competing groups. The missing human types are, in part, the casualties of climate.

The key point is that climate does not change evenly. It changes in pulses. It can shift rapidly enough to disrupt food webs and water availability within the span of a few generations. When rainfall patterns change, grasslands can shrink, forests can expand, deserts can advance, and rivers can move. Coastal zones can be transformed by sea-level rise. Cold phases can lock landscapes into ice and tundra; warm phases can flood corridors with vegetation or disease. For large, mobile animals, these shifts are navigable, but for human populations that rely on specific resource zones, rapid shifts can be devastating. A group specialized to a particular ecological niche can find that niche disappearing.

Extinction, in this context, is not simply dying out everywhere at once. It is often a process of fragmentation. A once-connected population breaks into smaller pockets as habitats fragment. Those pockets become isolated, and isolation reduces genetic diversity. Reduced diversity can make a population more vulnerable to disease, to further environmental stress, and to demographic collapse. A few harsh winters, a prolonged drought, a decline in game, or a change in plant availability can push a small isolated group past a tipping point. The group does not need to be wiped out by catastrophe; it can simply fail to reproduce enough to sustain itself. Over time, the population becomes a ghost—gone as a distinct lineage, even if some individuals were absorbed into others.

This matters because many hominin lineages likely lived in narrow ecological bands. Some were adapted to cold steppe environments, with bodies suited to conserving heat and diets suited to high-protein resources. Others may have thrived in woodland mosaics. Others may have been specialized to island environments. When climate shifted, these niches shifted. A lineage that thrived in one climate phase could struggle in the next. The same landscape could flip between abundance and scarcity over tens of thousands of years. This cycling creates a recurring selection filter that can repeatedly eliminate lineages or push them into refuges where they become more vulnerable.

Refuge zones are central to understanding extinction. During harsh climate phases, populations often retreat into regions where resources remain stable—coastal areas, river valleys, certain mountain foothills, or ecological “edges” where multiple habitats meet. These refuges can sustain life when surrounding regions become inhospitable. But refuges also concentrate populations. When multiple groups retreat into the same refuge, competition intensifies. The refuge becomes a crucible where survival depends not only on adaptation to environment but on negotiation, conflict, and interbreeding with other groups. In such crucibles, a lineage can disappear not by dying out completely but by being absorbed into others—or by being pushed out and failing elsewhere.

This is where climate and interaction intertwine. Climate collapse can set the conditions for extinction, but extinction often occurs through combined pressures: environment narrows options while other populations expand into the same spaces. A group that could survive climate stress alone might not survive climate stress plus competition. The stronger the bottleneck, the more decisive interaction becomes. Over long periods, this could produce the pattern we see in both fossils and genetics: lineages that were once widespread become restricted, then disappear as distinct groups, leaving only traces.

Sea-level change adds another layer. Lower sea levels exposed land bridges and coastal plains that may have served as migration routes and rich habitats. As sea levels rose, those landscapes were drowned. Populations living on those plains could be displaced inward, pressured into new territories, or isolated on newly formed islands. Some may have adapted; others may have declined. The drowning of coastal zones is therefore not only an archaeological problem; it is a demographic force. It can force sudden migration and increase conflict in inland refuges. It can also isolate groups, producing small populations vulnerable to extinction through drift and resource limitation.

Disease is another underappreciated dimension of climate-driven extinction. Climate shifts can alter the distribution of pathogens and parasites. Warmer, wetter conditions can expand disease zones. New environments can bring new exposures. When populations move into unfamiliar regions or come into contact with other groups, disease exchange can occur. Small, isolated populations are especially vulnerable to novel pathogens. While we should be cautious about over-claiming the specifics of ancient disease without direct evidence, the general principle is sound: ecological change and contact can create biological stress beyond simple food scarcity.

What emerges from this is a picture of deep time as a dynamic, unstable arena. In that arena, hominin populations are not static "species" waiting to be classified. They are living groups responding to shifting pressures. Some endure and spread. Some fragment and vanish. Some interbreed and dissolve into others. The fossil record captures a few snapshots, often from periods of stability or from particular preservation environments. Genetics captures remnants of ancestry that survived through absorption. But both point toward the same reality: climate collapse repeatedly rearranged the human world.

This reality complicates the way we talk about extinction narratives. It is tempting to treat the disappearance of certain lineages as a triumph of Homo sapiens—a story of superiority and conquest. There may be elements of competition, certainly, but climate forces caution. Superior technology does not guarantee survival if the ecological base collapses. Likewise, a lineage can disappear even if it is well-adapted, if its habitat shrinks and its population falls below sustainable size. Extinction is often a story of bad timing and constrained options as much as a story of capability.

This is also where the volume's larger theme becomes sharper. The "missing chapters" are not missing because evolution is simple; they are missing because evolution is harsh. Climate change does not merely sculpt species slowly; it can break populations abruptly. It can force contact. It can force mixing. It can force collapse. In that world, "lost human types" are an expected outcome. The more we understand climate volatility, the more we should expect that many lineages existed briefly or regionally and then vanished before leaving clear fossil footprints.

The next subsection turns to the most important implication of this climate-driven reshuffling: disappearance does not always mean erasure. Extinction can occur through absorption. Populations can vanish as distinct groups while persisting as ancestry within survivors. In that sense, the human story is not only a history of loss but a history of incorporation. Climate may have driven many lineages into contact, and contact may have turned extinction into integration.

When most people hear the word extinction, they imagine finality. A lineage exists, then it disappears, leaving nothing behind but fossils and silence. That image is partly true for many species, but it is not the only way extinction can happen—especially among closely related human populations capable of interbreeding. In the human story, extinction often looks less like erasure and more like absorption. A group can vanish as a distinct population while surviving as fragments inside another population. Its language may be lost, its identity dissolved, its territory taken, yet its biology persists in scattered inheritance. The group is gone and not gone at the same time. This is one of the most important conceptual shifts genetics has forced upon us, and it reframes what "lost human types" really means.

Absorption begins with contact. Climate change compresses populations into refuges. Migration corridors open and close. Resources shift. Groups meet. Sometimes they compete violently. Sometimes they avoid each other. Sometimes they exchange mates. But whenever interbreeding occurs across distinct populations, boundaries blur. If the contact is sustained, the smaller population can be absorbed into the larger one, leaving its descendants embedded in the dominant group's gene pool. Over generations, the absorbed population's unique genetic signature becomes diluted but not necessarily erased. Some segments persist as recognizable ancestry, especially if they confer advantage or if the admixture event was substantial.

This dynamic makes intuitive sense in demographic terms. Imagine two populations of humans living adjacent to each other. One is larger and more connected, with broader territory or stronger social networks. The other is smaller and more isolated. If the smaller population experiences stress—climate deterioration, resource shortage, disease—and begins to decline, its members may seek mates or alliances with the larger group. Or the larger group may expand into the smaller group's territory, bringing contact by proximity. Either way, interbreeding can occur. The smaller population's distinctiveness erodes, not because its members vanish instantly, but because their descendants become part of a larger, mixed population. Their identity as a separate "people" disappears, but their bloodline continues.

This is a key reason the old notion of a branching family tree fails. A tree assumes splits without reconnection. Absorption assumes reconnection. It assumes that population boundaries are real but porous. It assumes that divergence does not necessarily lead to permanent separation. In deep time, with hominins living in overlapping regions, experiencing repeated expansions and contractions, absorption could have been common. Instead of a world where lineages were neatly replaced, we may be looking at a world where lineages were repeatedly braided.

The genetic evidence for absorption is straightforward at the level of principle. Modern humans carry ancestry from archaic populations. That alone means some lineages did not simply vanish; they were incorporated.

But absorption is not only about one famous admixture event. It is about a repeating demographic pattern. Many lineages could have contributed small amounts of ancestry to others and then disappeared as distinct groups. This is exactly how ghost populations become plausible: they are the genetic footprints of absorbed groups whose fossil identities are not yet clear. The more we accept absorption as normal, the more we should expect that the human story includes many invisible incorporations.

Absorption also changes how we think about "success" in evolution. If one population absorbs another, is that a victory for the absorbing group? In one sense, yes: the absorbing group becomes the surviving identity. But in another sense, it is not victory; it is integration. The absorbing group is altered by the absorbed group. It becomes a composite. Over time, what survives is not a pure lineage but a blended one. This reframes the very idea of modern humans as a single coherent type. We are not the product of a single lineage that eliminated all others. We are the product of a lineage that incorporated others, sometimes through peaceful contact, sometimes through asymmetrical pressures, and sometimes through the harsh constraints of climate that forced proximity.

The process is not necessarily symmetrical. Absorption can be uneven, with one population contributing far more ancestry than the other. It can also be selective. Certain genetic variants may persist because they confer survival advantage—immunity-related traits, environmental tolerance, metabolic adaptations—while other variants fade. In that case, absorption acts like a filter. The absorbed population leaves behind a biased sample of itself: not a full inheritance, but the parts that either survived by chance or were favored by selection. This is one reason genetic traces can feel like "upgrades" in popular storytelling, even though that framing can be misleading. A retained variant may have provided an advantage in a particular context, but it does not imply that the absorbed population was globally superior. It implies that, in one ecological arena, one inherited trait mattered.

Absorption also has cultural implications. A population that is absorbed biologically may still leave cultural traces—tool styles, symbolic motifs, ritual practices—that are adopted by the dominant group. But these cultural traces can be difficult to detect because culture can spread without genes, and genes can spread without culture. A group could be biologically absorbed while its cultural identity disappears quickly. Or a group could be biologically absorbed while some of its cultural practices persist, reinterpreted within a new social order. Over long timescales, these cultural incorporations become almost impossible to disentangle from the broader flow of innovation. Yet the principle holds: absorption is not only genetic; it is a process of integration at multiple levels, with identity dissolving into a larger composite.

This is where the chapter's theme becomes philosophically sharp. What does it mean to call a human type "lost" if it survives inside us? The answer is that what is lost is not necessarily biological contribution, but distinctness. A population can be lost as a named, bounded people even while it remains present as ancestry. This is similar to how historical peoples can vanish as languages and identities while their descendants continue under new names. It is not unusual in human history; it is simply occurring on a deeper timescale. Genetic absorption is the prehistoric version of cultural assimilation—except the evidence is stored in DNA rather than in documents.

Absorption also helps explain why the human record can feel simultaneously rich and empty. It can feel rich because we find evidence of multiple lineages. It can feel empty because those lineages do not always appear as long, continuous streams. They flicker. They appear in one region, disappear in another. They leave traces in genomes without leaving clear fossil trails. This makes sense if absorption was common. A lineage could persist for a long time but rarely fossilize. It could leave genetic traces through contact and then dissolve. The record would capture only fragments: a jaw here, a tooth there, a ghost signal in genomes, an ambiguous trait in a skull. The lineage would be real, yet it would not be easily definable.

All of this sets up the final subsection's synthesis. If extinction often occurs through absorption, then humanity is not a single lineage that replaced others but a composite species—a mosaic assembled through repeated contact, mixing, and survival through integration. This does not erase the uniqueness of Homo sapiens; it reframes it. Our uniqueness may lie not in isolation but in permeability—in our capacity to expand, adapt, and incorporate. The next subsection will gather the chapter's threads into that broader conclusion without turning it into a chapter-ending wrap-up, but as a natural step forward: the recognition that "humanity" is not one clean origin story, but the living result of many vanished peoples.

To call humanity a composite species is not to indulge metaphor. It is to describe what the evidence increasingly demands when fossils, genetics, and deep-time ecology are placed side by side. The old image of a single, clean line—one ancestral population steadily refining into modern humans—cannot survive contact with what we now know. The human story is braided. It is woven from multiple lineages, shaped by repeated encounters, and punctuated by disappearances that were often not total erasures but absorptions. The living human is, in that sense, an inheritance from a crowded past.

This composite nature begins with a simple demographic reality: humans evolved in populations, not as a singular entity. Populations split, spread, and sometimes reconnect. In deep time, geographic separation does not always mean permanent separation. Climate shifts repeatedly changed what "separated" even means—opening corridors, closing them, pushing groups into refuges where contact became unavoidable. Under these conditions, the presence of admixture is not surprising; it is expected. If multiple hominin groups occupied overlapping regions and were compatible enough to interbreed, then some degree of genetic exchange would be part of the landscape. The surprise, in hindsight, is not that mixing happened. It is that we once taught a story as if it did not.

The composite species idea also clarifies a tension in human identity. We tend to treat "modern humans" as a defined type with a singular essence, but the evidence points to a mosaic of inherited traits and histories. Some of our inherited variants may come from archaic populations. Some may come from regional lineages that are no longer definable. Some may reflect adaptations acquired through long local histories and then spread through later migrations. The body becomes a map of multiple histories layered together. This does not make humanity less coherent; it makes humanity historically honest. A composite can still be a coherent form. It simply carries multiple sources.

This perspective changes how we interpret the disappearance of other human types. If extinction is often absorption, then "lost" does not always mean gone. It means the loss of distinctness. It means the collapse of boundaries between populations. A lineage becomes lost when it can no longer be identified as a separate demographic stream, even if its genetic contribution persists. The world of deep time may therefore have been filled with populations that existed for long spans, interacted, and then dissolved into others. We are left with the survivors not as pure lines but as composites that contain fragments of the vanished.

There is also a psychological consequence to this realization. Humans have long imagined themselves as singular—created once, chosen once, separate from all others. Even secular modern narratives often preserve this singularity, portraying Homo sapiens as the one winner that replaced all competitors. A composite story is more unsettling. It suggests our identity is inherited from others we can barely name. It suggests we are the descendants not only of triumph but of integration. It suggests that "human" is not a discrete package delivered fully formed, but an evolving coalition of lineages that merged and remerged as history unfolded.

In evolutionary terms, this coalition can be seen as a strategy. A population that interbreeds with neighboring groups can incorporate adaptive variants and expand its ecological flexibility. This does not imply conscious intent; it is simply what gene flow can do over time. When environments shift rapidly, flexible populations have an advantage. Flexibility can come from behavioral innovation, but it can also come from genetic diversity—especially if admixture introduces variants suited to local pathogens or extreme environments. The composite nature of humanity, then, may have been one mechanism by which our lineage became resilient: not by remaining isolated and pure, but by being permeable enough to incorporate what worked.

At the same time, we must be careful not to turn this into a triumphalist fantasy. Composite does not mean superior in every dimension. It means historically entangled. It means that survival came with cost, and that many lineages vanished as distinct peoples. It means that the past was not a clean tournament but a complex demographic ecology. Some groups were absorbed because they were small and vulnerable. Some disappeared because their habitats collapsed. Some likely perished through combined pressures of climate and competition. The composite human is therefore not the proof of a noble evolutionary plan. It is the residue of survival in a harsh and shifting world.

This composite framing also helps us interpret why certain questions remain so difficult. If the human past included many populations, some of which left only faint traces, then we should expect persistent uncertainty. We will continue to find fossils that do not fit neatly. We will continue to detect genetic signals that do not map cleanly to known lineages. We will continue to revise the story as new evidence emerges. The composite model does not eliminate mystery; it normalizes it. It says: complexity is not an anomaly; it is the baseline.

In the larger arc of this volume, the composite species idea does something even more important. It provides a scientifically grounded bridge to the book's broader theme of missing chapters without requiring speculative leaps. When myths speak of "others," when legends describe mighty ones, when traditions remember teachers who were not fully human, the disciplined response is not to equate mythic beings with specific hominin populations. The disciplined response is to recognize that the intuition of plurality is not wrong. The past did contain multiple kinds of humans. The present does carry traces of them. Cultural memory does preserve the sense that humanity was once more diverse than it appears now. The composite species model allows us to hold that intuition without turning it into literal myth-history.

And it sets the stage for what follows. The next chapter shifts from lost human types to evolution after catastrophe—bottlenecks, survival against odds, and the ways cultural memory can reset even when genes persist. If humanity is composite, then catastrophe does not simply prune a tree;

it reshapes a web. It can erase populations, force absorption, and leave survivors carrying fragments of many lineages into a world that no longer remembers their names. That is the heart of the missing chapters: not that we lack evidence of complexity, but that the complexity itself was produced by cycles of loss and integration.

CHAPTER 12 — EVOLUTION AFTER CATASTROPHE

If the human story is braided, catastrophe is one of the forces that tightens the braid. In deep time, populations do not simply expand across landscapes and keep expanding. They crash. They narrow. They pass through constrictions so severe that, in a strict statistical sense, survival begins to look improbable. These constrictions—bottlenecks—are moments when the number of breeding individuals falls dramatically, reducing genetic diversity and increasing vulnerability. For many species, bottlenecks of sufficient severity are death sentences. Yet humans, again and again, appear to have passed through narrow gates and emerged on the other side. This is not a claim that we were invincible. It is a claim that our lineage's existence implies survival through episodes that could easily have ended it.

A bottleneck is not simply "a hard time." It is a demographic collapse with evolutionary consequences. When numbers shrink, the gene pool shrinks. Rare variants are lost. Harmful mutations can accumulate more easily in small populations. Inbreeding becomes harder to avoid. The population becomes more fragile in the face of disease, climate variability, and competition. The world does not need to deliver a single dramatic blow; it only needs to keep conditions harsh long enough that births cannot replace deaths. In that state, extinction becomes a matter of arithmetic. If the bottleneck persists, the population disappears.

The phrase "should have killed us" captures the intuition that humans faced multiple overlapping hazards: climate instability, habitat shifts, resource collapses, and competition with other hominin groups. Each hazard alone might be survivable. Combined, they can be fatal. Consider the logic of the Pleistocene world. Glacial cycles repeatedly reshaped ecosystems. Regions that were once green became cold steppe or arid zones. Rivers shifted. Coastlines moved. Game distributions changed. Plant communities reorganized. For a small population dependent on a particular seasonal rhythm, these changes could be catastrophic. Food scarcity leads to malnutrition, which weakens immunity, which increases mortality, which reduces births. A population can enter a downward spiral. In such spirals, survival is not guaranteed by intelligence alone.

Bottlenecks also intensify the role of chance. In large populations, random events average out. In small populations, random events can dominate. A single drought, a harsh winter, an epidemic, or a conflict over a refuge zone can remove a large fraction of a group. If the group is already small, that fraction might be the difference between recovery and collapse. This is why bottlenecks can feel like existential roulette. When the numbers are low enough, the future hinges on the survival of a few families. From the outside, looking backward, the fact that we exist at all can feel astonishing—not because the universe intervened, but because the arithmetic was unforgiving.

The bottleneck concept also reframes how we think about "human origins" as a single event. In reality, what survives to become "modern humanity" is shaped by repeated filtering. Bottlenecks act as filters. They reduce diversity. They amplify certain lineages. They eliminate others. They can make one regional population disproportionately influential simply because it survived while others did not. Over time, the survivors become the ancestors of later expansions, giving the impression of a singular origin when what actually happened may have been repeated narrowing and re-expansion. The origin is not a single birth; it is a series of narrow survivals.

This is one reason the composite species model and the bottleneck model fit together. When populations are large and dispersed, there is room for diversity—multiple human types, multiple lineages, multiple regional adaptations. When catastrophe strikes and populations contract, diversity collapses. Some lineages disappear entirely. Others are absorbed into the survivors. The survivors carry fragments of what was, but much is lost. This produces exactly the pattern we see: a past that was crowded becomes a present that seems singular, with ghost traces of what vanished.

The bottleneck idea also helps explain why modern humans can be globally successful yet still carry signs of vulnerability. A lineage that survives bottlenecks is not necessarily the "best" in some abstract sense. It is the lineage that happened to navigate a particular sequence of hazards with enough flexibility, innovation, and luck. Surviving bottlenecks can produce a population that is resilient behaviorally but genetically reduced in certain ways. It can also produce a lineage that relies heavily on culture—on learned strategies—because genetic adaptation is slower in small populations. When numbers are low, rapid biological evolution is constrained by limited variation, but cultural adaptation can still move quickly. In this way, bottlenecks may have pushed humanity further toward being a cultural species: a species whose survival depends on knowledge, cooperation, and behavioral flexibility more than on specialized biology.

There is also a social dimension to bottlenecks that is easy to overlook. When populations shrink, social bonds become more critical. Cooperation becomes survival. Sharing food, protecting children, maintaining knowledge of landscapes, and caring for the injured can determine whether a group persists. In small groups, losing one skilled hunter or one midwife can matter more than losing a random individual in a large population. The group's survival may depend on a few key roles and a few key relationships. Bottlenecks therefore select not only for physical resilience but for social resilience—the capacity to maintain cohesion under stress.

This brings us to the deeper reason bottlenecks "should have killed us." Humans are not solitary animals. We require social learning and collective support. If a bottleneck becomes too severe, it can break the chain of cultural transmission. Children may not learn essential skills if elders die. Knowledge of seasonal patterns may vanish if the group is displaced. Tool traditions may be interrupted. If a group loses enough carriers of knowledge, it can become culturally as well as biologically endangered. In that sense, bottlenecks threaten humanity twice: by shrinking the gene pool and by breaking the memory chain.

And yet, the human story suggests that the memory chain often held. It held enough to allow recovery. It held enough to allow populations to expand again when conditions improved. This does not require heroic myth. It requires a realistic appreciation for how culture can function as a survival technology. A group that remembers where water persists during drought, that remembers which plants are edible during lean seasons, that maintains fire-making skills, that knows how to coordinate hunting and sharing, has a margin of resilience that can make the difference in a bottleneck. The margin is not infinite, but it is real.

The final complication is that bottlenecks are rarely clean. They do not happen to a single population in isolation. They happen within a world of other hominin groups, predators, and competitors. A bottleneck may reduce one population, making it vulnerable to absorption by another. Or a bottleneck may reduce multiple populations simultaneously, creating a landscape where refuges become shared and mixing increases. Catastrophe can therefore be a driver of hybridization as much as a driver of extinction. When the world narrows, contact becomes more likely. This is one reason "evolution after catastrophe" is not simply a story of survival; it is a story of reshaping ancestry.

The next subsection confronts the obvious follow-up: if bottlenecks were so severe, why did humans survive when other hominins did not? The answer will not be a single magic trait. It will be a cluster—mobility, cooperation, cultural flexibility, and the ability to occupy diverse habitats—working together under the relentless pressure of changing worlds.

To ask why humans survived when other hominins did not is to ask a question that tempts simplistic answers. One might want a single decisive advantage—language, creativity, superior weapons, or some spark of consciousness that made Homo sapiens inevitable winners. But deep time rarely rewards single-cause stories. Survival is usually an outcome of many interacting factors: ecology, demography, culture, and contingency. If modern humans outlasted other lineages, it was not necessarily because we were "better" in every dimension. It may have been because we were flexible across more dimensions at once, and because our social and cultural systems could respond quickly to an unstable world.

Begin with flexibility. In a volatile climate, specialization is risky. A population highly adapted to a narrow niche can thrive when that niche remains stable, but it can collapse when conditions swing. Humans appear to have been unusually good at shifting strategies—changing prey, changing tools, changing settlement patterns, changing social organization. This kind of flexibility is not an abstract virtue. It is a survival mechanism in a world where rainfall patterns, vegetation zones, and animal migrations could reorganize over surprisingly short timescales. A group that can reconfigure its behavior can ride environmental change rather than be crushed by it.

This flexibility is amplified by cultural transmission. Biological evolution is slow relative to rapid climate shifts. Genes can adapt a population over thousands of years, but a drought or a cold snap can transform a habitat within decades. Culture can move faster. A new hunting technique, a new tool design, a new seasonal route, a new method of storage—these can spread within a few generations, sometimes within a single generation. If Homo sapiens had a greater capacity to accumulate and transmit complex learned behaviors, then humans would have possessed an adaptive layer that does not require genetic change. That layer can buffer bottlenecks. It can allow small populations to survive conditions that might otherwise push them past the tipping point.

Social structure is another decisive layer. Humans are cooperative beyond what most species can sustain at scale. Cooperation is not simply friendliness. It is organized sharing, coordinated action, and mutual support that extends beyond immediate kin in many contexts. In harsh environments, this matters because it reduces variance. When hunting is unpredictable, sharing smooths the risk. When illness strikes, caregiving preserves valuable individuals who would otherwise die. When a group faces scarcity, cooperation allows pooling and rationing. These dynamics can be the difference between a population that collapses and one that endures.

Cooperation also intersects with information. Humans are not only cooperative with food; they are cooperative with knowledge. A group that shares information about landscapes—water sources, seasonal plants, dangerous zones—builds a collective mind that outperforms individual experience. In a changing climate, the collective mind can update faster than any single person can. This produces a kind of distributed intelligence. A small population with strong information sharing can punch above its demographic weight. It can also preserve knowledge through stress: if one carrier of knowledge dies, others still hold parts of the map.

Mobility is another advantage, but it is not just physical movement. It is the ability to treat space as a resource. A population that can move long distances can track shifting habitats and avoid localized disasters. It can relocate when rivers fail, when prey declines, when cold intensifies. Mobility also enables contact. Contact enables exchange—tools, mates, ideas. This exchange can refresh genetic diversity and introduce innovations that help survival. In a fragmented world, mobility creates network resilience. Populations linked by movement and exchange are less likely to become isolated pockets that collapse silently.

Technology matters as well, but again not as a single miraculous invention. The advantage is not that humans had "better tools" in some absolute sense. It is that humans could innovate toolkits and integrate them into social practice. Tools are not merely objects; they are behaviors and traditions. A spear point is a concept taught and refined. Fire use is not just ignition; it is fuel management, cooking, protection, and landscape interaction. Clothing and shelter construction are not just items; they are systems that extend habitats into colder zones and reduce seasonal vulnerability. A population that can maintain these systems robustly can survive in environments that would otherwise be lethal.

Demography adds another layer. Small populations are fragile, but populations that can maintain connectivity between groups—through mating networks, seasonal aggregation, and alliances—can reduce that fragility. If Homo sapiens maintained wider mating networks, it could have mitigated inbreeding and retained more genetic diversity through bottlenecks. Even modest differences in connectivity can produce large long-term consequences. A lineage that stays connected across a region has a better chance of persistence than one that breaks into isolated, vulnerable pockets.

This is not a covert way of declaring other hominins "inferior." Many other hominins were remarkably adapted to their environments and survived for long periods. Their disappearance may reflect not weakness but the combined pressures of climate instability, demographic contraction, and competition in refuge zones. If a lineage was less able—or less positioned—to maintain wide connectivity, it could be more vulnerable when refuges became crowded. If a lineage relied on a narrower ecological niche, it could be more vulnerable when the niche shrank. If a lineage faced repeated population crashes, it could simply run out of demographic resilience over time. Extinction can be the cumulative result of many near-survivals.

Competition likely played a role, but competition is not always direct warfare. It can be competitive exclusion through resource use. Two populations can compete for the same prey, the same shelter-rich zones, the same water sources. A group with slightly more flexible technology or broader diet can gradually outcompete another without a single dramatic battle. Over centuries, the less competitive group can decline, fragment, and be absorbed. In this scenario, survival is not decided by one event but by persistent asymmetry in resilience.

Hybridization itself may also be part of the survival story. If humans interbred with other hominins, then humans did not simply replace them; humans incorporated them. This may have provided adaptive benefits in certain environments, but even if it did not, it had demographic consequences. Hybridization can bring populations into closer integration rather than forcing strict boundaries that lead to repeated conflict. It can also mean that the disappearance of other lineages was not purely extinction but transformation into composite ancestry. Humans may have "survived" partly because the human category expanded through incorporation—because the surviving population absorbed others rather than remaining isolated.

What emerges, then, is not a single reason but a cluster that reinforces itself. Cultural flexibility supports mobility. Mobility supports contact. Contact supports exchange and genetic diversity. Cooperation supports survival through scarcity. Knowledge sharing supports innovation. Innovation supports the occupation of varied habitats. Varied habitats reduce vulnerability to any single ecological collapse. Together, these traits form a resilience package. Under repeated bottlenecks, resilience is what matters.

The next subsection turns from survival mechanisms to a different consequence of catastrophe: memory. Even when humans survived biologically, catastrophe can reset culture. A bottleneck can wipe out specialists, break traditions, erase accumulated knowledge, and leave survivors with a reduced cultural repertoire. This creates an uneven human story—moments of brilliance followed by regression, followed by reinvention. Understanding selective memory and cultural reset is essential if we want to understand why the archaeological record sometimes looks like sudden leaps rather than steady progress.

Catastrophe does not only prune populations. It prunes knowledge. A bottleneck can be survived biologically and still function as a cultural amputation. The survivors may carry the genome forward, but they do not necessarily carry the full library of skills, stories, and specialized techniques their ancestors once held. This is one of the most important and least intuitive truths about human history: culture is fragile. It can accumulate over long spans, but it can also collapse quickly when the social conditions that sustain it are disrupted. In that sense, catastrophe produces selective memory. It does not simply erase the past. It filters it, leaving a biased remainder that later generations mistake for the whole.

A cultural system is not stored in any one person. It is distributed across a community. One person knows how to knap a particular tool form. Another knows where the best stone source lies. Another knows how to predict seasonal winds. Another knows the medicinal plants. Another carries songs and rituals that encode social cohesion. Another remembers ancestral routes and water holes. A functioning culture is the coordination of these knowledges. But when a population shrinks, this distribution becomes a vulnerability. Lose a few key individuals, and entire traditions can vanish. Not because they were unimportant, but because their carriers were too few.

This is why small populations can experience cultural drift as well as genetic drift. In a large population, knowledge can be redundant; multiple people know the same craft, and the loss of one does not destroy it. In a small population, redundancy may be limited. Skills can become bottlenecked into single lineages. If those lineages break, the skill disappears. Over time, the population may simplify its technology and ritual life—not because it becomes less intelligent, but because it no longer has the social capacity to maintain complexity. Complexity requires infrastructure: time to teach, roles that can specialize, stable resource flows that allow experimentation and practice. Catastrophe attacks all three.

A "cultural reset" is therefore not a mystical event. It is a practical consequence of survival under stress. When food is scarce, long apprenticeships become difficult. When people are migrating constantly, heavy tool traditions may be abandoned. When a group is hiding in a refuge, elaborate ceremonial life may shrink to what can be carried and performed under pressure. When conflict is constant, innovation can narrow to survival needs. And when a population is reduced, social institutions that require scale—specialist classes, formal councils, complex exchange networks—may collapse. The culture that emerges after the crisis may be leaner, more portable, and less differentiated. It may preserve only a core set of practices that are essential for immediate survival and group cohesion.

Selective memory is not only about what is lost; it is also about what is kept. Certain forms of knowledge are more likely to survive because they are embedded in daily life. Fire-making survives because it is constantly used. Basic hunting strategies survive because they are constantly practiced. Kinship rules survive because they organize reproduction. But more specialized knowledge—astronomical tracking beyond what is needed for seasons, complex ceremonial architecture, long-distance navigation systems, sophisticated craft sequences—can be more vulnerable. These systems often require stable conditions and specialist roles. If the specialists die or the conditions collapse, the knowledge may fade rapidly. Later generations may retain fragments—stories of the old ways, ritual gestures without full explanation, mythic references to knowledge that once existed.

This creates an archaeological and psychological illusion. When a culture re-expands after a bottleneck, it may "rediscover" complexity. Tools become more sophisticated again. Symbolic art flourishes again. Long-distance trade networks reappear. Later observers may interpret this as a sudden cognitive leap, as if humans became smarter overnight. But what may be happening is not a sudden invention of intelligence;

it is a rebuilding of cultural complexity after a reset. The capacity was always there, but the infrastructure of transmission had been damaged. Once conditions stabilize and populations grow, complexity can reaccumulate—sometimes surprisingly quickly.

This dynamic helps explain why the record can look punctuated. Deep-time narratives often swing between "gradualism" and "revolution." Either we imagine slow incremental progress, or we imagine abrupt bursts of innovation. Cultural reset suggests a third pattern: progress interrupted by collapse. A group can innovate, build complexity, and then lose it. Later groups can reinvent or reassemble similar complexity. The underlying human cognitive capacity may be stable, but the cultural expression of that capacity can oscillate dramatically depending on demography and environment.

Selective memory also shapes myth. After catastrophe, survivors often carry a sense that something great was lost. They may remember a time when life was different—more ordered, more abundant, closer to the sacred. They may preserve legends of teachers, golden ages, forbidden knowledge, and civilizing gifts. In a disciplined reading, these myths can be understood as cultural memories of reset. They encode the feeling that knowledge once existed in fuller form and then disappeared. They also encode the moral interpretation of collapse: the gods withdrew, the world was punished, humanity fell from a higher state. This is not necessarily literal history. It is memory shaped into meaning—a way of explaining why the present is diminished relative to an imagined past.

Importantly, selective memory is not neutral. What survives is shaped by identity and power. In post-catastrophe worlds, the groups that become dominant often define the story of the past. They preserve certain myths because those myths justify their authority. They forget other myths because those myths belong to vanished lineages. Cultural reset can therefore be accompanied by narrative reset: the rewriting of origin stories to fit new realities. A surviving group may reinterpret inherited motifs—teachers, gods, ancestors—in ways that legitimize new leadership. In this way, catastrophe does not only reduce culture; it reorganizes it.

There is also a cognitive dimension to selective memory. Under prolonged stress, humans narrow attention. Survival demands focus on immediate threats and resources. Long-term planning can shrink. Risk-taking decreases. Curiosity becomes costly. In such conditions, societies may become more conservative and less experimental. When stability returns, curiosity and experimentation can expand again. This can contribute to the punctuated pattern of innovation: it is not that humans gain or lose intelligence, but that the social and ecological conditions that allow intelligence to express itself in complex forms expand and contract.

All of this reframes the concept of "evolution after catastrophe." Evolution is not only genetic change. It is also the evolution of culture through cycles of accumulation and reset. A bottleneck can compress a population, and that compression can compress culture. The survivors' world becomes a smaller world. Their knowledge becomes a smaller knowledge. And yet the human capacity to rebuild remains. When conditions improve, humans can reconstruct complexity—sometimes drawing on fragments of old knowledge, sometimes reinventing from scratch, sometimes integrating new ideas through contact.

The next subsection builds directly on this: the survivors' advantage. If catastrophe filters both genes and culture, then those who pass through the filter may be unusually resilient—not because they are chosen by fate, but because the bottleneck selects for certain traits and because the experience itself reshapes behavior. Survivors inherit a world reduced in memory but sharpened in adaptation. That sharpened edge may be one of the engines that drives rapid re-expansion and the next phase of human transformation.

A catastrophe does not create heroes in the romantic sense, but it does create survivors in the evolutionary sense. When a population passes through a bottleneck, the individuals and groups that remain are not random samples of what existed before. They are the ones who endured a specific set of pressures—ecological instability, scarcity, displacement, conflict, disease, and social strain—long enough to reproduce. That endurance can leave a signature. It can reshape the gene pool in subtle ways, but more immediately, it can reshape behavior, social norms, and cultural strategies. The survivors' advantage is not that survivors become "better humans" in some moral sense. It is that survival filters for resilience, and resilience—once established—can become a platform for expansion.

The most obvious advantage is adaptive flexibility. Under stable conditions, a population can afford specialization. Under unstable conditions, specialization becomes fragile. Bottlenecks favor those who can switch strategies: who can eat broader diets, move across varied terrains, and adjust to new resource patterns. Over time, this selects for behavioral versatility—an ability to treat uncertainty as normal. The survivors inherit a habit of adaptation. They build cultures that assume change. They preserve practices that are portable and robust. Their lives teach them that the world can flip, and therefore the best strategy is not to depend on any single ecological promise.

This advantage is amplified by cognitive and social selection. Under scarcity, groups that cooperate effectively survive better than groups that fracture. Cooperation reduces risk, smooths distribution, and preserves vulnerable individuals whose knowledge or reproductive capacity matters. Over time, the survivors are more likely to come from communities with strong cohesion, reliable sharing norms, and effective conflict management. This is not to say that bottlenecks always produce peaceful societies. They can also produce harshness. But they do tend to reward groups that can maintain collective function under stress. In that sense, the survivors' advantage is often institutional: better coordination, better mutual support, better systems for distributing food and care.

There is also an informational advantage. When the world is unstable, knowledge of landscapes becomes more valuable. Knowing where water persists during drought, where shelter exists in cold phases, how to read seasonal signals, how to store food, how to manage fire and insulation—these become survival determinants. The survivors are those whose knowledge systems worked under pressure. Even if culture resets in complexity, the core survival library that remains is often highly refined. It is tested knowledge. It has been selected by reality. When conditions improve, populations carrying such tested knowledge can expand rapidly because they are already practiced in risk management.

This is where bottlenecks can create a paradox. They reduce genetic diversity and cultural complexity, but they can increase efficiency in the remaining repertoire. The population becomes smaller and leaner, but the strategies that survive tend to be those that performed well. When the environment stabilizes or opens new habitats, these lean strategies can scale quickly. A population that has learned to survive in a narrow refuge can spread outward with confidence when corridors reopen. In that expansion, the survivors' advantage becomes visible as rapid dispersal.

Demography turns this into momentum. Bottlenecks often end when conditions improve—climates shift toward stability, resources return, habitats expand. When that happens, populations that endured the constriction can rebound. If the survivors maintain social cohesion and sufficient reproductive capacity, growth can be rapid relative to the small baseline. Over generations, a population can go from endangered to expansive. This demographic rebound is one reason the archaeological record can appear punctuated: after periods of scarcity and invisibility, human presence suddenly becomes more visible—more sites, more tools, more art—because there are simply more humans on the landscape.

The survivors' advantage can also include network resilience. Populations that endure catastrophe often do so by maintaining connections—alliances between groups, exchange routes, seasonal aggregation points. These connections allow mates to move, information to spread, and resources to be shared. Under stress, networks matter. A group that is isolated is vulnerable; a group that can draw on wider social ties is more resilient. Therefore, survivors may disproportionately come from populations with stronger network behaviors. When expansion becomes possible again, those same networks facilitate rapid spread and innovation diffusion.

This advantage does not mean that survivors are the most "advanced." It means they are the most resilient under particular conditions. This distinction matters because it avoids a common trap: turning survival into proof of superiority. Evolution does not reward abstract excellence; it rewards fit to circumstance. A population can be highly adapted to one environment and still vanish when the environment collapses. Another population can survive because it had access to refuges, because it was more connected, because it was more flexible, or because it was lucky.

The survivors' advantage is therefore conditional and historical. It reflects the pressures of the bottleneck, not a universal hierarchy of worth.

But there is another dimension to the survivors' advantage that ties directly into this volume's theme: psychological and mythic inheritance. A population that has passed through catastrophe often carries a memory of rupture—stories of floods, fires, long winters, the withdrawal of abundance, the loss of an earlier order. Even when selective memory compresses details, the emotional imprint can be strong. Such communities may develop rituals and myths that encode vigilance: the sense that the world can end, that stability is fragile, that sacred order must be maintained. These myths can become organizing principles. They can shape identity, morality, and social cohesion. In this way, catastrophe can produce not only demographic survivors but cultural survivors: communities whose worldview is built around the reality of discontinuity.

This worldview can, in turn, reinforce resilience. A society that expects disruption may store food differently, maintain mobility differently, preserve sacred knowledge differently, and prioritize cohesion differently. The mythic memory of catastrophe becomes a behavioral asset. It is not that myths predict disasters; it is that myths can encode practical lessons in emotionally compelling forms. A story about a "world reset" can be a way of teaching risk management across generations. Even if the story is symbolic, the behavior it supports can be adaptive.

The survivors' advantage also helps explain why, after bottlenecks, certain narratives of origin become so persistent. If a population rebounds after near-extinction, it may interpret its survival as a new beginning. It may frame itself as recreated, renewed, or chosen. This leads naturally into myths of "second creation"—stories where humanity is made again after a catastrophe, or where a few survivors repopulate the world, or where the gods reset the human project. These myths, as we will see next, are not necessarily literal records of bottleneck events, but they are psychologically and structurally aligned with what bottlenecks do: reduce the world to a few, then begin again.

Across cultures separated by oceans and millennia, a strangely consistent motif appears: humanity is not created only once. Something goes wrong. The world collapses. A prior humanity is destroyed, punished, washed away, burned, or otherwise erased. Then a smaller remnant survives—or a creator remakes the human line—and the world begins again. These are the myths of second creation: narratives of reset, renewal, and reauthorization after catastrophe. They are among the most enduring stories humans tell, and they fit with uncanny precision to the demographic logic of bottlenecks. This does not mean the myths are literal historical chronicles of specific events. It means they are culturally appropriate ways of remembering a reality that the genome and the archaeological record also imply: human history has been repeatedly narrowed and restarted.

Second creation myths are not merely flood stories, though floods are one of their most dramatic forms. They are reset stories. The mechanism varies—water, fire, ice, darkness, war among gods, moral collapse—but the structure repeats. There is an earlier world with its own order. That order fails. A cleansing event eliminates most life or most humans. The surviving remnant becomes the seed of the new world. Often, the new humanity is changed—made more mortal, more limited, more obedient, or more constrained. The myth is not only explaining disaster; it is explaining why the present human condition feels diminished relative to an imagined past.

From a disciplined standpoint, the most important thing to notice is what these myths are doing psychologically and socially. They offer a narrative form that can hold the trauma of discontinuity. Catastrophe threatens the sense that the world is stable and meaningful. A second creation myth restores meaning by turning catastrophe into a structured moral event. The world ended because something was out of order—violence, arrogance, disobedience, divine conflict, imbalance. The world began again because order was restored—through a remnant, through a covenant, through a redesigned humanity. In this way, second creation myths transform random suffering into interpretable history.

They also serve as boundary markers. After a crisis, survivors must rebuild identity. Who are we? Why do we live differently now? Why are certain rules sacred? Why do we practice certain rituals? A second creation story provides answers. It says: we are the people who began again. Our customs are the safeguards that prevent another reset. Our rituals maintain the contract between humans and the forces that govern the world. Our moral code is the condition of survival. Whether or not any specific detail is literally true, the story becomes a stabilizing framework for post-crisis societies.

This is where the second creation motif aligns with the concept of cultural reset. After a bottleneck, a population may lose complexity and rebuild from a reduced base. The world may literally feel newly made because the old social structures have collapsed. Networks have broken. Specialists are gone. Skills are lost. Landscapes are altered. The survivors must improvise. In such circumstances, the idea of a "new humanity" is not abstract. It is lived experience. A second creation myth can be understood as the symbolic expression of that experience: a way of saying, "we are not the same as before."

Some second creation myths go further and describe the new humanity as intentionally redesigned—more constrained, more limited, more mortal. This resonates with the observation that catastrophe filters. Bottlenecks reduce genetic and cultural diversity. The post-bottleneck population is, in a real sense, narrower. It contains less of what once existed. A myth that says the gods shortened human lifespan, reduced their power, or changed their nature can be read as a cultural interpretation of this narrowing. The myth is not measuring genomes. It is expressing a perception: something was lost, and we live with the consequences.

There is another subtle pattern in second creation stories: the survival of knowledge is often precarious. The remnant carries seeds, tools, sacred objects, or divine instructions. Sometimes the remnant must rebuild civilization from almost nothing. Sometimes a teacher figure appears again to reintroduce law, agriculture, metallurgy, or ritual. This is a mythic dramatization of how culture behaves after catastrophe. If knowledge chains break, the post-crisis world requires re-teaching. The myth preserves this reality by giving it an archetypal form: the gift of civilization reissued after reset.

In the context of this volume's theme—missing chapters of human evolution—second creation myths can be approached as memory structures rather than evidence claims. They are not data points proving specific floods or comet impacts. They are cultural forms that make sense in a world where bottlenecks and collapses were real. They preserve the sense that history contains discontinuities. They also preserve the sense that the human condition is not simply the natural unfolding of progress but the result of recoveries from near-ending events. That sense is scientifically compatible with what we see in demographic reconstructions and the fragmentary archaeological record, without requiring any single myth to be taken as literal reportage.

This disciplined reading also explains why second creation motifs often come packaged with moral warnings. Catastrophe is framed as consequence, not accident. The purpose is not to preserve geology; it is to regulate behavior. A society that believes the world can be reset will invest in rituals and laws meant to maintain order. The myth becomes a technology of cohesion. It warns against internal fracture because fracture is deadly when resources are scarce. It warns against hubris because hubris destabilizes hierarchy. It warns against violating sacred boundaries because boundaries preserve group identity. In this way, second creation myths function as survival psychology encoded in story.

At the same time, these myths can be exploited by power. If a ruling class claims to inherit authority from the post-reset covenant—if it claims to be the guardian of the second creation—then its legitimacy becomes cosmic. The myth can justify hierarchy as necessary to prevent chaos. In some contexts, the idea of a second creation is not only a communal memory; it is a political tool. It makes the present order feel like the only barrier between society and annihilation.

All of this leads naturally to the final subsection of the chapter: civilization after near-extinction. If bottlenecks and resets were real forces shaping human survival, what happens when populations recover and begin building complex societies again? How does near-extinction shape the drive toward organization, storage, monuments, and centralized power? How do the survivors rebuild meaning, and how do myths of reset influence the shape of post-crisis civilization? The answer will tie together demography, cultural memory, and the recurring human impulse to make order durable—perhaps because we have learned, in our bones and stories, how easily order can be lost.

Near-extinction does not leave humanity in a neutral state. It leaves humanity with scars—demographic, cultural, and psychological—and those scars can shape what comes next. When a bottleneck ends and populations begin to expand again, the world they expand into is often changed: coastlines shifted, habitats rearranged, familiar routes broken, old networks dissolved. The people who survive are not simply continuing a stable trajectory. They are rebuilding in a world that has proven itself capable of collapse. Civilization after near-extinction, then, is not merely the return of complexity. It is complexity built under the shadow of fragility.

The first driver of post-bottleneck civilization is demographic rebound. When conditions improve—climates stabilize, resources recover, corridors reopen—human numbers can grow. Growth changes everything. It increases encounter rates between groups. It pushes people into new territories. It intensifies competition over fertile zones. It creates the need for more formal social coordination. Small bands can manage disputes through personal relationships; larger communities need rules that scale. Growth also makes storage more important. When more mouths depend on a landscape, seasonal abundance must be captured and held. Storage is one of the thresholds between mobile survival and settled life. It transforms the relationship to time. It turns the future into a managed resource.

But the critical point is that this rebound happens with a reduced cultural inheritance. After reset, many communities carry lean toolkits and simplified ritual repertoires. Yet as populations grow, culture can reaccumulate rapidly. Humans are not only capable of innovation; they are compelled toward it when density increases. Dense populations produce specialization. Specialization produces craft refinement. Refinement produces surplus and new forms of exchange. Exchange produces networks that carry ideas faster than any single group could invent them. In this way, civilization can appear to "burst" into view after a near-extinction: not because intelligence suddenly increased, but because demographic and ecological conditions allowed cultural complexity to expand again.

This is where the psychology of catastrophe becomes relevant. Survivors do not simply seek abundance; they seek stability. A world that has collapsed teaches a lesson: what is not secured can be lost. This lesson can express itself as an intensified human desire to anchor life in structures that feel durable—stored grain, fortified settlements, codified laws, sacred calendars, and monuments. Monumentality, in particular, can be read as an attempt to make order physical. A monument is not only art or piety. It is a statement that continuity matters. It is a declaration that the group will persist. In a post-catastrophe world, that declaration carries emotional weight.

The drive toward sacred order can intensify as well. After near-extinction, people often interpret survival as meaningful. They seek reasons. They form narratives of covenant, punishment, rescue, and renewed obligation. These narratives become embedded in ritual. Ritual becomes a mechanism for maintaining cohesion and for managing anxiety about collapse. The sacred calendar becomes a way to coordinate society and to align labor with seasonal cycles. In practical terms, rituals can also function as social technologies: they gather people, redistribute resources, confirm alliances, and reinforce shared identity. The line between sacred practice and social management blurs because both serve stability.

This makes the emergence of priesthoods and administrative classes more understandable. In small bands, specialized roles exist, but they are informal and flexible. In larger settled communities, specialization hardens into institutions. Someone must track time for planting. Someone must manage communal storage. Someone must adjudicate disputes. Someone must organize labor for irrigation and construction. Over time, these roles become centralized, and those who hold them gain authority. Civilization after near-extinction often becomes civilization after vulnerability: societies build institutions to reduce the risk of chaos. Yet those institutions can also create new forms of inequality and control.

The paradox is that institutions meant to prevent collapse can become brittle. Storage and hierarchy can produce wealth concentration. Concentration can produce conflict. Conflict can destabilize the very order the institutions were meant to secure. Post-catastrophe civilization therefore carries a built-in tension. The impulse to make life stable pushes toward centralized power and fixed settlement, but centralization can create new vulnerabilities—dependency on irrigation systems, reliance on stored surpluses, susceptibility to disease in dense populations, exposure to raids and conquest. The shadow of near-extinction never fully leaves; it is replaced by new forms of risk generated by the solutions.

This tension helps explain why early civilizations often wrapped themselves in mythic narratives of origin, order, and divine mandate. If society is fragile, then legitimacy must feel strong. If the world can collapse, then authority must appear cosmic. This is where second creation myths and origin myths connect to state formation. A people who see themselves as survivors of a reset may be especially receptive to narratives that define rules as sacred and institutions as necessary. The memory of collapse becomes fuel for obedience, cohesion, and monumental investment. In that sense, near-extinction can shape not only demography but ideology.

It also shapes the way the past is remembered. After a reset and rebound, a society may look back on a broken prehistory and interpret it as a sequence of ages—golden age, fall, flood, renewal. These age structures appear in many traditions. They can be read as cultural metaphors for demographic cycles: periods of abundance and growth, followed by collapse and narrowing, followed by recovery. Again, this does not mean the myths are literal demographic records. It means they mirror the rhythm of human survival in a volatile world. They encode a deep intuition: history is not a straight ascent; it is a series of climbs after falls.

In the framework of this volume, civilization after near-extinction is one of the keys to the "missing chapters" thesis, grounded without exaggeration. The missing chapters are missing not only because evidence is fragmentary, but because catastrophe and reset can interrupt cultural continuity. Knowledge can be lost and later reinvented. Populations can vanish and be absorbed. Networks can dissolve and later re-form. When we see sudden blooms of complexity in the archaeological record, we should consider the possibility that these blooms are not purely beginnings—they may be rebuildings. Civilization may not be the first rise; it may be the rise after survival.

This perspective does not demand a hidden advanced global civilization. It does not require secret technologies. It requires only what the evidence already suggests: that humans endured severe bottlenecks, that culture can reset, and that recovery can be rapid under the right conditions. From there, the human story becomes more honest and more dramatic: not a smooth march toward progress, but a repeated struggle to reassemble meaning and order in a world that repeatedly threatened to erase both.

CHAPTER 13 — SUPPRESSION, RIDICULE, AND CONTROL

There is a peculiar asymmetry in how modern culture treats questions about human origins. We celebrate discovery, praise curiosity, and build institutions around the ideal of open inquiry. Yet certain lines of questioning—especially those that complicate the standard narrative of human evolution—trigger a reflexive tightening of the room. The tone shifts. The questioner is asked not only for evidence but for motives. The topic becomes associated with fringe thinking before the argument is even heard. This is not because mainstream science is a conspiracy, and it is not because the questions are necessarily correct. It is because origins are not merely scientific territory. They are identity territory. When a question threatens identity, it becomes taboo even in cultures that claim to love truth.

The taboo begins with the way origin stories function in human societies. Every society carries an implicit narrative about what humans are and how we came to be. In modern secular societies, evolutionary biology often plays this role. It is not only a scientific framework; it has become a cultural foundation story. It tells us we are natural beings produced by long processes, not special creations outside nature. It tells us our minds emerged gradually, our moral instincts were shaped by selection, and our place in the world is continuous with other life. For many people, this story is emotionally stabilizing. It is the substitute for older religious creation accounts. When a foundation story becomes stabilizing, it becomes protected. Challenges are felt as attacks, not as proposals.

This protection intensifies because human origins questions have high symbolic stakes. If the narrative changes, it does not change only a timeline. It changes what people think about human uniqueness, morality, and meaning. It changes how we interpret race, difference, and equality. It changes how we understand intelligence and hierarchy. It changes the boundary between science and spirituality. Even when researchers insist they are discussing fossils and genes, the public hears implications about identity and ethics. Because the stakes feel large, the threshold for acceptable speculation becomes unusually low. People demand certainty in a domain where certainty is structurally impossible.

Another reason the topic becomes taboo is historical contamination. Human origins has been abused. Pseudoscientific claims have been used to justify racism, eugenics, colonial domination, and social hierarchy. This history created a legitimate ethical sensitivity: when people start talking about "different human types" and "hybrids" and "lost lineages," it can sound like a revival of dangerous ideas. Even when the modern evidence is real—multiple hominin lineages, admixture, ghost ancestry—the cultural memory of past abuses makes the conversation fragile. Institutions respond by building protective norms: strict language, strict boundaries, and a quick suspicion of narratives that sound like biological hierarchy. In that climate, even good questions can become socially risky.

Taboo also arises from the problem of associative stigma. Certain topics attract sensationalism—giants, hidden civilizations, non-human teachers, suppression claims—and sensationalism corrodes credibility. When a field has repeatedly been exploited by loud, uncareful voices, professionals become defensive. They learn to treat certain framing as a warning sign. This is understandable, but it can create collateral damage. A legitimate question can be dismissed because it resembles a pattern used by illegitimate narratives. The result is a social sorting mechanism: not only the content of a claim matters, but the cultural "vibe" of the claim.

Over time, taboo is enforced less by direct censorship than by reputation management.

There is also an epistemic reason these questions become taboo: the evidence is complex and easily misinterpreted. Human origins research spans multiple disciplines—genetics, paleoanthropology, archaeology, climatology, cognitive science—and each has its own methods and uncertainties. A compelling narrative can be constructed from partial evidence, and it can sound persuasive even when it overreaches. Professionals therefore develop a cautious posture toward broad syntheses, especially those that mix mythic motifs with scientific data. They worry, often rightly, that the synthesis will collapse into storytelling. That worry translates into a norm: stay within your lane. But the cost of this norm is that interdisciplinary questions—often the most interesting ones—become socially discouraged.

Taboo is strengthened further by the sociology of expertise. In any mature scientific field, communities maintain quality control through peer review, methodological standards, and shared assumptions. This is necessary, but it can also create "default narratives" that feel safer than exploring anomalies. When a topic sits at the intersection of uncertain evidence and high public interest, experts become more conservative. They have seen how quickly nuance is lost outside the field. So they simplify. They defend the core narrative. They treat deviations as threats to public understanding. Over time, this creates an environment where anomalies are acknowledged privately but minimized publicly, and where speculative questions are treated as reputational hazards rather than as exploratory tools.

None of this requires bad faith. It can emerge from ordinary incentives. Scientists are human. They have careers, reputations, and limited time.

They have to choose which questions are worth pursuing. In a world where resources are finite, "safe" projects that fit within accepted frameworks are more likely to be funded, published, and praised. "Risky" projects—especially those that could be misunderstood by the public—can feel like professional self-harm. That risk calculus creates taboo without any central authority enforcing it. The taboo is an emergent property of the system: a collective preference for stability over uncertain exploration.

But the deepest reason these questions are taboo is that they sit at the boundary between what science can currently answer and what humans desperately want answered. We want a complete story. We want clean lines. We want definitive origins. Yet the record is fragmentary. The genome holds ghosts. Fossils are sparse. Cultural memory is symbolic. The gap between desire and evidence creates tension. To reduce the tension, societies adopt a simplified narrative and defend it. Complexity feels like vulnerability. Uncertainty feels like weakness. So the questions that highlight uncertainty—questions about unknown hominins, sudden cognitive shifts, deep-time discontinuities—become taboo because they expose the limits of our knowledge in the very domain where we most crave certainty.

This is the tone and logic that shapes everything that follows in this chapter. The issue is not that inquiry is literally forbidden. It is that inquiry is socially managed. Certain questions are framed as unserious. Certain framings are treated as disqualifying. Certain ideas are punished through ridicule rather than debated on their merits. Understanding this social atmosphere matters, because it influences what is studied, how findings are communicated, and how the public interprets the boundaries of legitimate curiosity. The next subsection examines the mechanism that turns taboo into practical control: funding, reputation, and academic risk.

Taboo becomes real not when someone is formally forbidden to ask a question, but when asking it carries a cost. In modern research culture, the most effective forms of control are rarely explicit. They are structural. They operate through funding priorities, publication incentives, career timelines, and the fragile economy of reputation. A scientist does not need to be censored to be discouraged. It is enough that certain questions are associated with professional danger, while other questions are associated with steady rewards. Over time, whole landscapes of inquiry become underexplored—not because nobody is curious, but because curiosity has to pay rent.

Funding is the first gate. Modern research is expensive. Field excavations require permits, teams, equipment, and years of labor. Ancient DNA work requires specialized labs, contamination control, sequencing costs, and sophisticated computation. Even theoretical synthesis requires time—time bought through grants, institutional support, and secure employment. Funding agencies, whether public or private, do not fund "curiosity" in the abstract. They fund projects that can be justified as methodologically sound, socially beneficial, and likely to produce publishable results. These requirements are reasonable, but they create a conservative bias. Projects that challenge dominant narratives, or that involve uncertain outcomes, are harder to justify. If the proposal sounds speculative, it is likely to be rejected, regardless of whether the underlying question is meaningful.

This does not mean funding agencies are malicious. It means they are risk managers. They are accountable to boards, taxpayers, donors, and institutions. They prefer projects that promise clear deliverables. A project aimed at mapping an archaeological site with well-understood methods appears safer than a project framed around searching for "unknown hominin contributions" in an understudied region where preservation is poor. Even if the second project might yield transformative insight, it is easier to dismiss as too uncertain. In a system where success is measured by predictable outputs, the safest path becomes the dominant path.

Reputation is the second gate, and it is often more powerful than money. In science, credibility is a form of currency. It determines who gets hired, who gets invited, who gets cited, who gets trusted with resources. Reputation is built slowly and can be damaged quickly, especially in fields that attract public sensationalism. Human origins research is one such field. It sits at the crossroads of identity, ideology, and popular fascination. That makes it vulnerable to misrepresentation. A researcher who engages publicly with questions that resemble fringe narratives can be grouped with those narratives, regardless of the researcher's actual rigor. The fear of being misunderstood becomes a deterrent.

This fear is not paranoia. Scientists have seen how easily complex findings can be distorted. A cautious statement about gene flow can become "scientists admit humans are hybrids of unknown species" in the public imagination. A speculative remark about missing evidence can become "they're hiding the truth." The researcher then faces a dilemma: either avoid the topic altogether, or spend enormous energy clarifying nuance and defending credibility. Many choose avoidance, not because they do not care about truth, but because they have limited bandwidth and a career to sustain.

Career structure intensifies this. Academic timelines are unforgiving. Graduate students and early-career researchers are evaluated on publications, grants, and the ability to produce results on schedule. Risky projects can be career disasters. A young researcher cannot easily afford a five-year field effort that might yield ambiguous fossils or no publishable DNA. They cannot afford to be labeled speculative when they need letters of recommendation and job offers. So they choose safer questions, safer sites, safer methods, and safer framings. The system selects for caution at precisely the stage when innovation is most needed.

Publication incentives reinforce the same pattern. Journals and reviewers are gatekeepers of legitimacy. They demand methodological clarity, strong evidence, and alignment with existing scholarship. Again, this is necessary for quality control. But it can also create a subtle bias against research that does not fit within established categories. If a finding suggests a complex admixture scenario that challenges standard interpretations, reviewers may demand extraordinary evidence. If a paper proposes a new model that integrates multiple disciplines, reviewers may reject it as too broad or insufficiently specialized. The result is a field that can be simultaneously open to new data and resistant to new frameworks.

Even the language researchers use becomes shaped by these pressures. Scientists learn to frame results in ways that will not trigger unnecessary controversy. They adopt cautious terminology, avoid dramatic metaphors, and emphasize continuity with accepted models. This is not deception; it is professional survival. Yet it also means that the public-facing narrative can lag behind the complexity of the data. Researchers may privately acknowledge uncertainty, missing populations, and unresolved questions, while publicly emphasizing the stable core story. The gap between private nuance and public simplicity can then feed mistrust, because outsiders sense that complexity is being downplayed and interpret that as suppression.

Institutional politics adds another layer. Anthropology and genetics sit within universities, museums, and research institutes that have their own reputational concerns. Donors may not want controversy. Museums may prefer coherent exhibits. Universities may worry about public misunderstanding. There is a constant incentive to present a clean story that can be communicated to wide audiences without confusion. Complexity does not photograph well in an exhibition hall. A tidy narrative does. This creates an institutional preference for clarity over exploration, which can indirectly discourage lines of inquiry that would complicate the public story.

The irony is that, at the level of data, the field has become more complex than ever. Genetics has revealed ghost ancestries and admixture. Fossils continue to surprise. Archaeology shows that humans dispersed earlier and more widely than once assumed. The scientific reality is already a web. Yet the public narrative often remains closer to a ladder, because ladders are easier to explain and safer to defend. Researchers who emphasize the web risk being associated with sensationalism, even when they are merely following evidence.

None of this means that mainstream science is suppressing the truth. It means the ecosystem of science has built-in incentives that shape what is asked, what is funded, what is published, and what is emphasized. Certain questions become "high-risk, low-reward." Others become "low-risk, high-reward." Over decades, these incentives create blind spots. They do not forbid inquiry, but they make inquiry uneven. This is how academic risk can function like control without anyone needing to conspire.

The next subsection looks at an even sharper mechanism: ridicule. Funding and reputation pressures can discourage work quietly, but ridicule can shut down conversation openly. Skepticism is essential to science, but when skepticism is used as a social weapon—when it becomes a tool for policing boundaries rather than evaluating evidence—it can distort the entire landscape of inquiry. Understanding that weaponization is essential if we want to see why certain questions remain taboo even when the evidence suggests complexity.

Skepticism is one of the noblest tools in human thought. It is the discipline of doubt, the refusal to accept claims without adequate support, the habit of testing ideas against evidence and logic. Without skepticism, science collapses into storytelling. Yet skepticism has a shadow form. It can be turned from a method into a posture, from a tool into a weapon. When skepticism is weaponized, it stops being a way to evaluate claims and becomes a way to manage status—who is allowed to speak, which questions are permitted, and what kinds of curiosity are treated as respectable. In the debates surrounding human origins, this weaponization is not an abstract possibility. It is a pattern that shapes what gets explored and how.

The first sign that skepticism has become weaponized is when it targets the question rather than the evidence. A genuine skeptic asks, "What evidence supports this? What alternative explanations exist? What predictions follow?" A weaponized skeptic asks, "Why are you asking this? Who else asks this? What kind of person believes this?" In other words, the focus shifts from content to character and association. The question becomes guilty by proximity. If a topic has been discussed by sensationalists, then any mention of it is treated as contamination. This is an efficient social tactic because it avoids the hard work of analysis. It collapses nuance into stereotype.

The second sign is asymmetry of standards. Healthy skepticism applies similar evidential expectations across comparable claims. Weaponized skepticism applies extreme standards to disfavored ideas and relaxed standards to favored ones. In practice, this can look like an insistence that anomalies must meet an impossible burden of proof while the mainstream narrative is treated as self-evident even when it rests on inference and partial data. This does not mean mainstream models are wrong. It means that the cultural defense of mainstream models can slip into unfairness. When skepticism becomes a gatekeeping tool, it is less about truth and more about maintaining stability.

In human origins, the conditions for weaponization are unusually strong. The field sits at the intersection of real uncertainty and intense public interest. There are genuine anomalies: ghost ancestries, unexpected dispersals, ambiguous fossils, abrupt-looking cultural shifts. There are also many overreaches, misuses, and sensational claims. In that environment, skepticism is both necessary and easily corrupted. It becomes tempting to protect the field by policing discourse rather than by refining models. The skeptic's role can morph into that of a cultural immune system, attacking anything that resembles infection. But immune systems can also attack the body itself if they become too aggressive. Legitimate questions can be treated as pathological.

Ridicule is one of the primary weapons. Ridicule is not argument. It is social punishment. It signals to observers that a question is not worthy of serious engagement. This can be effective because most people fear being associated with foolishness more than they fear being wrong. Ridicule creates a chilling effect: even those who have reasonable questions choose silence rather than risk social humiliation. In academic environments, where reputation is currency, ridicule can be more damaging than disagreement. A researcher can survive criticism; they may not survive being branded as unserious.

Another weapon is the strategic conflation of categories. When skepticism is weaponized, it becomes common to bundle distinct ideas together and dismiss them as a single "fringe package." A careful discussion of admixture and unknown contributors can be lumped together with claims of deliberate suppression. A conversation about missing data can be lumped together with claims of hidden advanced civilizations. A symbolic reading of myth as cultural memory can be lumped together with literal claims of non-human genetic engineering. Once bundled, the entire package can be rejected without discrimination. This is epistemically lazy, but socially efficient. It prevents the need to draw fine distinctions.

The weaponization of skepticism also relies on rhetorical traps. One common trap is demanding absolute proof before allowing inquiry to proceed. But inquiry exists precisely because absolute proof is not yet available. If a skeptic insists that only fully proven claims deserve attention, then entire frontiers of science would never be explored. In practice, research often begins with anomalies, weak signals, and partial patterns. It advances by testing, refining, and sometimes discarding. Weaponized skepticism collapses this process by treating uncertainty as disqualification rather than as the natural condition of exploration.

Another trap is framing curiosity itself as dangerous. Because human origins has a history of misuse—racism, eugenics, pseudoscience—some skepticism adopts an implicit moral posture: certain questions should not be asked because they could be misused. Ethical caution is legitimate, but it becomes weaponized when it functions as a blanket veto rather than as a demand for careful framing. The result is a paradox: the field becomes less able to address complexity precisely because it fears how complexity might be misinterpreted. But avoiding complexity does not prevent misuse. It simply leaves the public vulnerable to unregulated storytellers who will fill the gap.

There is also an institutional incentive behind weaponized skepticism: it preserves narrative coherence. Public communication often rewards clean stories. A clean story is easier to teach, easier to defend, and less likely to be twisted. Skepticism can therefore be used to enforce simplicity. Anomalies become "noise." Alternative interpretations become "distractions." Interdisciplinary syntheses become "speculation." The field protects its public image by narrowing the range of acceptable discourse. Again, this does not require malicious intent. It can arise from well-meaning attempts to prevent confusion. But the effect is still a form of control: inquiry is guided toward what is safe to communicate, not only toward what is true.

The tragedy of weaponized skepticism is that it ultimately weakens the very thing it claims to protect. Science thrives when skepticism is paired with curiosity—when doubt is used to refine questions, not to punish them. If skepticism becomes primarily a social weapon, then it drives complexity into the shadows. Researchers avoid high-risk questions. The public senses that uncomfortable topics are being dodged. That perception can feed mistrust and amplify conspiratorial interpretations. In other words, weaponized skepticism can produce the very fringe culture it fears by leaving a vacuum where nuanced inquiry should be.

A healthier posture is not naïve openness to every claim. It is disciplined openness: the willingness to ask difficult questions while maintaining clear standards for evidence. It is the ability to separate anomalies from fantasies without dismissing anomalies as fantasies. It is the ability to critique without contempt. It is the recognition that the human story is complex and that complexity should not be treated as a threat.

This leads directly into how narratives are policed—not through censorship, but through boundary maintenance, language control, and social reinforcement of "acceptable" stories. The next subsection examines those mechanisms: the quiet ways communities decide what is sayable, which interpretations are publishable, and how the public story of human origins is kept coherent even when the underlying evidence becomes increasingly web-like.

Narratives are rarely policed by a single authority. They are policed by ecosystems—by incentives, norms, gatekeepers, and the quiet human tendency to protect shared stories from disruption. In a field like human origins, where evidence is partial and public interest is intense, policing does not usually look like censorship. It looks like boundary maintenance. It looks like deciding which questions are "serious," which framings are "responsible," which metaphors are "dangerous," and which uncertainties are "too confusing" to foreground. Over time, these decisions shape what the public thinks the science says, and they shape what researchers feel safe emphasizing even when the underlying data is more complex.

The first mechanism is language control. Words determine the emotional temperature of a topic. Terms like "anomaly," "unknown," and "missing" can be treated as invitations to curiosity, but they can also be treated as triggers for sensationalism. Because of that risk, communities often develop preferred phrases that minimize interpretive openings. Instead of saying "unknown hominin," one might say "unsampled population." Instead of "ghost ancestry," one might say "unmodeled gene flow." These phrases are technically cautious, but they also narrow the imaginative space. They make the topic sound like a minor statistical adjustment rather than a profound implication about how many human groups once existed. This is not dishonest; it is a form of narrative risk management. It is how a community keeps its story from being hijacked.

A second mechanism is the control of framing. The same evidence can be presented as revolutionary or as incremental. A cautious field often prefers incremental framing because it reduces controversy. New findings are described as "refining" rather than "replacing." Uncertainty is treated as "future work" rather than as a challenge to current models. Competing interpretations are downplayed to preserve coherence. This is especially common when communicating to the public, because the public tends to interpret scientific disagreement as evidence that "nobody knows anything." To avoid that, institutions present a unified narrative. The cost is that the public receives a simplified version of reality, and the simplified version becomes a cultural expectation that then pressures researchers to conform.

The third mechanism is gatekeeping through peer review and editorial culture. Peer review is essential, but it is also a social process. Reviewers are trained within existing frameworks and may unconsciously reward papers that align with those frameworks. They may be harsher on work that proposes broad synthesis or that challenges established interpretations. Again, this is not necessarily ideological suppression; it is how communities enforce methodological norms. But in effect, it can narrow the space of publishable storytelling. Certain models become the default. Alternatives must fight uphill. A young researcher learns quickly which interpretations produce smooth reviews and which interpretations invite endless demands for proof.

A fourth mechanism is credential boundary enforcement. Not all critiques of narratives are treated equally. A question asked by a leading expert is treated as provocative. The same question asked by an outsider is treated as suspicious. This difference is not always unfair—expertise matters—but it can become a form of narrative policing when it shuts down inquiry based on status rather than on evidence. Outsiders are more likely to be dismissed as sensationalists, even when they raise legitimate issues. Insiders are allowed to be curious because they are trusted to be careful. This creates a closed loop: only trusted voices can explore the edges, and the edges remain inaccessible to broader discourse.

A fifth mechanism is public relations and institutional storytelling. Museums, universities, and media outlets often prefer stable narratives because they are easier to communicate and less likely to attract controversy. A museum exhibit is not built for constant revision. A documentary script needs a coherent arc. Press releases often highlight certainty and minimize nuance. Over time, these public narratives become canonical in the popular mind. Once canonical, they become difficult to change. Researchers may privately discuss complexity, but public-facing institutions continue to present the simplified story because it has become part of their identity and brand. Narrative policing here is not enforced by reprimand; it is enforced by inertia.

A sixth mechanism is social reinforcement within academic communities. Researchers learn quickly which topics cause colleagues to raise eyebrows. They learn which conference questions are met with supportive engagement and which are met with cold silence. They learn which words trigger eye-rolls. This social feedback shapes behavior long before any formal punishment occurs. A scientist who wants to stay in good standing internalizes the boundaries. They self-edit. They choose safer framings. Over decades, this self-editing becomes culture. Narrative policing becomes invisible because it is embedded in what feels "normal."

Importantly, narrative policing also happens through the management of uncertainty. Every scientific field contains uncertainty, but not every field faces the same public pressure. In human origins, uncertainty is often treated as threatening because it can be exploited by ideologues and sensationalists. So institutions try to contain it. They emphasize what is known and treat unknowns as minor. But when the unknowns accumulate—ghost ancestries, ambiguous fossils, missing regions—the containment becomes harder. The public senses the gap. Some people respond by mistrusting institutions, imagining suppression. Others respond by clinging more tightly to the simplified narrative. The tension increases, and narrative policing becomes more aggressive, not less, because the stakes feel higher.

This is the central irony: the more complex the evidence becomes, the more some institutions feel compelled to present a simple story. Simplicity becomes a protective shell. But shells can crack. When the public eventually encounters complexity through genetics news, new fossil finds, or alternative storytelling, the contrast between the official narrative and the messy reality can feel like deception, even when it is just communication strategy. That perception can fuel exactly the fringe narratives that institutions are trying to prevent.

A healthier approach would treat complexity as normal and train the public to live with uncertainty without panic. But that requires cultural maturity, and it requires institutions willing to risk short-term confusion for long-term trust. Until then, narrative policing will remain a dominant force: a set of subtle pressures that shape which questions are allowed to be asked in public, which questions are explored quietly, and which questions are treated as socially radioactive.

This brings us to the cost. When inquiry is narrowed and certain questions are discouraged, something is lost—not only potential discoveries, but the scientific imagination itself. The next subsection examines that cost directly: what is lost when inquiry is closed, when ridicule replaces debate, and when the desire for narrative stability overrides the field's duty to explore uncertainty honestly.

When inquiry is closed, the loss is not always visible. There is no single experiment that fails in public. No siren announces the questions that were never asked. The loss is quieter: a narrowing of imagination, a thinning of intellectual courage, a gradual reduction in the range of models the culture permits. Over time, this narrowing shapes not only what we know, but what we believe is knowable. In a field as consequential as human origins—where the evidence is fragmentary and the stakes are existential—closing inquiry does more than delay discovery. It distorts the story we tell about ourselves.

The first loss is the loss of anomaly sensitivity. Science advances not only by confirming patterns, but by noticing what does not fit. An anomaly is not automatically a revolution, but it is a signal that the current model may be incomplete. If anomalies are routinely treated as nuisances—errors, noise, distractions—then the field becomes less sensitive to the very data that could refine it. This is especially damaging in human evolution because the record is already broken. When the archive is fragmentary, every fragment matters. If the culture trains itself to dismiss odd fragments quickly to preserve narrative coherence, then it risks missing the deeper patterns hidden in those fragments.

The second loss is the loss of alternative hypotheses. A healthy scientific culture does not cling to a single story; it maintains competing models and tests them against emerging evidence. Competing models are not indulgences. They are protections against premature closure. When inquiry is socially policed, alternative hypotheses become socially dangerous, and researchers stop proposing them. This creates an intellectual monoculture. In a monoculture, even accurate models become brittle because they are not constantly challenged and strengthened. A brittle model can persist for decades not because it is correct in every detail, but because it has become the only socially acceptable way to think.

The third loss is the loss of interdisciplinary synthesis. Human origins sits at the intersection of genetics, paleoanthropology, archaeology, ecology, and cognitive science. None of these disciplines alone can tell the full story. Genetics can infer contact and ancestry but cannot reconstruct lived culture. Archaeology can reconstruct behavior but often cannot identify biological lineages. Fossils can show form but rarely show social life. Ecology can describe pressures but not the exact choices populations made. The full narrative emerges only when these disciplines converse. Yet interdisciplinary work is among the most socially risky, because it crosses boundaries and invites criticism from multiple sides. If inquiry closes, the field retreats into specialization, and the story fractures into separate conversations that never quite integrate. The result is a public narrative that feels coherent, but a scientific reality that remains siloed and incomplete.

The fourth loss is the loss of public trust. This may sound counterintuitive, because narrative policing is often justified as a way to protect public understanding. But trust does not ultimately depend on simplicity; it depends on honesty. When the public senses that certain questions are dismissed with ridicule rather than addressed with clarity, suspicion grows. When people hear that the evidence is complex but are offered only simplified stories, they look elsewhere for explanations. That "elsewhere" is often filled with confident storytellers who are not constrained by rigor. In that environment, closing inquiry does not eliminate fringe narratives; it fertilizes them. It creates the impression that official narratives are maintained by social pressure rather than by evidence, and that impression can be corrosive even when the official narrative is broadly correct.

The fifth loss is the loss of ethical clarity. Ironically, avoiding certain questions because they could be misused can make misuse more likely. If professionals refuse to discuss complexity openly, then complexity becomes the territory of those who are willing to misuse it. The public then encounters the topic through ideologically driven filters—racist distortions, conspiracy framings, sensational hybrid fantasies. A responsible scientific culture should not deny complexity; it should teach how to handle it without moral collapse. That includes distinguishing between biological variation and moral worth, between ancestry and hierarchy, between explanation and justification. Closing inquiry avoids the teaching responsibility and leaves the cultural field open to exploitation.

The sixth loss is the loss of intellectual humility. When inquiry closes, the field begins to treat its current story as if it were more complete than it really is. This can happen even to careful scientists, because institutional narratives harden. Over time, the community confuses the map with the territory. It forgets that the fossil record is partial, that genetic models are inferential, that large parts of the world are under-sampled, that many lineages may have existed without leaving clear signatures. The field becomes less comfortable admitting uncertainty. Yet uncertainty is not weakness in a fragmentary domain; it is accuracy. Losing the habit of humility is losing the habit of truth.

The seventh loss is the loss of human meaning-making in its healthiest form. Humans will always tell stories about origins. If science refuses to address certain origin questions openly, people will not stop asking them; they will simply ask them in less disciplined spaces. The mythic impulse will fill the vacuum. This is not a condemnation of myth; myth can preserve insight and memory structures. But without disciplined conversation, myth can harden into literalism and become a substitute for evidence. Closing inquiry therefore risks creating a split culture: science as a guarded priesthood of acceptable statements, and a public sphere of unregulated narratives. That split is unstable.

In the context of this volume, what is lost when inquiry is closed is precisely what the book is trying to recover: the right to treat human evolution as complex, hybridized, discontinuous, and partially unknown without collapsing into either naive certainty or unmoored fantasy. The missing chapters of human evolution are missing in part because the archive is broken. But they are also missing because certain kinds of curiosity are socially punished. The result is an origin story that is too smooth for the evidence we already possess.

The final subsection of this chapter is therefore not a call to believe every anomaly. It is a call to reclaim the right to ask. To separate skepticism from ridicule. To separate methodological rigor from narrative policing. To make room for disciplined exploration of the unknown. If the human story is a web rather than a ladder, then we need a culture of inquiry that can tolerate webs—messy, interconnected, and unresolved—without demanding that everything be made simple before it is allowed to be said.

If inquiry can be narrowed without anyone issuing a formal ban, then the right to ask must be reclaimed without waiting for anyone to grant permission. This reclamation is not rebellion against science. It is loyalty to science's deepest purpose: to follow evidence where it leads, to admit uncertainty honestly, and to treat questions as the engines of discovery rather than as social threats. Reclaiming the right to ask does not mean abandoning standards. It means separating rigor from intimidation, and skepticism from contempt. It means restoring a cultural atmosphere in which curiosity is not automatically treated as subversion.

The first step is to normalize uncertainty. In human origins, uncertainty is not a temporary failure; it is the default condition of working with a broken archive. Fossils are sparse. DNA preservation is uneven. Archaeological sites cluster by chance and geography. Models are inferential. If a culture demands complete certainty before it allows questions, then it will permanently forbid itself from exploring its own history. Reclaiming the right to ask begins with a simple confession: we do not know everything, and we should not pretend we do. This confession does not weaken science. It strengthens it, because it aligns public expectations with the reality of evidence.

The second step is to distinguish question from claim. Many taboos arise because a question is treated as an assertion. If someone asks whether unknown hominin contributions exist, they may be accused of claiming that "mystery beings" shaped humanity. If someone asks whether myths preserve memory structures, they may be accused of claiming myths are literal history. If someone asks whether cultural resets occurred, they may be accused of claiming advanced lost civilizations.

The right to ask requires the ability to explore possibilities without being forced into conclusions. A question is a tool for mapping uncertainty. It should not be treated as a declaration of belief.

The third step is to practice disciplined curiosity. Disciplined curiosity is curiosity that is willing to be corrected. It is curiosity that seeks the strongest counterarguments. It is curiosity that welcomes falsification. It is curiosity that values precision over drama. This is the middle path between timid silence and reckless speculation. A disciplined culture can ask bold questions without collapsing into fantasy, because it treats evidence as the final arbiter and treats models as provisional. When disciplined curiosity becomes normal, taboo loses its grip. Questions can be asked, examined, and—if necessary—discarded without social punishment.

The fourth step is to restore fairness in skepticism. Skepticism should interrogate methods, not motives. It should challenge claims, not shame people. It should apply standards consistently. This requires a cultural shift, because weaponized skepticism often feels like virtue: it signals intelligence, seriousness, and immunity to gullibility. But seriousness is not measured by how quickly one can mock. It is measured by how carefully one can evaluate. Reclaiming the right to ask means rebuilding the ethic of critique: firm, precise, and free of contempt. In a healthier environment, disagreement becomes an engine of refinement rather than a mechanism of exclusion.

The fifth step is to protect space for synthesis. Human origins demands synthesis because the evidence is distributed across domains. Genetics without archaeology becomes abstract. Archaeology without genetics can misattribute populations. Fossils without ecology can misread adaptation. Myth without discipline becomes literalism. Yet synthesis is often punished because it is harder to police. It crosses boundaries and exposes uncertainty. Reclaiming inquiry means legitimizing the integrative imagination—the willingness to hold multiple archives in view at once and to build models that can be tested across disciplines. This does not require reducing everything to one story. It requires building bridges where evidence permits and admitting gaps where it does not.

The sixth step is to separate the ethical fear of misuse from the intellectual fear of complexity. It is true that human origins can be misused. It has been misused. But the solution to misuse is not silence; it is responsible framing and education. A mature culture can hold complex truths—multiple lineages, admixture, regional variation—without turning them into moral hierarchies. It can teach that biological difference is not moral rank. It can teach that ancestry does not justify domination. It can teach that the existence of lost human types is a testament to deep-time diversity, not a license for modern prejudice. Reclaiming the right to ask includes reclaiming the responsibility to teach nuance rather than hiding complexity.

The seventh step is to recognize that public trust depends on transparency, not on narrative control. People can tolerate complexity when it is communicated honestly. What they struggle to tolerate is the sense that complexity is being managed through ridicule and dismissal. If institutions want trust, they must invite the public into the process of uncertainty: how models are built, how they are tested, why some interpretations are rejected, and what remains unknown. This does not mean giving equal weight to every claim. It means showing the difference between disciplined speculation and unfounded assertion, and explaining why that difference matters.

In the context of this book, reclaiming the right to ask is not the same as claiming the right to conclude. The volume's core thesis insists that modern humans emerged from a complex, hybridized past and that many chapters remain misunderstood or missing. That thesis is strengthened, not weakened, by careful separation between established evidence and interpretive framing. Genetics supports complexity. Fossils support plurality. Ecology supports discontinuity. Myth supports memory structures and identity narratives. But no single archive licenses limitless claims. The right to ask must be paired with the discipline to stop where evidence stops, and to mark clearly where hypothesis begins.

This chapter's function is therefore to clear a path forward. It is to explain why the conversation becomes socially charged, and to insist that charge should not dictate truth. The next chapter turns back toward reconstruction: building a model of human origins that can hold hybridization, missing populations, and the dialogue between science and myth without surrendering to either dogma or fantasy. To do that, we must be able to ask questions in full daylight.

CHAPTER 14 — RECONSTRUCTING THE MISSING CHAPTERS

A new model of human origins does not begin by throwing out the old one. It begins by admitting what the old one was built to do: provide a coherent story under conditions of limited evidence. For decades, that coherence was achieved by simplification. A main line was drawn, side branches were minimized, and the public narrative emphasized clarity: modern humans emerged, spread, and replaced others. Even when the science itself contained caveats, the story presented to most people was tidy enough to memorize. But the evidence now refuses tidiness. Genomes have revealed interbreeding and ghost ancestry. Fossils have exposed unexpected diversity. Archaeology has shown early dispersals and complex behaviors appearing in uneven bursts. Ecology has clarified the volatility of the world in which these populations lived. The old model was not "wrong" in the way a failed myth is wrong. It was incomplete in the way an early map is incomplete: it shows the coastlines and hides the interior.

So what does a new model look like if it is built to reflect complexity rather than suppress it? It looks less like a ladder and more like an ecosystem. It treats humanity not as a single river running cleanly from source to sea, but as a watershed—multiple streams branching, merging, disappearing underground, resurfacing, and feeding into a shared downstream flow. In this model, Homo sapiens is real and distinct, but it is not solitary. It is a lineage embedded in a larger human landscape. Our species is not the only actor; it is the survivor that carries traces of other actors within it.

The first pillar of this model is plurality. The Pleistocene world was not waiting for modern humans to arrive. It already contained multiple hominin populations. Some were widely distributed. Some were regional.

Some likely remained isolated long enough to develop distinctive traits. Some may have been transitional forms that existed for long spans without fitting neatly into our categories. The plurality model says: assume diversity unless proven otherwise. It treats the existence of unknown or poorly defined populations as expected under conditions of fragmentary fossils and uneven sampling. The burden is no longer on the idea of plurality to justify itself. The burden shifts to any model that insists the world was simple.

The second pillar is permeability. Population boundaries mattered, but they were not sealed. Interbreeding was not an exotic exception; it was part of how hominins interacted when they came into contact and were biologically compatible. This does not mean constant mixing across all boundaries. It means that when climate shifts forced populations into shared refuges or opened corridors between regions, gene flow could occur. The result is not a clean pattern of replacement, but a pattern of incorporation and absorption. In this model, extinction often means loss of distinctness rather than total disappearance. Populations vanish as names and forms, but pieces of them persist.

The third pillar is volatility. The world in which these populations lived was not stable enough to support a smooth narrative. Glacial cycles, sea-level changes, habitat reorganizations, and resource collapses repeatedly disrupted populations. Bottlenecks narrowed the human line. Refuges concentrated interaction. Expansions followed contractions. The model therefore treats discontinuity as normal. It expects gaps in the record not only because preservation is uneven, but because populations genuinely rose and fell. It also expects cultural resets: knowledge systems can collapse when populations shrink, and complexity can reaccumulate quickly when conditions improve. The archaeological record's "bursts" of behavior are not necessarily sudden evolutions of intelligence; they may be rebounds after disruption.

The fourth pillar is multi-layered inheritance. In the old simplified story, inheritance was treated primarily as genetic descent. In the new model, inheritance operates on at least three channels at once: genes, culture, and ecology. Genes carry ancestry and certain adaptive variants. Culture carries survival knowledge and social organization. Ecology shapes the pressures that determine which genes and cultures persist. These channels interact. Cultural flexibility can allow survival without genetic change. Genetic diversity can support biological resilience in new environments. Ecological volatility can filter both. The new model insists that human origins cannot be explained by any single channel. It is an entangled process.

The fifth pillar is asymmetry. Human history is not evenly distributed across space. Some regions preserve evidence better. Some regions were ecological crossroads. Some regions served as refuges. Some regions were under-sampled. The model therefore rejects the assumption that the best-known evidence is automatically the most representative. It anticipates that major chapters may be missing simply because they occurred in places where fossils and DNA do not preserve well or where research has been limited. This is not a license for fantasy. It is a methodological warning against overconfidence.

From these pillars, a new narrative emerges—one that remains grounded but no longer overly clean. Modern humans are not the result of a single, isolated lineage that climbed alone. We are the downstream expression of a networked past. Our lineage formed within a wider field of human diversity, experienced repeated crises, interacted with neighboring populations, incorporated some of them, and then expanded through a world reshaped by catastrophe. In this model, the question "Where did we come from?" becomes less like identifying a single birthplace and more like tracing a braided history: multiple origins converging into a survivor population that carried fragments forward.

This framing changes how we interpret "missing chapters." Missing chapters are not only unknown because someone refused to look. They are unknown because the archive is broken and because the process itself was messy. If humanity is composite, then some ancestral contributors may never be tied to clear fossil identities. If volatility repeatedly reset populations, then cultural sophistication may have risen and fallen more than once. If permeability allowed absorption, then the disappearance of "other humans" may be partly a story of integration rather than replacement. The missing chapters are not mere blanks. They are the places where our old narrative demanded simplicity and therefore failed to ask the right questions.

Yet a new model must remain disciplined. It must clearly separate what is established from what is interpretive. Plurality, permeability, and volatility are strongly supported as general principles by the convergence of fossils, genetics, and climate science. The precise number of lost populations, the timing and location of every admixture event, and the degree to which cultural resets occurred in specific regions—these remain open. The model is not a finished map. It is a better orientation: a way to think that matches the evidence's complexity and keeps inquiry honest.

The next subsection deepens one of these pillars: hybridization not as embarrassment, but as creative force. If gene flow and absorption shaped the human line, then hybridization is not a footnote. It is one of the engines of our survival story. And understanding that engine may be the key to reconstructing why modern humans became what they became.

For much of modern history, hybridization carried a stigma. In popular imagination, mixing implies impurity, and impurity implies weakness. Even within scientific culture, hybridization was often treated as an awkward complication—a messy side effect in what people preferred to imagine as clean evolutionary divergence. But the genomic revolution has forced a reversal. Hybridization is not a minor footnote to the human story. It is one of the mechanisms by which the human story became possible. In a volatile world crowded with related populations, gene flow was not merely a leak in the system. It was part of the system's creative capacity.

To call hybridization a creative force is not to romanticize it. It is to recognize what gene flow does in evolutionary terms. It moves variation between populations. It can introduce alleles that have already been tested by selection in one environment into another population that encounters similar pressures. It can increase genetic diversity, which can be a buffer against disease and environmental change. It can break the isolation that makes small populations vulnerable to inbreeding and drift. In short, hybridization can provide resilience. In a world where climates oscillated, habitats shifted, and bottlenecks narrowed populations, resilience could mean survival.

This is especially relevant for Homo sapiens because our success appears tied not to a single specialized niche, but to a capacity to inhabit many. Humans expanded across deserts, forests, coastlines, steppes, and high-altitude plateaus. That expansion required flexibility, and flexibility is not only cultural. It also has biological components: immunity, metabolism, oxygen regulation, skin pigmentation, and more. Some of these traits can evolve within a lineage, but hybridization can accelerate the process by introducing useful variants already present in neighboring populations. Hybridization does not create traits out of nothing; it redistributes what exists. But redistribution can change the speed and direction of adaptation.

At the demographic level, hybridization is also a way out of extinction traps. Small populations, especially those isolated in refuges, face a brutal arithmetic. Low numbers increase inbreeding, inbreeding increases genetic load, and genetic load reduces reproductive success. A population can spiral downward. Contact with a neighboring population and the exchange of mates can interrupt that spiral. It can restore diversity and increase viability. In such cases, hybridization is not a curiosity; it is a lifeline. And over deep time, lifelines determine which lineages persist.

Hybridization is also a mechanism of cultural contact. When populations interbreed, they do not exchange only genes. They exchange behaviors. Even if the cultural exchange is asymmetric—one group absorbing another—there can be borrowing of tool traditions, hunting strategies, survival knowledge, ritual forms, and social practices. Over time, the merged population becomes culturally composite as well as genetically composite. This complicates any attempt to draw sharp lines between "our culture" and "their culture" in deep time. The human story is not simply the spread of one culture. It is the emergence of blended cultural systems in contact zones, refined by selection pressures and demographic expansion.

Yet hybridization also has boundaries. It is not a universal blending of everything with everything. Populations can interbreed and still remain distinct if gene flow is limited. Hybridization can occur in pulses rather than continuously. It can be shaped by geography, by social behavior, by mate preferences, and by competition. It can produce mixed offspring that are fully viable, or it can be constrained by reduced fertility in certain crosses, depending on how divergent populations are. In the human case, the fact that admixture occurred at all indicates a level of compatibility, but compatibility does not imply seamless integration. Hybridization is a complex phenomenon, not a simple merging.

This complexity matters because it prevents the topic from becoming ideology. Hybridization is not a moral symbol; it is a biological process. It does not imply that "mixing is always good" or "mixing is always bad." It implies that, under certain conditions, gene flow can be adaptive, and under other conditions it can be neutral or even detrimental. A disciplined model recognizes this and avoids turning biology into social metaphor. The creative force of hybridization is not a slogan. It is a conditional reality: in volatile environments and small populations, diversity can be life.

In the new model of human origins, hybridization changes what we mean by "who we are." It suggests that Homo sapiens is not a sealed category that developed in splendid isolation. It suggests that modern humanity is, at least in part, an emergent outcome of contact. Our genome contains inherited segments from other hominin groups. Our ancestry includes ghosts—contributors we can detect but not fully define. Our biological inheritance is therefore a mosaic of multiple human histories. Hybridization is one reason that mosaic exists.

But the creative force of hybridization is not only genetic. It is also narrative. It reshapes how we tell the human story. A replacement narrative implies winners and losers, with one lineage triumphing through superiority. A hybridization narrative implies integration, absorption, and composite survival. It suggests that what survived is not only the strongest, but the most connected, the most flexible, and the most capable of incorporating what it encountered. It implies that our lineage's strength may have been partly ecological and social: the ability to move across landscapes, to form networks, and to persist through instability by not remaining isolated.

This is where hybridization becomes a bridge between science and certain mythic intuitions—without collapsing one into the other. Many mythologies speak of unions between humans and "others," producing exceptional lineages. A disciplined reading does not equate mythic beings with archaic hominins. But it can recognize that the idea of mixed origins resonates with a real biological truth: humans are not the product of one isolated line. Hybridization was part of our deep ancestry. The mythic motif may be symbolic, but the biological reality is concrete: we carry inherited fragments from other humans who once walked the world.

The next subsection takes the model into more speculative territory and must therefore be handled with strict care: the question of non-human intelligence in prehistory. The goal will not be to assert what cannot be proven. The goal will be to define the boundary between evidence and interpretation, and to explore why this idea appears so persistently in myth and modern imagination, while maintaining the discipline that separates hypothesis from established history.

The question of non-human intelligence in prehistory is one of those topics that attracts heat before it attracts light. It triggers immediate associations—aliens, secret histories, engineered humanity—and because those associations have been used irresponsibly, the entire question is often treated as unserious by default. Yet the persistence of the motif across cultures, and the psychological power it holds in modern imagination, makes it worth addressing carefully. Not as a claim, not as a conclusion, but as a boundary question: what can evidence support, what can interpretation suggest, and why does the human mind keep returning to the idea that "others" once shaped the human world?

Begin with what is established. Prehistory, as we can responsibly reconstruct it, is full of non-human intelligences in a broad sense. Humans lived alongside animals with complex cognition—wolves, corvids, elephants, dolphins, primates—and interacted with them in ways that influenced culture. The domestication of the dog, for example, represents a long co-evolutionary partnership between two intelligent species. This is not the "non-human intelligence" people usually mean when they invoke Watchers or sky teachers, but it matters because it grounds the conversation. Humans have always been shaped by relationships with other minds. We are a species that lives in cognitive ecologies.

But the question in this subsection is narrower and more charged: did something beyond the known spectrum of human and animal cognition intervene in human prehistory in a way that altered the trajectory of culture or biology? Here we must be disciplined. There is no mainstream scientific evidence that a technologically advanced non-human species engineered Homo sapiens, taught metallurgy, or directed civilization. There is evidence of human ingenuity emerging through complex social and ecological processes. There is evidence of cultural accumulation and spread. There is evidence of hybridization among hominins. But there is not evidence, in the strict sense, of an external intelligence directing the process.

So why does the idea persist? One reason is that human history contains genuine discontinuities that are emotionally difficult to hold. The archaeological record can look punctuated. Symbolic behavior appears in bursts. Complex cultural systems seem to flare and vanish. Monuments and sophisticated practices can appear in contexts that feel sudden.

Even when these patterns can be explained by demography, preservation bias, and cultural accumulation, they still feel mysterious. Human beings, confronted with mystery, often prefer agency-based explanations. We would rather imagine an actor than accept the slow, emergent complexity of many interacting factors. “Someone did it” is a cognitively satisfying narrative shape.

Another reason is the tension between the simplicity of public narratives and the complexity of the evidence. When people are taught a clean ladder story and then encounter real data that looks web-like—multiple hominins, admixture, missing populations, ancient migrations—the gap can feel like hidden truth. Into that gap, the imagination pours. Non-human intelligence becomes a placeholder for what the simplified story cannot accommodate. It becomes an explanatory magnet for everything that feels anomalous. This is not evidence for the idea; it is a diagnosis of why the idea thrives.

A third reason is mythic memory structure. Many cultures preserve motifs of teachers from beyond the human community: beings who descend, instruct, impose law, teach agriculture, teach metallurgy, teach writing, and then withdraw. These motifs appear in forms shaped by local environments and religious systems, but the structure is recognizable: knowledge arrives from an “other.” In a disciplined reading, this does not require literal extraterrestrials or engineers. It can reflect how humans interpret the arrival of new knowledge systems that originate elsewhere—brought by migrants, traders, conquerors, or specialist elites whose skills seem magical to outsiders. To a community that has never seen a certain craft or a certain astronomical calendar, the bearers of that knowledge can appear non-human in status. Myth then encodes that social reality in symbolic form.

This connects to a more subtle, historically plausible mechanism: elite differentiation. Many early societies developed priestly or specialist classes who held knowledge of calendars, navigation, metallurgy, ritual, and law. These specialists were often treated as mediators between worlds—between gods and humans, between sky and earth. Over time, myth could retroject this social structure backward, turning human specialists into semi-divine teachers, or teachers into beings who were never fully human. The psychological function is clear: it legitimizes authority by placing its origin beyond ordinary humanity. The myth says: our order was founded by beings above us, therefore our institutions are sacred.

There is also a cognitive and emotional reason the motif persists: it externalizes the burden of human creativity. Human civilization is astonishing. The capacity to build complex symbolic systems, to manipulate matter, to carve monuments, to map the skies—these achievements can feel disproportionate to our animal origins. The mind searches for an explanation that matches the magnitude of the outcome. “We did it ourselves” can feel too small, even when it is true. Non-human teachers offer a narrative that flatters the wonder of civilization by making it a gift. It preserves awe by externalizing origin.

Yet to engage this question responsibly, we must outline what kinds of evidence would actually matter if one were to test such a hypothesis. It would not be mythic similarity alone, because humans converge in symbolism. It would not be the existence of complex ancient structures, because humans can build them. It would not be gaps in the record, because gaps are normal. It would require material signatures that cannot plausibly be generated by known human technologies and social processes in the relevant contexts—clear artifacts, reproducible traces, or unambiguous anomalies that withstand alternative explanations. In the absence of that, the idea remains an interpretation, a modern myth of its own: a way of explaining wonder and discontinuity, not a conclusion supported by the current evidence.

However, the question still performs an important function in this book. It forces us to confront the limits of our models. When people reach for non-human intelligence, they are often responding to real feelings: that the story is incomplete, that complexity is under-acknowledged, that discontinuities exist. The responsible move is not to mock the impulse. The responsible move is to redirect it toward disciplined questions: What unknown populations existed? What cultural networks spread knowledge? What bottlenecks reset societies? What specialist classes preserved astronomy and ritual? What mythic motifs might encode social memory rather than literal events? In other words, the impulse can be used as a compass—even if the destination it points to is symbolic rather than literal.

That brings us naturally to the next subsection: science and myth in dialogue. If myth cannot be treated as literal evidence, it also cannot be dismissed as meaningless noise. Myth is a human archive of perception, trauma, identity, and social memory. The challenge is to read it with discipline, letting science anchor what can be anchored, and letting myth illuminate what science alone cannot easily capture: the felt experience of discontinuity, the meaning humans assign to catastrophe, and the symbolic memory of a world that once contained more kinds of humans than the present admits.

Science and myth are often treated as enemies because they make different kinds of claims. Science aims to describe what happened using testable methods and material evidence. Myth aims to encode meaning—identity, morality, memory, and cosmic order—through symbolic narrative. When they are forced into competition, both are damaged. Science becomes defensive, narrowing itself to what can be proven immediately. Myth becomes literalized, turned into pseudo-history to compete with scientific authority. The dialogue this book requires is a different relationship: science as anchor, myth as archive of human perception and social memory. Not a fusion where myth becomes "evidence" in the scientific sense, but a conversation where each corrects the other's blind spots.

The first step in that dialogue is to clarify what myth is good at preserving. Myth does not preserve dates the way a stratigraphic layer does. It does not preserve genomes the way DNA does. But it can preserve patterns of experience: floods and droughts, migrations and displacements, encounters with strangers, the rise and fall of orders, the terror of catastrophe, the loss of knowledge, the legitimacy of elites, the boundary between human and other. These are not trivial. They are the psychological and social realities that shape how communities survive and rebuild. Even when the mythic narrative is cosmological—gods, monsters, world ages—the underlying structure often maps to real human experiences. The myth is a compressed memory in symbolic form.

This matters because the scientific archive of prehistory is incomplete in precisely the places where myth is most active. Science struggles to recover the inner life of extinct communities: what they feared, what they revered, how they interpreted catastrophe, how they legitimized authority, how they experienced encounters with unfamiliar peoples. Archaeology can infer behavior, but it rarely captures meaning directly. Genetics can infer contact, but it cannot describe how contact felt. Myth can. Not with literal precision, but with emotional truth. A disciplined dialogue therefore uses science to constrain what could have happened, and myth to illuminate how humans made sense of what happened.

The second step is to recognize that myth is often born at moments of transition. Stable societies tell stories, but societies under stress generate the most powerful myths—because transition demands explanation. When worlds change—through climate collapse, migration, conflict, or social reorganization—communities need narratives that stabilize identity. They need to know who they are in the new order. They need to explain loss. They need to justify authority. They need to encode survival lessons. Myth is a technology for doing that. This is why motifs like floods, second creations, sky teachers, and ages of decline recur so widely. They are not random fantasies. They are narrative forms suited to a world where discontinuity is real.

In the context of human evolution, the dialogue becomes especially interesting because science has revealed plurality and hybridization. The past truly contained multiple human types. The present truly contains traces of them. Yet most modern people experience humanity as singular. That mismatch creates psychological space for "others" in myth—giants, watchers, semi-divine hybrids, strange builders, beings who are not fully human. A disciplined approach does not declare that these myths describe Neanderthals or Denisovans. It recognizes that the motif of "others" becomes culturally plausible in a world that actually had others. Myth may preserve an intuition of deep-time diversity without preserving the taxonomy.

The dialogue also helps us reinterpret "civilizing gifts." Myths often claim that crucial knowledge—agriculture, metallurgy, astronomy, writing—was given by external beings. Science, meanwhile, sees these as cultural developments emerging over time through human innovation and exchange. The bridge is not to literalize the myth or to dismiss it, but to consider what social reality the myth encodes. Knowledge often does arrive from outside a community: through migrants, traders, conquerors, or specialist elites. To those who receive it, the bearers can feel superhuman. Their skills can seem magical. Myth then records the social shock of innovation as divine instruction. This reading keeps the myth meaningful without making it scientifically literal.

Another key domain for dialogue is catastrophe. Science shows climatic volatility, sea-level change, and repeated demographic bottlenecks. Myth preserves world-reset stories—floods, fires, darkness, the collapse of ages. The temptation is to treat myth as direct geological evidence. That is a mistake, because myth compresses and moralizes. But it would also be a mistake to ignore the convergence of structure: both archives tell us that discontinuity matters. Science says populations shrink and rebound; myth says worlds end and begin again. The scientific statement is about demography and ecology. The mythic statement is about meaning and moral order. Both can be true in their own domains. In dialogue, they enrich rather than cancel.

A disciplined dialogue also protects against a common error: using myth to fill every gap in evidence. Gaps exist because preservation is uneven and because the past was genuinely complex. Myth can inspire hypotheses, but it cannot substitute for material confirmation. The correct posture is to let myth pose questions and let science test what can be tested. Myth can suggest that humans remember encounters with "others," but science must determine what kinds of "others" were biologically and culturally plausible in a given time and place. Myth can suggest that knowledge was lost and reissued, but archaeology must determine whether cultural resets and rebuildings occurred in that region.

Dialogue is not a shortcut. It is a way of expanding the question space responsibly.

This approach also changes how we interpret the modern hunger for hidden histories. People often turn to mythic literalism because they feel the official story is too small—too linear, too sterile, too dismissive of mystery. A science-and-myth dialogue offers a third way. It allows a story that is both grounded and profound: humanity emerged through plurality, contact, catastrophe, and recovery. We are a composite species shaped by a volatile world. Our myths preserve the emotional and social memory of discontinuity, even when they do not preserve literal facts. This story has enough depth to satisfy the sense that something is missing without requiring invented evidence.

The next subsection addresses the final piece of the reconstruction: why this story was fragmented. Fragmentation is not only a function of broken bones and degraded DNA. It is also a function of bottlenecks, cultural resets, absorption, and narrative policing. The human story is fragmented because the process itself produced fragmentation—populations collapsing, knowledge chains breaking, identities dissolving into others, and later societies rewriting origins to stabilize themselves. To reconstruct the missing chapters, we must understand fragmentation not as an accident, but as a force within the story.

The missing chapters of human evolution are missing for more than one reason, and the temptation is to reach for a single explanation—poor preservation, academic conservatism, or the slow pace of discovery. But fragmentation is not just an external problem of evidence. It is also an internal feature of the story itself. Human history was fragmented by the forces that shaped it: climate volatility, demographic collapse, population absorption, cultural reset, and the later rewriting of origins into stable myths. When we ask why the story is hard to reconstruct, we are not only asking why the archive is incomplete. We are asking why the human past produced discontinuity in the first place.

Start with the simplest fragmenter: preservation. Bones do not fossilize reliably. Most bodies vanish without trace. Even when bones survive, they often survive as fragments—teeth, a jaw, a partial skull. DNA preservation is even more selective. It depends on temperature, humidity, soil chemistry, and time. Whole regions of the world are hostile to ancient DNA, and many of the places where humans lived longest are precisely the places where DNA degrades fastest. Archaeological sites, too, are not evenly preserved. Coastlines have moved. River valleys have shifted. Sites have been buried, eroded, drowned, or destroyed. This is the basic tragedy of prehistory: the more ancient the event, the more likely its physical traces have been erased.

But if preservation were the only problem, the story would be incomplete but not necessarily discontinuous. We would expect smooth trends with missing data points. Instead, the record often looks punctuated: abrupt appearances, mysterious disappearances, sudden cultural blooms, and long silences. That pattern suggests a deeper fragmenter: population dynamics. The Pleistocene world did not support stable, continuously expanding populations. It produced cycles—expansion, contraction, refuge, re-expansion. Each contraction did not merely reduce numbers; it restructured ancestry. Lineages that might have persisted for tens of thousands of years could vanish suddenly through local extinction. Other lineages could be absorbed into neighbors. Bottlenecks could narrow the human stream so sharply that vast diversity was lost, leaving only a few surviving branches to carry the future. In other words, the story is fragmented because the populations that lived it were repeatedly broken and reassembled.

Absorption is a particularly powerful fragmenter because it erases identities without erasing genetic contribution. If a small population is absorbed into a larger one, the absorbed group can disappear as a distinct demographic reality. Their language, their rituals, their social boundaries vanish. Their bones may never be found. Yet fragments of their ancestry persist in the genome of the absorbing population. This creates a strange archaeological silence: the biology can be present without a clear cultural label, and the genetic trace can survive without a corresponding fossil signature. It is fragmentation through integration. The very act of survival—through blending—can erase the clear outline of who contributed.

Cultural reset fragments the story in a different way. Even if populations survive biologically, knowledge chains can break. Specialists die. Traditions collapse. Innovations vanish. After a bottleneck, a society may rebuild with a reduced cultural repertoire. Later observers might see a gap: a region once had complex symbolic artifacts, then it appears simple, then complexity reappears later. This can be misread as a sudden invention of intelligence, when it may be a cycle of loss and reaccumulation. Cultural reset fragments continuity, creating the impression that humans “jump” forward when, in reality, they may be recovering from collapse.

This is where the emotional and symbolic record becomes part of fragmentation. When a society experiences collapse, it does not preserve the event as a neutral report. It preserves it as meaning. Catastrophe becomes moralized. Loss becomes explained through divine order. Recovery becomes framed as a new creation. Over time, these symbolic interpretations replace the literal memory of what happened. The story fragments into mythic ages: before the flood, after the flood; before the fall, after the fall. These age structures encode discontinuity, but they do so in a way that compresses detail and reorganizes chronology. Myth preserves the feeling of rupture while discarding the specific sequence of events. The result is a cultural archive that is rich in structure but fragmented in time.

Later civilizations then add another layer of fragmentation: deliberate simplification. As societies grow complex, they need coherent origin narratives. Coherent narratives stabilize identity and authority. They define who belongs, who rules, and what obligations are sacred. In creating coherence, societies often compress complexity. They merge multiple ancestral streams into one lineage. They turn migrations into single heroic journeys. They turn long processes into founding moments. They claim divine sanction to legitimize institutions. This is not necessarily deceit; it is social necessity. But it means that when we later look back through the lens of those civilizations, we are not seeing prehistory directly. We are seeing prehistory edited into a usable past.

This editing can also be intensified by conquest and cultural dominance. When one group conquers another, it can absorb not only genes but narratives. It can overwrite local origin stories with imperial ones. It can reinterpret local gods as versions of its own. It can rename places and reframe ancestry. Over centuries, the memory of earlier peoples becomes ghostly—not because nothing happened, but because the surviving narrative belongs to the winners. In that sense, fragmentation is also a politics of memory. The story we inherit is often the story that had the power to reproduce itself through institutions, texts, and monuments.

Finally, there is modern fragmentation: the fragmentation created by disciplinary boundaries and public storytelling. Genetics speaks one language, archaeology another, mythology another. Each archive is partial. When these archives are not integrated carefully, they produce competing fragments—stories that cannot easily fit together. Add to this the social pressures discussed in the previous chapter—taboo, ridicule, narrative policing—and the modern picture becomes even more fractured. Complexity is real, but it is not always communicated. Unknowns are real, but they are often minimized. The public, sensing the gaps, fills them with whatever narratives are available. Fragmentation multiplies.

Put together, these forces explain why the missing chapters feel both tantalizing and elusive. We are not trying to reconstruct a clean, continuous film. We are trying to reconstruct a story that was repeatedly torn, edited, rewritten, and partially erased by the very processes that made us. The archive is broken because the world was unstable. The lineages are unclear because absorption dissolved boundaries. The cultural record is uneven because resets erased knowledge chains. The myths are symbolic because humans preserved meaning rather than chronology. The later narratives are simplified because societies need coherence. And the modern discourse is cautious because the topic is socially charged.

The next subsection takes this understanding and turns it into a final interpretive frame for the reconstruction: humanity as an inherited project. Not a claim of external engineering, but a recognition that we inherit ourselves from many hands—biological and cultural—across deep time. If the story is fragmented, it is also layered. And those layers are what we must learn to read.

To say that humanity is an inherited project is to use a phrase that can easily be misunderstood. It does not mean, in the strict evidentiary sense, that an external intelligence engineered our species or directed our fate. It does not mean that prehistory is a secret laboratory. It means something both more grounded and, in its own way, more profound: we are the outcome of layered inheritance. We inherit ourselves across deep time through genes, through culture, through selective pressures, through contact and absorption, through catastrophe and recovery. We did not simply "become" in isolation. We were shaped, revised, narrowed, diversified, and reassembled—again and again—by forces both inside and outside any single population's control. In that sense, humanity is not a finished product. It is a cumulative work.

The inherited project begins with biology. Modern humans carry genetic traces of other humans—lineages that disappeared as distinct peoples but persist as fragments within our genomes. This alone dismantles the idea that humanity is a single, sealed creation. Our bodies are composite archives. They contain layers of deep ancestry, some mapped to known groups, some detectable only as statistical signals, some likely still hidden because sampling is incomplete. If we are a project, it is a project built from many contributors, some unnamed, some unknowable. The "we" of modern humanity is not the product of one ancestral stream. It is a convergence.

But inheritance is not only genetic. Culture is an inheritance system with its own evolutionary dynamics. A tool tradition, a ritual, a way of organizing kinship, a method of storing food—these are not random behaviors. They are transmitted, selected, and refined. A group that preserves effective cultural practices survives better; its practices spread. Over generations, culture becomes a second heredity, layered over biology. Culture can move faster than genes, adapt faster than genes, and sometimes compensate for genetic limitations. In a volatile world, this cultural inheritance can be the difference between extinction and endurance. If the human story is a project, culture is one of its main builders.

Catastrophe adds another layer. Bottlenecks do not simply reduce population size; they reshape the project by filtering which lineages and which cultural practices survive. A bottleneck is a brutal editor. It cuts out diversity, breaks knowledge chains, and leaves survivors with a narrowed inheritance. But it also leaves behind resilience-tested strategies. After catastrophe, the project continues, but it continues under new constraints. The survivors rebuild from what remains. They reinvent what was lost. They merge with others. They create new forms of order because old forms have failed. In that sense, humanity's inheritance is not smooth accumulation. It is accumulation through rupture.

Contact and absorption are also project dynamics. When populations meet, they do not exchange only genes; they exchange worlds. They exchange ways of living. They transmit techniques and symbols. They sometimes dominate and overwrite, and sometimes integrate and blend. Over time, contact zones become engines of novelty. New identities form at the edges where older identities dissolve. This is one way a project evolves: not by isolated refinement, but by synthesis under pressure. A composite species is a composite culture as well—an inheritance of encounters.

The phrase "inherited project" also highlights something that the simplified evolutionary story often hides: the role of continuity across discontinuity. Humans repeatedly face reset. And yet, something persists. A remnant survives. A fragment of knowledge remains. A symbolic memory continues to be told. A technique is carried forward in the hands of a few. This persistence is not always obvious in the material record, because the record is broken. But it is visible in the human capacity to rebuild complexity quickly after collapse. That capacity itself may be one of our deepest inherited traits—not a single gene, but a systems-level inheritance: social learning, cooperation, abstraction, and the ability to transmit meaning.

Myth belongs here, not as literal history, but as one of the ways the project remembers itself. If the human line has been repeatedly narrowed and restarted, it makes sense that cultures would preserve stories of second creation, of lost ages, of teachers who reintroduce order, of covenants established after catastrophe. These stories do not need to be factual in every detail to be meaningful. They can encode the psychological truth that humans have lived through rupture, that order is fragile, and that knowledge can be lost. Myth, in this framing, becomes part of the inheritance system: it transmits caution, cohesion, and identity across generations.

This is also why the project metaphor has moral weight without requiring supernatural claims. If humanity is inherited, then our present condition is not inevitable. It is contingent. We are here because many others were not. We carry fragments of those who vanished. We carry the outcomes of survival filters we did not choose. Recognizing this can produce humility. It undermines triumphalist narratives that treat modern humans as the obvious winners of a clean competition. It suggests instead that we are the survivors of a complex and often brutal history—survivors who carry the past within us, whether we acknowledge it or not.

At the same time, the inherited project points forward. If we are shaped by layered inheritance, then the project continues. We are not only inheritors; we are contributors. Our choices shape the cultural inheritance of future humans. Our technologies reshape selection pressures. Our institutions reshape cooperation and conflict. Our narratives reshape identity. In a post-catastrophe framing, this matters because we now live in a world where our actions can create bottlenecks of our own making—through ecological disruption, war, and systemic risk. The story of deep-time vulnerability is not only ancient history; it is a warning embedded in our own inheritance.

Chapter 14 has therefore reconstructed the missing chapters not as a catalogue of secret events, but as a model that can hold complexity honestly. Human origins, in this model, are plural, permeable, volatile, and layered. Hybridization is not a scandal; it is part of our creative resilience. Myth is not a rival to science; it is a human memory archive shaped by rupture and recovery. And the story is fragmented not only because evidence is missing, but because the process itself produced fragmentation—through bottlenecks, absorption, and narrative rewriting. To inherit humanity is to inherit all of that.

The final chapter now turns from reconstruction to reflection: who are we really, once we take this complex model seriously? What does genetics force us to admit, what has myth been telling us all along, and what does it mean to be a species that may have forgotten the depth of its own origins?

CHAPTER 15 — WHO ARE WE REALLY?

For most of the twentieth century, the deep story of human origins was reconstructed from bones and stones. Fossils offered shapes and proportions, hints of brains and bodies. Tools offered traces of behavior. But both archives were sparse, uneven, and vulnerable to interpretation. Genetics has changed that—not by replacing the older evidence, but by adding a new kind of witness. The genome is not a narrative written in sentences, but it is a record written in inheritance. It preserves traces of ancestry that fossils may never reveal. And what genetics now forces us to admit is that the old, tidy story cannot survive contact with this witness. The human line was not clean. It was not solitary. It was not purely linear. Modern humanity is, in a literal biological sense, a mosaic.

The first admission is plurality. The genome confirms what fossils had begun to suggest: multiple human populations existed in parallel, for long spans, across wide geographies. The older popular story implied a mostly singular march: one line evolving steadily into us. Genetics refuses that simplification. It shows that there were neighboring lineages close enough to interbreed, distinct enough to be meaningfully different, and present enough to leave traces in living people. If the past contained such neighbors in places we have sampled well, the most reasonable inference is that the past contained even more neighbors in places we have sampled poorly. Genetics does not merely add a branch or two to the family tree; it pushes us toward a new image altogether. The ancestral landscape was crowded.

The second admission is permeability. Modern humans carry inherited segments that did not originate within the main Homo sapiens lineage. This means contact was not only conflict or avoidance. It was, at least sometimes, intimacy. The boundaries between populations were real, but they were not sealed. When populations met—through migration corridors, refuge zones, or overlapping ranges—gene flow occurred. The result is a deeper truth than any moral metaphor: species boundaries in deep time are not always the hard walls we imagine. The human past operated more like a network of partially compatible populations whose separations were shaped by geography and time, and whose contacts were shaped by opportunity and pressure.

The third admission is that extinction is not always total erasure. In a purely fossil-based imagination, an extinct lineage disappears like a candle blown out—gone, leaving only bones if we are lucky. Genetics reveals a different reality: a lineage can vanish as a distinct population while leaving living traces through admixture. Some ancestral contributors are dead as peoples but alive in fragments. Their identities dissolved into a larger survivor population. Their languages and rituals may be lost. Their bones may never be recovered. But their genetic signatures persist as quiet inheritance. This is a profound correction to the replacement narrative. It suggests that the human story is not only a story of one group eliminating another. It is also a story of incorporation.

The fourth admission is the existence of ancestry we can detect but cannot fully name. The genome contains signals that point to "ghost populations"—contributors inferred from patterns of variation rather than directly identified through fossils or sequenced remains. This is not mystical. It is a logical outcome of statistical inference applied to deep ancestry. Yet the implication is unsettling for tidy narratives: there were humans—or human-like populations—whose existence is real enough to leave genetic traces but whose faces we have never seen, whose bones we have not yet found, and whose cultural worlds are entirely unknown to us. Genetics forces humility here. It tells us that our catalog of human diversity is incomplete, and that the map of ancestry includes hidden terrain.

The fifth admission concerns time. Genetics undermines the assumption that human evolution is best told as a sequence of neatly separated eras. The genome is a palimpsest: layers written over layers. Some signals are ancient and diffuse. Others are more recent and regional. The boundaries between "then" and "now" blur as inherited segments travel through generations and across continents. This makes the human story less like a timeline of discrete steps and more like a long, overlapping process where multiple histories continue to echo in the present. We do not simply descend from the past; we carry it.

The sixth admission is that adaptation can arrive through borrowing as well as through internal mutation. This does not mean humans collected traits like trophies. It means that when a population already adapted to a harsh environment interbred with another population entering that environment, useful variants could move. In this sense, genetics suggests that survival was sometimes collaborative at the level of inheritance. The human project—if we use that metaphor—was not built solely by one lineage inventing everything within itself. It was also shaped by exchange. This makes sense in a world of volatility, where adaptation speed mattered and where small populations faced extinction pressures. Gene flow could be a form of resilience.

The seventh admission is that modern human identity cannot be mapped cleanly onto ancient biological categories. People often want to imagine ancient populations as stable "races" or fixed types. Genetics undermines this impulse. Deep time was not composed of neatly bounded groups that correspond to modern categories. It was composed of shifting populations, expansions, contractions, isolations, and contacts. The boundaries moved. The mixtures changed. The surviving patterns we see today are the end of a long process of rearrangement. Genetics therefore forces us to admit that our present-day categories are not windows into deep time. They are recent outcomes layered onto a far older and more complex demographic reality.

All of these admissions converge on one central point: modern humans are not the product of a single clean evolutionary line. We are the survivors of a complicated ancestry in which multiple lineages contributed, some through long separation, some through brief contact, some through absorption, and some through histories we can only infer. This does not reduce human dignity; it enlarges human reality. It means our species is not a solitary achievement but a composite continuity. It means the human "we" has always been wider than any single lineage, and that the present carries ghosts of past diversity within it.

This genetic reality also changes how we interpret the emotional hunger for missing chapters. People sense that the story has been simplified. Genetics confirms that it has. The simplification may have been pedagogically useful, but it is no longer adequate. If we want to know who we really are, we must accept that our ancestry includes lost populations, hybridization, and a demographic history shaped by collapse and recovery. The genome does not give us a myth. It gives us a constraint: the past was more complex than the story we inherited.

The next subsection turns to a different witness: myth. If genetics forces us to admit biological plurality and permeability, myth has been telling a parallel story in symbolic language—stories of others, hybrids, teachers, falls, resets, and hidden inheritance. The task is not to collapse myth into genetics, but to ask what myth has been preserving all along about how humans remember complexity.

If genetics is a record written in inheritance, myth is a record written in meaning. It does not preserve the past as a laboratory transcript. It preserves the past as a human experience—compressed, dramatized, moralized, and shaped into forms that can survive across generations. That is why myth has often been dismissed as useless for reconstructing origins. It is not "accurate" in the modern sense. Yet the deeper question is not whether myth functions like a scientific report. It is whether myth preserves recurring intuitions about the human condition that align with what science now reveals: plurality, contact, discontinuity, and the sense that something about our origins is layered rather than singular. In that sense, myth has been telling a story all along—not the story of events, but the story of how humans have always felt their own complexity.

The first thing myth tells us is that humanity is not alone in the old world. Across cultures, the prehistoric landscape is crowded with "others": giants, sages, spirits, builders, ancestors who are different in scale or power, beings who stand at the edge of the human category. These "others" are not uniform. Sometimes they are benevolent; sometimes they are dangerous; often they are both. The crucial point is structural: myth rarely imagines early humanity as the only intelligent presence in the world. It imagines a layered landscape in which humans share space with non-ordinary beings. In a disciplined reading, we do not need to literalize these beings to see what the motif preserves. It preserves the intuition that the past contained more kinds of persons than the present.

Science now gives that intuition a grounded form. The Pleistocene world did contain other humans—different populations, different lineages, different bodies, different adaptive histories. Most modern people live in a world where humanity appears singular, and so the idea of "others" feels fantastic. But in deep time, "others" were real. Myth may encode the cognitive memory of plurality as symbolism: the sense that the world was once crowded with beings not identical to us. It is not a fossil catalogue, but it is a pattern of remembrance.

The second thing myth tells us is that boundaries are porous. Myths are filled with unions across categories: gods with humans, spirits with mortals, kings descended from divine lines, heroes born from mixed ancestry. Hybrid children are often depicted as exceptional—stronger, wiser, more luminous, more capable. Again, a disciplined reading does not equate mythic gods with archaic hominins. But it recognizes that myth persistently imagines human origins as mixed. It imagines ancestry as layered and blended, not purely contained within one kind of being. This resonates with what genetics forces us to admit: we are not the result of a single clean line. Hybridization is not an invention of modern storytellers; it is a deep motif of human imagination.

The third thing myth tells us is that history is not smooth. Myths are structured around ruptures: falls, floods, fires, world ages ending and beginning, second creations, the withdrawal of gods, the collapse of an earlier order. These stories are not simply entertainment. They are cultural tools for remembering discontinuity. They preserve the idea that the world humans inhabit has been reset—that the present is not continuous with the past in an unbroken line. This is striking because it matches the demographic and ecological reality of deep time: bottlenecks, refuges, expansions, cultural resets. Myth is not giving us dates, but it is preserving a truth about structure: the human story contains breaks.

The fourth thing myth tells us is that knowledge feels inherited rather than purely discovered. Many traditions claim that crucial arts were "given" by teachers: agriculture, metallurgy, law, astronomy, writing, navigation, medicine. Even when the myth frames these teachers as divine, the motif is consistent: civilization's foundations are received, not invented. This can be read literally, but it can also be read socially. Knowledge often enters a community through contact—through migrants, traders, specialists, or elites who carry practices developed elsewhere. To those who receive it, the knowledge feels like a gift from beyond their world. Myth then encodes this experience in cosmic terms. The story is not "we gradually learned," but "it came to us."

This motif is especially powerful after catastrophe and reset. When societies rebuild after collapse, the reappearance of knowledge can feel like restoration rather than innovation. The mythic teacher becomes the figure who reintroduces order. This aligns with cultural reset dynamics: knowledge can be lost and later recovered or reinvented. Myth preserves the felt truth of that process: the sense that what is most valuable in human life—order, technique, sacred timing—does not always originate in the present moment but descends from a deeper past.

The fifth thing myth tells us is that authority is often justified by origins that are not purely human. Kings and elites frequently claim descent from gods, from sky beings, from founders who were more than ordinary mortals. This is not a scientific claim; it is a political technology. It legitimizes hierarchy by placing it beyond dispute. Yet it also reveals something about how humans think about exceptional capability. Societies often interpret unusual skill, knowledge, or power as evidence of non-ordinary ancestry. In a world where populations mixed, where newcomers arrived with unfamiliar technologies, where specialist classes emerged, it is easy to see how mythic genealogy could become the symbolic language of social differentiation. Myth is telling us that humans repeatedly experience difference within humanity and then explain that difference through origin stories of "others."

The sixth thing myth tells us is that humans fear forgetting. Many myths are haunted by loss: lost golden ages, lost languages, lost wisdom, forbidden knowledge, broken covenants. These are not incidental themes. They reflect the reality that culture can vanish. A society that has lived through rupture becomes obsessed with preservation. It builds rituals to maintain memory. It tells stories to keep identity intact. Myth itself becomes a memory technology designed to survive when writing does not. When the book argues that humanity may have forgotten parts of its own origins, it is not merely projecting modern anxiety onto the past. Myth shows that humans have long carried the intuition that forgetting is part of our condition—that we inherit fragments and live among gaps.

Taken together, myth has been telling a coherent meta-story: the past was crowded, boundaries were porous, history was punctuated by resets, knowledge felt inherited, authority claimed non-ordinary origins, and forgetting haunted the human condition. None of this proves that myth is literal history. But it does suggest that myth is not random. It preserves durable structures of experience that make sense in a world where humanity was diverse, where populations mixed, where catastrophes narrowed societies, and where memory had to be carried through disruption.

This sets the stage for the next subsection, which focuses on a particularly potent mythic motif: the Watchers. Whether interpreted symbolically or more literally by different traditions, the Watchers represent the archetype of non-human instructors and boundary-crossers—figures who bring knowledge, alter the human condition, and then vanish or are punished. The question will not be "Were they real?" in a simplistic sense, but "What does the Watchers motif preserve about human anxieties, memories, and the felt discontinuity between what we are and what we imagine we once were?"

The Watchers are less a single story than a recurring human shape: the idea that knowledge entered the human world from beings who were not fully human, and that this gift carried consequences. Across cultures, the names change—angels, sages, sky people, civilizers, shining ones—but the structure persists with remarkable durability. A group of "others" arrives or descends. They see humanity from above, as if from a different domain. They teach arts that transform the human condition. Their presence disrupts boundaries—between worlds, between classes, between nature and culture. Then a rupture follows: punishment, exile, withdrawal, collapse. What remains is not simply civilization, but a wound in memory. Humans inherit the gifts and the fear that the gifts were not earned. This is the legacy of the Watchers motif: it frames civilization as both elevation and risk, and it frames human origins as a boundary-crossing event whose full truth is partially hidden.

In a disciplined reading, the Watchers do not need to be literal extraterrestrials or engineered overseers to have explanatory power. Their power is psychological and historical. The Watchers motif concentrates several deep anxieties into one narrative: the anxiety that human knowledge appears too sudden; the anxiety that elites possess secret instruction; the anxiety that catastrophe follows hubris; and the anxiety that our story contains missing pages. The Watchers are a way of expressing that the human condition is not merely natural continuity, but a complex inheritance that feels, in some sense, borrowed.

One reason this motif persists is that it provides a narrative model for the asymmetry of knowledge. In many early societies, certain classes held specialized understanding—calendar systems, metallurgy, ritual law, navigation, medicine. To those outside the specialist circle, such knowledge can appear superhuman. It controls time, transforms stone into metal, predicts seasons, and commands the invisible forces of social order. The Watchers become the mythic origin of this asymmetry. They explain why knowledge is not equally distributed: it came from beyond ordinary humanity and was entrusted to a few. In this way, the Watchers motif legitimizes hierarchy while also making it morally precarious. If knowledge is a gift, it can be revoked. If it is stolen, it can be punished.

Another reason is that the motif expresses the felt discontinuity between animal origins and cultural achievement. Humans are biological creatures, yet we live in symbolic worlds. We carve matter into meaning. We track the sky and encode time. We transmit complex knowledge across generations. When people confront the scale of this transformation, it can feel like a leap rather than a slope. Even when science can explain the processes—demography, cultural accumulation, selection—the intuition of leap remains powerful. The Watchers externalize that leap. They give it an agent. They make the transformation intelligible to the story-making mind: knowledge did not simply emerge; it arrived.

But the motif is never purely celebratory. The Watchers bring both gifts and corruption. The story often includes forbidden arts, violence, sexual transgression, or the destabilization of a previous moral order. This duality is essential. It encodes the ambivalence humans feel about their own powers. Metallurgy yields tools and weapons. Astronomy yields calendars and omens. Writing yields memory and bureaucracy. Law yields order and control. Every civilizing gift has a shadow. The Watchers motif gathers that shadow into a moral drama: power expands faster than wisdom, and the world responds with catastrophe.

In the framework of this volume, the Watchers motif can be approached as a memory structure for real dynamics in deep time. Consider what science now forces us to admit: humans encountered other human populations, mixed with them, absorbed them, and survived through composite inheritance. That alone can generate a sense of “others” in the deep past. Add to it the reality of migration waves and cultural transmission: new knowledge systems often arrive with newcomers. Add to it catastrophe and reset: societies rebuild and remember earlier worlds as lost ages. In such a world, it is not surprising that myths imagine teachers who are not fully human and eras that end violently. The Watchers become the symbolic integration of these realities: contact with others, the arrival of knowledge, and the trauma of discontinuity.

The “watching” itself is important. To watch is to observe from a distance, to see patterns humans within the system cannot see. The Watchers are often framed as beings who know the sky, who track time, who perceive cosmic order. This mirrors the historical role of priest-astronomers and specialist classes who did exactly that—watch the heavens, regulate calendars, and thereby regulate society. The Watchers may be the mythic origin of the watchers: the sacred specialists whose authority depended on access to knowledge beyond the common eye. The motif therefore carries a social memory of how power and knowledge are intertwined.

At the same time, the Watchers motif can be read as a moral warning against the collapse of boundaries. Many versions of the story emphasize that the Watchers transgress: they cross from one realm to another, mingle where they should not, and thereby disrupt cosmic order. This transgression is followed by punishment and a world reset. In a post-catastrophe society, this is a potent lesson. It says: boundaries exist for a reason; when they are violated, chaos follows. Whether the boundary is social, sexual, ecological, or spiritual, the motif teaches restraint through fear of collapse. In that sense, the Watchers legacy is not only about origins. It is about governance—how societies regulate behavior by encoding risk into myth.

The Watchers also offer an explanation for missing knowledge. Many traditions carry the sense that humans once knew more—lived longer, understood deeper secrets, possessed arts now lost. The Watchers can be invoked to explain this: knowledge was given, misused, and then withdrawn. This maps to cultural reset dynamics: knowledge can indeed be lost through bottleneck, collapse, and migration. The myth turns that loss into a moral event, making forgetting legible. It also creates an enduring hunger: if knowledge was withdrawn, can it be recovered? The Watchers motif thus becomes a generator of “forbidden history” thinking. It encourages the sense that the past contains hidden instruction.

Yet discipline matters here more than anywhere. The Watchers motif is powerful precisely because it can seduce the mind into literalism. It can become a convenient explanation for every discontinuity. But the responsible approach keeps boundaries clear. Myth can preserve structure without preserving fact. The Watchers can be symbolic of contact, elite knowledge, and catastrophe memory without being literal non-human beings. And even when one entertains more speculative interpretations, the absence of material evidence must be acknowledged. The motif is an interpretive lens, not a proof.

Still, its legacy cannot be dismissed, because it reveals something about what humans keep returning to when they ask who they are. They return to the idea that we are inheritors of gifts we did not fully understand, that our power carries risk, and that our story is not entirely ours to narrate from the present alone. The Watchers, in mythic language, are the guardians of that uncomfortable intuition: that humanity's origins are complex, boundary-crossing, and partially forgotten.

The next subsection takes the final step toward synthesis: evolution as an open story. If genetics reveals a web and myth preserves the memory of rupture and otherness, then the human narrative cannot be closed into a finished doctrine. It remains open—revisable as evidence grows, and open in meaning as well, because what we make of our origins shapes what we become.

One of the quiet temptations of modern culture is to treat scientific narratives as if they were finished scriptures. Once a story becomes widely taught—once it is drawn in textbooks, repeated in documentaries, and embedded in the public imagination—it begins to feel complete. It develops a moral aura: to question it seems like attacking truth itself. Yet evolution is not a closed doctrine, and the human branch of evolution is least of all. The evidence is too fragmentary, the processes too complex, and the pace of discovery too disruptive for closure. If this volume has argued anything consistently, it is that the human story is not a smooth line but a web of interactions, collapses, absorptions, and recoveries. A web is not a final sentence. It is a living structure. Evolution, for us, must be understood as an open story.

To say the story is open is not to say anything can be true. It is to say that the model must remain revisable as the archive expands. In human origins, new discoveries routinely reframe what seemed settled. Genetic evidence can expose ancestry where fossils are absent. A single fossil can force reconsideration of timing and range. Archaeological sites can extend human presence earlier than expected. Even the meaning of "species" can shift depending on how much interbreeding is detected. These are not minor details. They change the shape of the narrative itself. An open story is one that expects this—one that treats revision not as embarrassment, but as a sign of a healthy relationship with evidence.

The openness also comes from the structure of the past. Deep time did not operate like a controlled experiment. It was a dynamic system with feedback loops: climate shifts reshaping habitats, habitats reshaping migration, migration reshaping contact, contact reshaping inheritance, inheritance reshaping adaptation. Cultural evolution ran alongside biological evolution, sometimes accelerating, sometimes buffering, sometimes collapsing and rebuilding. In such a system, no single explanatory lever is sufficient. A closed story prefers a single dominant mechanism: gradual improvement, linear progress, or one great dispersal. An open story accepts multiplicity: many mechanisms acting together, producing outcomes that are emergent rather than designed.

This openness changes how we should interpret the gaps. In a closed story, gaps are treated as minor holes in an otherwise complete picture. In an open story, gaps are signals. They are places where the archive is thin and therefore where confidence must be modest. They are also places where new chapters may exist that would alter how we understand the whole. The genome has already taught us this lesson. It revealed contributors we did not know to look for. That alone should make us cautious about declaring the story finished. If unknown ancestry can be detected without a fossil identity, then unknown chapters can exist without obvious artifacts. Openness is therefore not a philosophical preference. It is an epistemic requirement.

An open story also corrects the misuse of evolution as a narrative of moral progress. People often smuggle an assumption into the human evolution story: that evolution is a ladder toward improvement, and that modern humans represent the pinnacle. This is not science; it is mythology disguised as science. Evolution produces adaptation, not virtue. It produces fit to circumstance, not a universal march toward betterment. The open story refuses that moralizing structure. It recognizes that humans survived not necessarily because we were "best," but because we were resilient, flexible, and often lucky in a volatile world. It recognizes that many other human types may have been just as complex in their own niches, and that their disappearance may reflect vulnerability to catastrophe and demographic absorption rather than inferiority. Openness, here, is humility.

This humility has a second implication: modern humanity is not a monolith. If our ancestry is composite, then our present humanity is the downstream expression of many histories converging. That convergence undermines simplistic narratives about purity, singular origin, or essential types. It also undermines the desire to make one group the primary inheritor of humanity's "true" lineage. The open story is inherently pluralistic because the evidence is pluralistic. It forces us to accept that the human past was diverse and that modern humans carry that diversity as deep inheritance. In this sense, openness is not only about scientific revision; it is about resisting the psychological impulse to make origins into ideology.

Openness also reframes the relationship between science and myth. If evolution is open, then the boundary between what is known and what is interpreted must be actively maintained. Science anchors. Myth illuminates meaning. The two can speak without collapsing into each other precisely because the story remains open. When the story is treated as closed, myth becomes either a rival to be defeated or a secret truth to replace it. When the story is open, myth can be treated as a human archive—one that preserves patterns of rupture, otherness, and inherited knowledge in symbolic form. It can be taken seriously without being taken literally. The openness creates room for dialogue without surrendering evidence.

The open story also changes how we think about ourselves moving forward. If the past is a web and not a ladder, then the future is not guaranteed progress. Evolution does not promise improvement; it responds to pressures. Cultural evolution can produce astonishing achievements, but it can also produce collapse. The same cognitive and social capacities that allow civilization also allow self-destruction. An open story of origins therefore carries a warning: the human condition has always been precarious. Our line survived bottlenecks that could have ended it. Our civilizations have risen after near-extinction and then created new vulnerabilities. We should not treat our present stability as proof of permanence.

To live inside an open story is therefore to live with two truths at once. The first is that evidence matters and must constrain belief. The second is that humility matters because the archive is incomplete. We must be willing to revise our models as new data arrives, but we must also be willing to revise our cultural self-image as complexity becomes undeniable. We are not the endpoint of a clean ascent. We are the inheritors of a tangled past, built through contact, collapse, and recovery.

The next subsection draws the moral and practical consequence of this openness: why the past matters for the future. If our story contains bottlenecks, resets, and the loss of knowledge—if the human project has been repeatedly narrowed and rebuilt—then our future cannot be planned as if collapse is impossible. The past matters not as nostalgia, but as warning and instruction.

The past matters for the future because humanity is not built on permanence. It is built on survival through instability. That is the deepest lesson buried beneath genomes, fossils, and myths alike: the human story is a sequence of narrow escapes. Bottlenecks reduced our numbers. Climate shifts rearranged our habitats. Populations vanished. Others were absorbed. Knowledge chains broke and had to be rebuilt. When we look backward honestly, we do not see a species guaranteed to rise. We see a species that repeatedly came close to erasure and repeatedly recovered—not because the world is kind, but because humans are adaptive and because chance sometimes favored our persistence. In an age when human actions can shape planetary systems, this lesson is no longer academic. It is strategic.

The first reason the past matters is that it corrects the myth of safety. Modern people often live inside infrastructures that feel stable: electricity, supply chains, medicine, communication networks, global institutions. These systems create the illusion that collapse belongs to ancient times. But deep history reveals that collapse is not an anomaly. It is a recurring pattern in complex systems under stress. In prehistory, stress was climate and ecology. In history, stress includes conflict, disease, resource depletion, and institutional fragility. The fact that humans survived through repeated resets should not reassure us that we will always survive. It should warn us that survival is contingent and that complex systems can fail abruptly.

The second reason the past matters is that it reveals how narrow the margins can be. A bottleneck is not simply a dramatic event; it is a reduction in options. Diversity—genetic and cultural—is a reservoir of resilience. When diversity is lost, vulnerability increases. Past bottlenecks narrowed the human line, and the survivors rebuilt from reduced inheritance. This has direct relevance to the modern world. If modern civilization creates new bottlenecks—through ecological collapse, systemic conflict, or catastrophic technological failures—the future may inherit a reduced world. It may inherit fewer languages, fewer cultural strategies, fewer biological options. The past shows what that looks like: recovery is possible, but the cost is loss that cannot be fully restored.

The third reason the past matters is that it teaches us the value of redundancy. Prehistoric survival often depended on flexible strategies—multiple food sources, varied toolkits, mobility, social alliances across groups. A society that depends on one narrow resource or one narrow system is fragile. Modern civilization often moves in the opposite direction: it increases efficiency by reducing redundancy. This can be economically powerful, but it can be dangerous in a volatile world. The deep past suggests that resilience is not the same as efficiency. Resilience is the capacity to endure disruption, and that capacity often requires backup pathways and cultural adaptability.

The fourth reason the past matters is that it reframes human identity in a way that can either divide or unify. If the public clings to simplistic origin stories, identity becomes brittle. People seek purity myths, single-line descent, and hierarchy disguised as history. But genetics forces us to admit composite ancestry. Prehistory forces us to admit plurality and contact. When this is absorbed culturally, it can undermine the ideological misuse of origins. It can teach that humanity has always been mixed, that boundaries have always been porous, and that survival has often been cooperative as well as competitive. In a future where global coordination is essential, this reframe matters. A species that understands itself as composite may be less tempted by purity illusions.

The fifth reason the past matters is that it reveals a pattern of forgetting—and the costs of forgetting. Cultural memory is not guaranteed. Knowledge can disappear when populations shrink or when institutions collapse. The mythic obsession with lost ages and withdrawn knowledge can be read as a symbolic memory of this reality. In modern terms, the risk is not that we will forget everything. The risk is that we will assume our knowledge is permanent and therefore fail to protect it. A future shaped by crisis could lose technical systems that depend on global infrastructure. It could lose archives if energy systems fail. It could lose institutional continuity. The past teaches that knowledge must be actively preserved, not assumed.

The sixth reason the past matters is that it offers a realistic model of adaptation. Human survival has always depended on behavioral flexibility and social learning. When environments changed, humans migrated, innovated, and reorganized. They formed alliances, shared techniques, and built new identities. This is not romantic. It is hard-won. But it suggests that the human capacity for adaptation is real—and that it is most powerful when societies can coordinate rather than fragment. The past therefore serves as both warning and encouragement: collapse is possible, but so is recovery. The question is not whether humans can adapt; it is whether our social systems can adapt fast enough and fairly enough to prevent catastrophic narrowing.

The seventh reason the past matters is moral: it changes what we owe to the future. If we understand that our existence is the result of many lost lineages and many narrow escapes, then we may feel a responsibility to not squander the inheritance. The human project was not guaranteed.

It is an improbable continuity carried through deep time. To treat the future casually, to risk large-scale collapse through avoidable choices, is to betray that continuity. This is not a mystical claim; it is a sober recognition of contingency. We stand on a fragile chain of survival. We can choose to strengthen it or to break it.

This brings the chapter—and the volume—toward its final closure. If genetics forces complexity, and myth preserves symbolic memory of rupture, and history demonstrates the fragility of order, then the central question becomes unavoidable: what kind of species are we, really? Are we the triumphant endpoint of a clean evolutionary climb, or are we the composite survivors of a tangled past who repeatedly forget and rebuild? The concluding subsection answers that question in the tone the whole volume has been building toward: not as a final dogma, but as a portrait of humanity that can hold both wonder and vulnerability.

If the human story were simple, we would not need to keep retelling it. We would not keep arguing about it, revising it, defending it, and filling its gaps with legends. A simple story would settle into the mind like a finished map. But our story does not settle. It keeps moving beneath our feet, because it was never a single line to begin with. It was a field of lives—populations branching, meeting, mixing, vanishing, and reappearing through the narrow corridors left by catastrophe. The deepest conclusion of this volume is not a sensational claim about hidden rulers of prehistory. It is a grounded and unsettling truth: modern humanity is the survivor of a complex, hybridized, and partially forgotten past. We are a species that carries missing chapters in our bodies and in our myths, and we have spent much of modern time pretending the chapters do not exist.

Genetics forces the first part of this conclusion with cold clarity. We are not purely the product of one isolated lineage. The human genome is not a clean inheritance; it is a mosaic. It contains traces of other humans—lineages that once shared this planet with us, lineages that were absorbed rather than merely erased, lineages that sometimes left signatures without leaving identifiable bones. This is not a metaphor. It is biology speaking in the only language it has: inherited fragments. And once you accept this, the old story of linear progress becomes difficult to maintain. Humanity did not simply replace all others in a clean sweep. Humanity incorporated, blended, narrowed through bottlenecks, expanded again, and carried forward a composite inheritance. The "we" of modern humans is, in deep time, an accumulation of contributions and losses.

The fossil record and archaeology do not contradict this; they contextualize it. They show a world in which multiple human types existed and adapted, often under extreme environmental pressures. They show that behaviors we associate with "modernity" do not always appear as a smooth gradient. They flare in some regions, fade in others, and sometimes reappear later, suggesting that culture itself can be interrupted by collapse and rebuilt by survivors. The past does not look like a straight road; it looks like a landscape crossed by migrations, refuges, and resets. Our species is not simply the result of steady improvement. It is the result of endurance through volatility.

Myth, for its part, has been telling a parallel story, but in symbolic form. Across cultures, the past is crowded with "others"—giants, watchers, sages, semi-divine hybrids, builders who are not fully human. World ages end and begin again. Knowledge arrives as a gift and is sometimes withdrawn. Authority claims descent from beings beyond the ordinary.

A disciplined approach does not turn these motifs into literal evidence. It recognizes them as memory structures: ways humans encode discontinuity, contact, and the unease that civilization feels both sudden and precarious. Myth is not a fossil, but it is not meaningless either. It preserves what the scientific archive often cannot: the emotional signature of rupture, the social memory of asymmetrical knowledge, the cultural fear of forgetting.

When these archives are brought into dialogue—science anchoring, myth illuminating—another conclusion becomes visible: humanity's self-understanding has often been shaped by the desire for a clean origin story. We prefer a ladder because ladders are comforting. A ladder implies progress is natural and the endpoint is justified. A ladder implies we are the culmination. A web is harder to live with. A web implies contingency. It implies that our existence depends on many vanished lives and many accidents of survival. It implies that other human paths existed and ended, and that we carry their traces without knowing their names. It implies that our story could have gone differently. The web demands humility.

This is where forgetting enters. Forgetting is not merely a flaw in memory. It is a survival adaptation. In the aftermath of catastrophe, societies rebuild by narrowing their focus. They preserve what they can, and they let much of the past dissolve because it is too fragmented to hold. In the aftermath of absorption and migration, identities merge and older boundaries are lost. In the rise of complex civilizations, origin narratives are simplified to stabilize order and legitimize power. Even in modern science, narratives have often been streamlined to communicate clearly and defend against misuse. Layer by layer, complexity is compressed. Over time, the compression becomes normal. The species forgets—not necessarily by choice, but by structure.

Yet forgetting has a cost. When a species forgets its origins, it becomes vulnerable to myths of certainty. It can mistake its present stability for permanence. It can turn origin stories into ideology. It can treat complexity as a threat and discourage the very questions that could refine understanding. This volume has argued that the missing chapters of human evolution are missing not only because evidence is scarce, but because human culture repeatedly favors coherence over complexity. The result is a self-image that is too smooth for the reality written in our genes.

So who are we really? We are not simply the victorious endpoint of a clean evolutionary climb. We are the composite survivors of a turbulent world. We are the inheritors of other humans whose lives ended as distinct lineages but persist in fragments within us. We are builders of meaning who encode trauma and discontinuity in myths because literal memory breaks under the weight of collapse. We are a species capable of astonishing reconstruction after near-extinction, and also a species prone to creating new vulnerabilities through the very systems that give us power. We are not a finished story. We are an open one.

This conclusion does not close inquiry; it reopens it on better terms. It invites a future science that is more comfortable with uncertainty and more honest about complexity. It invites a public culture that can tolerate nuance without turning it into conspiracy. It invites a deeper humility about what it means to be human: not an isolated achievement, but a layered inheritance. And it leaves us with the quiet warning embedded in deep time: survival is never guaranteed, and forgetting is not harmless.

APPENDIX
Bibliography

Aiello, Leslie & Dean, Christopher. *An Introduction to Human Evolution.* Academic Press, 2002.

Ambrose, Stanley. *Late Pleistocene Human Population Bottlenecks.* University of Chicago Press, 2008.

Arsuaga, Juan Luis. *The Neanderthal's Necklace.* Four Walls Eight Windows, 2002.

Barham, Lawrence & Mitchell, Peter. *The First Africans.* Cambridge University Press, 2008.

Beaumont, Mark. *Population Genetics and Human Evolution.* Oxford University Press, 2010.

Berger, Lee R. *Almost Human: The Astonishing Tale of Homo naledi.* National Geographic, 2017.

Bergström, Anders et al. *Origins of Modern Humans.* Nature Publishing Group, 2021.

Bickerton, Derek. *Language and Human Evolution.* Harvard University Press, 2014.

Bocquet-Appel, Jean-Pierre. *When the World's Population Took Off.* Springer, 2011.

Boyd, Robert & Richerson, Peter. *Culture and the Evolutionary Process.* University of Chicago Press, 1985.

Brace, C. Loring. *The Stages of Human Evolution.* Prentice Hall, 1995.

Brown, Peter & Maeda, T. *Homo floresiensis: A New Human Species.* Nature Publishing, 2004.

Cann, Rebecca, Stoneking, Mark & Wilson, Allan. *Mitochondrial DNA and Human Evolution.* University of California Press, 1987.

Carroll, Sean B. *Endless Forms Most Beautiful.* W. W. Norton, 2005.

Cavalli-Sforza, Luigi Luca. *Genes, Peoples, and Languages.* University of California Press, 2001.

Cavalli-Sforza, Luigi Luca et al. *The History and Geography of Human Genes.* Princeton University Press, 1994.

Clark, Grahame. *World Prehistory in New Perspective.* Cambridge University Press, 1977.

Conkey, Margaret & Soffer, Olga. *Beyond Art: Pleistocene Image and Symbol.* Cambridge University Press, 1997.

Coolidge, Frederick & Wynn, Thomas. *The Rise of Homo sapiens.* Oxford University Press, 2009.

Darwin, Charles. *The Descent of Man.* Penguin Classics, 1871/2004.

Deacon, Terrence. *The Symbolic Species.* W. W. Norton, 1997.

Dennell, Robin. *The Palaeolithic Settlement of Asia.* Cambridge University Press, 2009.

Diamond, Jared. *The Third Chimpanzee.* HarperCollins, 1992.

Diamond, Jared. *Guns, Germs, and Steel.* W. W. Norton, 1997.

Dunbar, Robin. *Grooming, Gossip, and the Evolution of Language.* Harvard University Press, 1996.

Finlayson, Clive. *The Humans Who Went Extinct.* Oxford University Press, 2009.

Foley, Robert. *Humans Before Humanity.* Blackwell, 1995.

Foley, Robert & Gamble, Clive. *The Ecology of Social Transitions in Human Evolution.* Oxford University Press, 2009.

Fuentes, Agustin. *The Creative Spark.* Dutton, 2017.

Gamble, Clive. *Origins and Revolutions.* Cambridge University Press, 2007.

Gould, Stephen Jay. *Wonderful Life.* W. W. Norton, 1989.

Gould, Stephen Jay. *The Structure of Evolutionary Theory.* Harvard University Press, 2002.

Green, Richard et al. *A Draft Sequence of the Neanderthal Genome.* Nature Publishing, 2010.

Hawks, John. *The Evolution of Human Diversity.* University of Wisconsin Press, 2016.

Hublin, Jean-Jacques. *The Origin of Neanderthals.* Princeton University Press, 2009.

Hublin, Jean-Jacques & Richards, Michael. *The Evolution of the Human Diet.* Springer, 2009.

Johanson, Donald & Edgar, Blake. *Lucy: The Beginnings of Humankind.* Simon & Schuster, 1996.

Johanson, Donald & Wong, Kate. *Lucy's Legacy.* Harmony Books, 2009.

Jones, Steve. *The Language of the Genes.* HarperCollins, 1993.

Klein, Richard. *The Human Career.* University of Chicago Press, 2009.

Krause, Johannes & Trappe, Thomas. *A Short History of Humanity.* Random House, 2021.

Kuhn, Steven & Stiner, Mary. *What's a Mother to Do? The Division of Labor among Neanderthals and Modern Humans.* Current Anthropology, 2006.

Leakey, Richard. *Origins Reconsidered.* Anchor Books, 1993.

Leakey, Richard & Lewin, Roger. *People of the Lake.* Anchor Books, 1993.

Lewin, Roger & Foley, Robert. *Principles of Human Evolution.* Wiley-Blackwell, 2004.

Lewis-Williams, David. *The Mind in the Cave.* Thames & Hudson, 2002.

Mellars, Paul. *The Neanderthal Legacy.* Princeton University Press, 1996.

Mithen, Steven. *The Singing Neanderthals.* Harvard University Press, 2006.

Nielsen, Rasmus et al. *Genomic Evidence for Human Adaptation.* Annual Review of Genetics, 2007.

Nowell, April. *Growing Up in the Ice Age.* Oxbow Books, 2015.

O'Connell, James et al. *Hadza Hunting.* University of Utah Press, 2012.

Pääbo, Svante. *Neanderthal Man.* Basic Books, 2014.

Pagel, Mark. *Wired for Culture.* W. W. Norton, 2012.

Pinhasi, Ron & Stock, Jay. *Human Bioarchaeology of the Transition to Agriculture.* Wiley, 2011.

Pinker, Steven. *The Language Instinct.* William Morrow, 1994.

Ponce de León, Marcia & Zollikofer, Christoph. *Evolution of the Human Skull.* Cambridge University Press, 2013.

Potts, Rick. *Human Evolution and Environmental Change.* Smithsonian Press, 1996.

Reich, David. *Who We Are and How We Got Here.* Oxford University Press, 2018.

Richerson, Peter & Boyd, Robert. *Not by Genes Alone.* University of Chicago Press, 2005.

Roberts, Alice. *The Incredible Unlikeliness of Being.* Heron Books, 2014.

Rose, Jerome. *Bioarchaeology of the First Americans.* University Press of Florida, 2010.

Rosenberg, Karen & Trevathan, Wenda. *Birth, Obstetrics and Human Evolution.* Annual Review of Anthropology, 2002.

Scarre, Chris. *The Human Past.* Thames & Hudson, 2005.

Schwartz, Jeffrey. *Sudden Origins.* Wiley, 1999.

Shennan, Stephen. *Genes, Memes and Human History.* Thames & Hudson, 2002.

Shipman, Pat. *The Invaders.* Harvard University Press, 2015.

Silk, Joan. *The Evolution of Cooperation.* Annual Review of Anthropology, 2007.

Smith, Fred & Ahern, James. *The Origins of Modern Humans.* Wiley-Blackwell, 2013.

Somel, Mehmet et al. *Human Brain Evolution.* Nature Reviews Genetics, 2013.

Stanford, Craig. *The Hunting Apes.* Princeton University Press, 1999.

Stringer, Chris. *The Origin of Our Species.* Allen Lane, 2011.

Stringer, Chris & Andrews, Peter. *The Complete World of Human Evolution.* Thames & Hudson, 2005.

Tattersall, Ian. *Masters of the Planet.* Palgrave Macmillan, 2012.

Tattersall, Ian. *The Fossil Trail.* Oxford University Press, 1995.

Templeton, Alan. *Human Evolutionary Genetics.* Wiley, 2006.

Tomasello, Michael. *Origins of Human Communication.* MIT Press, 2008.

Tomasello, Michael. *A Natural History of Human Thinking.* Harvard University Press, 2014.

Toth, Nicholas & Schick, Kathy. *The Oldowan.* Indiana University Press, 2006.

Trinkaus, Erik. *The Neanderthals.* Cambridge University Press, 1983.

Van Schaik, Carel. *The Good Book of Human Nature.* Basic Books, 2016.

Varki, Ajit et al. *Denial: Self-Deception, False Beliefs, and the Origins of the Human Mind.* Twelve Books, 2013.

Walker, Alan & Shipman, Pat. *The Wisdom of the Bones*. Vintage, 1996.

Watson, James et al. *Molecular Biology of the Gene*. Pearson, 2008.

Wells, Spencer. *The Journey of Man*. Princeton University Press, 2002.

White, Tim. *Prehistoric Cannibalism at Mancos*. Princeton University Press, 1992.

Wolpoff, Milford & Caspari, Rachel. *Race and Human Evolution*. Simon & Schuster, 1997.

Wrangham, Richard. *Catching Fire*. Basic Books, 2009.

Wynn, Thomas & Coolidge, Frederick. *How to Think Like a Neanderthal*. Oxford University Press, 2011.

Zilhão, João. *Neanderthals and Modern Humans*. Cambridge University Press, 2006.

Suggested Reading

Harari, Yuval Noah. *Sapiens: A Brief History of Humankind.* Harper, 2015.

Reich, David. *Who We Are and How We Got Here.* Oxford University Press, 2018.

Stringer, Chris. *The Origin of Our Species.* Allen Lane, 2011.

Tattersall, Ian. *Masters of the Planet.* Palgrave Macmillan, 2012.

Pääbo, Svante. *Neanderthal Man.* Basic Books, 2014.

Wrangham, Richard. *Catching Fire.* Basic Books, 2009.

Mithen, Steven. *The Singing Neanderthals.* Harvard University Press, 2006.

Lewis-Williams, David. *The Mind in the Cave.* Thames & Hudson, 2002.

Roberts, Alice. *The Incredible Unlikeliness of Being.* Heron Books, 2014.

Wells, Spencer. *The Journey of Man.* Princeton University Press, 2002.

Diamond, Jared. *The Third Chimpanzee.* HarperCollins, 1992.

Gould, Stephen Jay. *Wonderful Life.* W. W. Norton, 1989.

Stringer, Chris & Andrews, Peter. *The Complete World of Human Evolution.* Thames & Hudson, 2005.

Johanson, Donald. *Lucy: The Beginnings of Humankind.* Simon & Schuster, 1996.

Finlayson, Clive. *The Humans Who Went Extinct.* Oxford University Press, 2009.

Gamble, Clive. *Origins and Revolutions.* Cambridge University Press, 2007.

Stanford, Craig. *The Hunting Apes.* Princeton University Press, 1999.

Coolidge, Frederick & Wynn, Thomas. *The Rise of Homo sapiens.* Oxford University Press, 2009.

Potts, Rick. *Human Evolution and Environmental Change*. Smithsonian Press, 1996.

Tattersall, Ian. *The Fossil Trail*. Oxford University Press, 1995.

Lewin, Roger & Foley, Robert. *Principles of Human Evolution*. Wiley-Blackwell, 2004.

Scarre, Chris. *The Human Past*. Thames & Hudson, 2005.

Van Schaik, Carel. *The Good Book of Human Nature*. Basic Books, 2016.

Pagel, Mark. *Wired for Culture*. W. W. Norton, 2012.

Fuentes, Agustin. *The Creative Spark*. Dutton, 2017.

Don't miss out!

Visit the website below and you can sign up to receive emails whenever Alistair Ravenhurst publishes a new book. There's no charge and no obligation.

https://books2read.com/r/B-A-IVYCF-KCOCJ

BOOKS 2 READ

Connecting independent readers to independent writers.

Did you love *The Hidden Chapters of Human Origins*? Then you should read *Megalithic Science, Architecture and Astronomy in the Ancient World*[1] by Alistair Ravenhurst!

[2]

Across every continent, ancient societies raised massive stone monuments—precisely aligned, carefully proportioned, and built to endure for millennia. These structures have long been described as ritual symbols or expressions of belief, yet their consistent orientation to celestial cycles and their disciplined architectural design tell a deeper story. *Megalithic Science* explores a compelling and evidence-based perspective: that many of the world's great stone monuments were not only symbolic, but functional—designed through systematic observation of the sky and an advanced understanding of space, time, and geometry long before the invention of writing.

---

1. https://books2read.com/u/meQZd9
2. https://books2read.com/u/meQZd9

This book examines how prehistoric builders used stone architecture as a scientific medium—encoding astronomical knowledge, calendrical systems, and spatial measurement directly into the landscape. Drawing on archaeology, archaeoastronomy, and architectural analysis, it reveals how ancient cultures tracked solar and lunar cycles, established precise orientations, and preserved knowledge through monumentality rather than text. Written for readers interested in ancient science, early architecture, and the true intellectual capabilities of prehistoric societies, *Megalithic Science* offers a rigorous, grounded reappraisal of humanity's earliest monumental achievements—and the forgotten scientific traditions that shaped them.

Read more at https://www.youtube.com/@AncientCivilization.

Also by Alistair Ravenhurst

Ancient Civilizations and Enigmas

Egypt Before Egypt, the World Before the Pharaohs

ANCIENT CIVILIZATIONS AND MYSTERIES

After the Flood, Cataclysm and the Forgotten Origins of Civilization

Megalithic Science, Architecture and Astronomy in the Ancient World

Skywatchers of Prehistory, the Hidden Origins of Ancient Shamanic Astronomy

Beneath the Sands: The Lost Worlds of Earth's Forgotten Deserts

Empire of the Ocean: The Lost Maritime Civilization of the Pacific

The Serpent Code: The Hidden Language of Ancient Myth

The Hidden Chapters of Human Origins

Watch for more at https://www.youtube.com/@AncientCivilization.

About the Author

Alistair Ravenhurst is an independent author and researcher whose work sits at the intersection of comparative mythology, ancient history, and archaeological interpretation. Trained in the close reading of mythic texts and historical traditions—and informed by archaeological method, site formation theory, and paleoenvironmental research—he investigates how human societies encode upheaval, migration, and cultural rupture into enduring narrative forms. His writing is characterized by a disciplined, evidence-minded approach: distinguishing between primary sources, scholarly consensus, and responsible inference while tracing the long-term continuity of motifs that appear across widely separated civilizations.

Ravenhurst's research interests include catastrophe memory and oral tradition, coastal settlement and submerged landscapes, early monumentality and calendrical systems, and the ways political authority is shaped by sacred time and ancestral origins. Drawing on scholarship in Quaternary climate history, geoarchaeology, and myth studies, he examines how environmental shocks can fragment material evidence while preserving cultural remembrance through story, ritual, and symbol. He writes for readers seeking academically grounded exploration with narrative momentum—books that treat the ancient past as a field of inquiry where the most enduring questions are not merely what happened, but how humanity remembered it, transmitted it, and rebuilt after it.

Read more at https://www.youtube.com/@AncientCivilization.

About the Publisher

AJ Carmichael is a renowned author and explorer known for his expertise in ancient history, ancient mysteries, and adventure. With a passion for uncovering the secrets of the past, AJ has dedicated his life to exploring the ancient cultures of the world.

AJ's journey began at a young age when he became fascinated with the ancient Egyptians and their mysterious pyramids. As he grew older, he began to delve deeper into studying ancient history, immersing himself in the cultures and customs of civilizations long gone.

After completing his studies, AJ set out on a journey to explore the ancient world firsthand. He traveled to Egypt, Mesopotamia, Greece, and Rome, studying the ruins and artifacts left behind by these ancient cultures. Along the way, he made many exciting discoveries, in addition to his explorations. He has published several bestselling books on ancient history and mysteries, including his 'Ancient Worlds and Civilizations' series. These works have earned AJ international acclaim and have established him as a leading authority on ancient history and adventure.

AJ continues to travel the world in search of new discoveries and adventures. His passion for ancient history and his thirst for adventure have taken him to some of the most remote and exotic locations on earth, and his work continues to inspire and captivate audiences around the world.

With his combination of knowledge, skills, and passion, his work continues to shed new light on past civilizations, and his discoveries and explorations have helped bring the ancient world to life for countless readers and enthusiasts.

www.ingramcontent.com/pod-product-compliance
Lightning Source LLC
LaVergne TN
LVHW060016210926
848481LV00013B/3727